INTERNATIONAL ORGANIZATION FOR CHEMICAL SCIENCES IN DEVELOPMENT

WORKING GROUP ON PLANT CHEMISTRY

CHEMISTRY, BIOLOGICAL AND PHARMACOLOGICAL PROPERTIES OF AFRICAN MEDICINAL PLANTS

Proceedings of the First International IOCD-Symposium
Victoria Falls, Zimbabwe, February 25-28, 1996

Edited by

K. HOSTETTMANN, F. CHINYANGANYA, M. MAILLARD and J.-L. WOLFENDER

Institut de Pharmacognosie et Phytochimie, Université de Lausanne, BEP, CH-1015 Lausanne, Switzerland and Department of Pharmacy, University of Zimbabwe, P.O. Box M.P. 167, Harare, Zimbabwe

UNIVERSITY OF ZIMBABWE PUBLICATIONS
1996

Foreword

The present book contains the Proceedings of the first International Symposium held under the auspices of the International Organization for Chemical Sciences in Development by its Working Group on Plant Chemistry. The success of this meeting is a great satisfaction for all of us who think that chemistry in its broadest sense, has a major role to play in the development of countries now and in the future. In the name of the IOCD, I wish to thank most warmly all those who have made it possible: the Organizing and Scientific Committees, the speakers and the participants as well as the sponsors. They have done a splendid job, with dedication and generosity. We all look forward to the future meetings which will pursue the action so masterfully initiated.

Jean-Marie Lehn, Nobel Laureate,

Professor of Chemistry at the College de France, Paris and at the University of Strasbourg, France.
Chairman of the International Organization for Chemical Sciences in Development

Foreword

Preface

The first symposium of the Working Group on Plant Chemistry of the International Organization for Chemical Sciences in Development (IOCD) was held at Victoria Falls, Zimbabwe from 25 to 28 February 1996. The aim of the symposium was to bring together scientists from African countries and other parts of the world in order to provide a forum for the exchange of ideas and give an update on the current status of research into African medicinal plants. The motivation for holding the meeting in Zimbabwe was to enable as many researchers as possible from neighbouring countries to attend. That this aim was achieved can be concluded from the fact that representatives from a total of 25 countries were present, including 13 African countries. A unique opportunity was thus provided for communication and the transmission of know-how between developing and industrialized countries. A platform was available for Africans themselves to present results of their own research. This volume is a compilation of contributions from the speakers giving plenary lectures and from those presenting short oral communications.

African medicinal plants have long provided important sources of healing drugs to local populations . The information obtained from ethnomedicine is now being put on a scientific basis and it is very important to investigate the pharmacological and phytochemical aspects of different preparations from vegetable sources. This type of research requires an interdisciplinary approach in which botanists, ethnobotanists, chemists, phytochemists, biologists and pharmacologists work together to unravel the mysteries. There is a certain urgency associated with this work since plant habitats are rapidly being changed or destroyed and it is essential to study as much of the disappearing plant material as possible before the opportunities are lost. Another factor is that with the decreasing number of traditional healers, the cumulation of their valuable know-how is progressively diminishing. This source of priceless data also needs to be exploited before it is too late.

This book provides articles on the evaluation of traditional remedies and also on the botany and chemotaxonomy of the medicinal plants involved. There are sections on the strategies employed for their study and details of isolation procedures for active compounds. The latest results concerning different biological activities of African plants and their constituents - anticancer, antiviral, fungicidal and antiparasitic activities, for example - are presented.

The authors would like to express their gratitude to the speakers who have contributed to this volume. Thanks are also due to the sponsors of the meeting and to the local organizers in Zimbabwe, with special emphasis on the University of Zimbabwe

Finally, it is hoped that this book will provide the stimulus for new research on the many African plants that remain to be studied and provide a catalyst for useful collaboration and cooperation at all levels.

Lausanne and *Harare,* September 1996

K. Hostettmann
F. Chinyanganya
M. Maillard
J.-L. Wolfender

Sponsorship

The Organizing Committee gratefully acknowledges the following companies and institutions for their generous support during the IOCD-Symposium on "Chemistry, Biological and Pharmacological Properties of African Medicinal Plants" held at Victoria Falls, Zimbabwe, 25th to 28th September, 1996. These contributions allowed the participation of a large number of African scientists.

Ciba-Geigy AG, Basle, Switzerland
The Commonwealth Foundation, London, United Kingdom
Galenica SA, Ecublens, Switzerland
Groupe LVMH Moët Hennessy-Louis Vuitton, Nanterre, France
John Wiley & Sons Ltd., Chichester, UK
OFAC SA, Geneva, Switzerland
L'Oréal, Clichy, France
Pharmaton SA, Lugano, Switzerland
Sandoz Pharma AG, Basle, Switzerland
Schaper & Brümmer GmbH, Salzgitter, Germany
Shaman Pharmaceuticals, South San Francisco, USA
Swiss Academy of Sciences, Bern, Switzerland
Swiss Agency for Development and Cooperation (SDC), Bern, Switzerland
The Third World Academy of Sciences, Trieste, Italy
US National Academy of Sciences and the American Chemical Society, Washington, USA
Varian International AG, Zug, Switzerland
Zeller und Söhne AG, Romanshorn, Switzerland
Zyma SA, Nyon, Switzerland

First published in 1996 by
University of Zimbabwe Publications
P.O. Box MP 203
Mount Pleasant
Harare
Zimbabwe

ISBN 0-908307-59-4

Cover photos.
African traditional healer and *Harpagophytum procumbens* (Pedaliaceae)
© K. Hostettmann

Printed by Mazongororo Paper Converters Pvt. Ltd., Harare

Contents

Contents

Contributors

B.M. Abegaz: Department of Chemistry, University of Botswana, Private Bag 0022, Gaborone, Botswana.

M.A. Abrahams: Department of Pharmacology, University of Cape Town, Observatory 7925, South Africa.

E.K. Adesogan: University of Ibadan, Ibadan, Nigeria.

C. Albrecht: Department of Pharmacology, Faculty of Medicine, University of Stellenbosch, Tygerberg 7505, South Africa.

C.F. Albrecht: Laboratory for Ecological Chemistry, University of Stellenbosch, Stellenbosch 7600, South Africa.

G. Alemayehu: Department of Chemistry, Addis Ababa University, P.O. Box 1176, Addis Ababa, Ethiopia.

O.O.G. Amusan: Chemisitry Department, University of Swaziland, Private Bag 4, Kwaluzeni, Swaziland.

C.C. Appleton: Department of Zoology and Entomology, University of Natal, Pietermaritzburg 3209, South Africa.

J.T. Arnason: Institutes of Chemistry and Biology, University of Ottawa, Ottawa, Ontario, Canada K1N 6N5.

D.V.C. Awang: MediPlant, P.O. Box 8693, Ottawa, Ontario, Canada K1J 3G1.

M.R. Boyd: Developmental Therapeutics Program, Division of Cancer Treatment, Diagnosis and Centers, National Cancer Institute, Bethesda, Maryland 20892, USA.

T. Brackenbury: Department of Zoology and Entomology, University of Natal, Pietermaritzburg 3201, South Africa.

G. Bringmann: Institut für Organische Chemie, Am Hubland, D-97074 Würzburg, Germany.

B.V. Burger: Laboratory for Ecological Chemistry, University of Stellenbosch, Stellenbosch 7600, South Africa.

I. Burger: Laboratory for Ecological Chemistry, University of Stellenbosch, Stellenbosch 7600, South Africa.

W.E. Campbell: Department of Chemistry, University of Cape Town, Observatory 7925, South Africa.

M.A. Christini: Office of Technology Development, National Cancer Institute, Bethesda, Maryland 20892, USA.

T. Clark: Department of Chemistry and Chemical Technology, University of Natal, Pietermaritzburg 3209, South Africa.

G.M. Cragg: Developmental Therapeutics Program, Division of Cancer Treatment, Diagnosis and Centers, National Cancer Institute, Bethesda, Maryland 20892, USA.

C.F. Curtis: London School of Tropical Medicine and Hygiene, Keppel Street, London, U.K.

E. Dagne: Department of Chemistry, Addis Ababa University, P.O. Box 30270, Addis Ababa, Ethiopia.

D. Diallo: Département de Médecine Traditionnelle, Institut National de Recherche en Santé Publique, B.P. 1746, Bamako, Mali.

S.E. Drewes: Department of Chemistry and Chemical Technology, University of Natal, Pietermaritzburg 3209, South Africa.

T. Durst: Institutes of Chemistry and Biology, University of Ottawa, Ottawa, Ontario, Canada K1N 6N5.

P.I. Folb: Department of Pharmacology, University of Cape Town, Observatory 7925, South Africa.

D.W. Gammon: Department of Chemistry, University of Cape Town, Observatory 7925, South Africa.

M. Gbeassor: Faculté des Sciences, Université du Bénin, Lomé, Togo.

D. Guénard: Institut de Chimie des Substances Naturelles du C.N.R.S., F-91198 Gif-sur-Yvette Cedex, France.

F. Guéritte-Voegelein: Institut de Chimie des Substances Naturelles du C.N.R.S., F-91198 Gif-sur-Yvette Cedex, France.

M. Horn: Department of Chemistry and Chemical Technology, University of Natal, Pietermaritzburg 3209, South Africa.

K. Hostettmann: Institut de Pharmacognosie et Phytochimie, BEP, Université de Lausanne, CH-1015 Lausanne, Switzerland.

B. Hvemm: Center for Development and the Environment, University of Oslo, Oslo, Norway.

L Kayonga: Department of Chemistry and Chemical Technology, University of Natal, Pietermaritzburg 3209, South Africa.

T. Kebede: Department of Chemistry, Addis Ababa University, P.O. Box 1176, Addis Ababa, Ethiopia.

H.K. Koumaglo: Faculté des Sciences, Université du Bénin, Lomé, Togo.

N. Lukwa: Blair Research Laboratory, P.O. Box CY573, Harare, Zimbabwe.

S. MacKinnon: Institutes of Chemistry and Biology, University of Ottawa, Ottawa, Ontario, Canada K1N 6N5.

D. Mahajan: Department of Chemistry, University of Botswana, Private Bag 0022, Gaborone, Botswana.

M. Maillard: Institut de Pharmacognosie et Phytochimie, BEP, Université de Lausanne, CH-1015 Lausanne, Switzerland.

J.M. Makinde: Chemsitry Department University of Swaziland, Private Bag, Kwazuleni, Swaziland.

A. Marston: Institut de Pharmacognosie et Phytochimie, BEP, Université de Lausanne, CH-1015 Lausanne, Switzerland.

C. Masedza: Blair Research Laboratory, P.O. Box CY573, Harare, Zimbabwe.

T.D. Mays: Office of Technology Development, National Cancer Institute, Bethesda, Maryland 20892, USA.

K.D. Mazan: Office of Technology Development, National Cancer Institute, Bethesda, Maryland 20892, USA.

J.D. Msonthi: Chemisitry Department, University of Swaziland, Private Bag 4, Kwaluzeni, Swaziland.

D.A. Mulholland: Natural Products Research Group, Department of Chemistry, University of Natal, Private Bag X10, Dalbridge, 4014 Durban, South Africa.

G.L. Mwaiko: NIMR, P.O. Box 4, Amani, Tanzania.

M.M. Nindi: Department of Chemistry, University of Botswana, Private Bag 0022, Gaborone, Botswana.

N.H.H. Nkunya: Department of Chemistry, University of Dar es Salaam, P.O. Box 35061, Dar es Salaam, Tanzania.

N.Z. Nyazema: Department of Clinical Pharmacology, P.O. Box A178 Avondale, Harare, Zimbabwe.

A.A. Onayade: Department of Community Health, Obafemi Awolowo University, Ile-Ife, Nigeria.

O.A. Onayade: Department of Pharmacognosy, Obafemi Awolowo University, Ile-Ife, Nigeria.

B.S. Paulsen: Institute of Pharmacy, University of Oslo, Oslo, Norway.

P. Potier: Institut de Chimie des Substances Naturelles du C.N.R.S., F-91198 Gif-sur-Yvette Cedex, France.

L. Van Puyvelde: Department of Organic Chemsitry, Faculty of Agricultural and Applied Biological Sciences, University of Gent, Coupure links, B-9000 Gent, Belgium.

S. Rodriguez: Institut de Pharmacognosie et Phytochimie, BEP, Université de Lausanne, CH-1015 Lausanne, Switzerland.

C.B. Rogers: Department of Chemistry, University of Durban-Westville, Private Bag X54001, Durban, South Africa.

E.A. Sausville: Developmental Therapeutics Program, Division of Cancer Treatment, Diagnosis and Centers, National Cancer Institute, Bethesda, Maryland 20892, USA.

B.M. Sehlapelo: Department of Chemistry, University of Cape Town, Observatory 7925, South Africa.

P.J. Smith: Department of Pharmacology, University of Cape Town, Observatory 7925, South Africa.

A. Sofowora: Department of Pharmacognosy, Obafemi Awolowo University, Ile-Ife, Nigeria.

C. Spatafora: Dipartimento di Scienze Chimiche, Università di Catania, Viale A. Doria 6, I-95125 Catania, Italy.

H.S.C. Spies: Laboratory for Ecological Chemistry, University of Stellenbosch, Stellenbosch 7600, South Africa.

C. Tringali: Dipartimento di Scienze Chimiche, Università di Catania, Viale A. Doria 6, I-95125 Catania, Italy.

L. Verotta: Dipartimento di Chimica Organica e Industriale, Universita degli Studi di Milano, via Venezian 21, I-20133 Milano, Italy.

P.G. Waterman: Strathclyde Institute for Drug Research, and The Phytochemistry Research Laboratories, Department of Pharmaceutical Sciences, University of Strathclyde, Glasgow G1 1XW, Scotland, U.K.

J.-L. Wolfender: Institut de Pharmacognosie et Phytochimie, BEP, Université de Lausanne, CH-1015 Lausanne, Switzerland.

1. African plants as sources of pharmacologically exciting biaryl and quaterary alkaloids

G. BRINGMANN

Institut für Organische Chemie, Am Hubland, D-97074 Würzburg, Germany

Introduction

African plants, in particular medical plants, constitute a rich, but still largely untapped pool of natural products. Two plant families that are of greatest interest to us, are the Ancistrocladaceae and the Dioncophyllaceae. These are very small but highly productive families of tropical lianas. They produce a unique class of natural products, the naphthylisoquinoline alkaloids (Bringmann and Pokorny 1995), such as dioncophylline A (**1**) and ancistrocladine (**2**). These compounds (Fig. 1.1) consist of a naphthalene and an isoquinoline moiety, linked by a biaryl axis, which, due to the steric demand of the two halves, gives rise to restricted rotation and thus to the phenomenon of atropisomerism and axial chirality. Further characteristic structural features are the methyl group at C-3 and the oxygen function at C-8, but these alkaloids are also interesting for their biogenetic origin, their pharmacological activities, and the taxonomic position of the plants that produce these compounds.

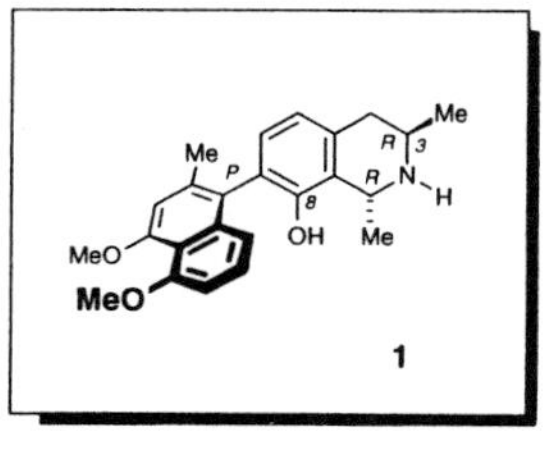

Dioncophylline A
from *Triphyophyllum peltatum*
(Dioncophyllaceae)

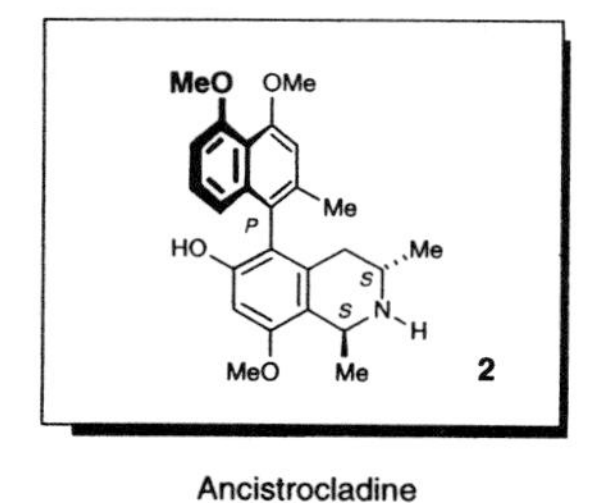

Ancistrocladine
from *Ancistrocladus heyneanus*
(Ancistrocladaceae)

Fig. 1.1. Two typical naphthylisoquinoline alkaloids from Ancistrocladaceae and Dioncophyllaceae.

'Normal', *i.e.* monomeric naphthylisoquinolines

Isolation, structure elucidation, and total synthesis

For these reasons, we have intensively been investigating the naphthylisoquinolines under most different aspects. For the isolation and structure elucidation, in particular for the unambiguous attribution of the absolute configuration at the stereocenters and the stereogenic axes, we apply a broad spectrum of efficient methods (see Fig. 1.2), among them the fruitful interplay between theoretical and experimental CD spectroscopy (Bringmann *et al.* 1993a), a novel oxidative degradation procedure to give simple known amino acids (Bringmann *et al.* 1991a; Bringmann *et al.*, *Phytochemistry,* in press), and, finally, the first total synthesis (Bringmann *et al.* 1986; Bringmann and Jansen 1989, 1991; Bringmann and Reuscher 1989a, 1989b; Bringmann *et al.* 1990a, 1990b, 1992a, 1995) of these natural products, as developed in our lab.

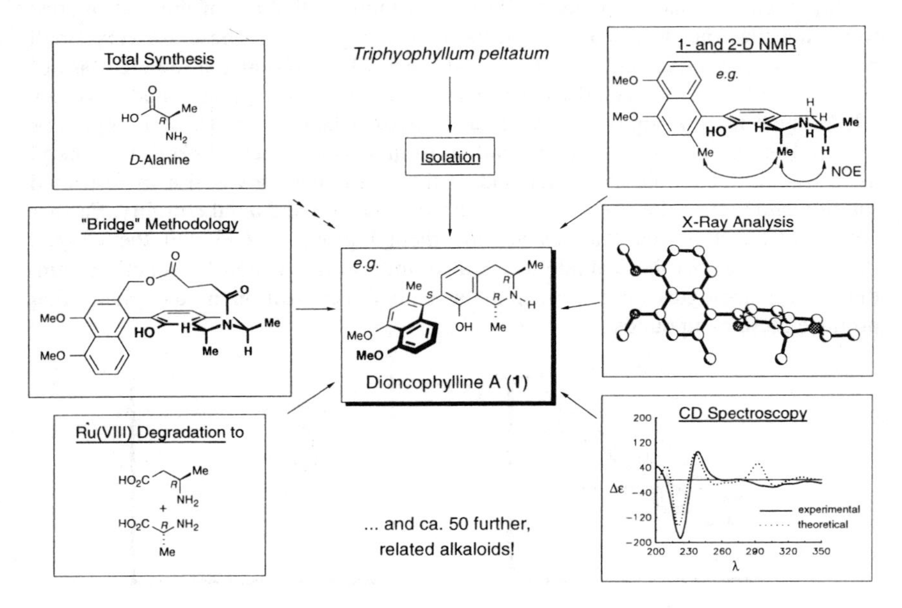

Fig. 1.2. Isolation and structure elucidation of naphthylisoquinoline alkaloids.

By these methods, we have established the structures of dioncophylline A (**1**) and a whole series of more than 50 further related compounds (Bringmann *et al.* 1991b, 1992b, 1992c, 1992d, 1993b; Bringmann and Pokorny 1995) - Fig. 1.3 shows a small selection of some of those alkaloids that are produced by the West-African species *Ancistrocladus abbreviatus*, among them dioncophylline A (**1**), ancistrocladine (**2**) and another interesting alkaloid, named ancistrobrevine B (**3**), which will be seen again below (see Fig. 1.9).

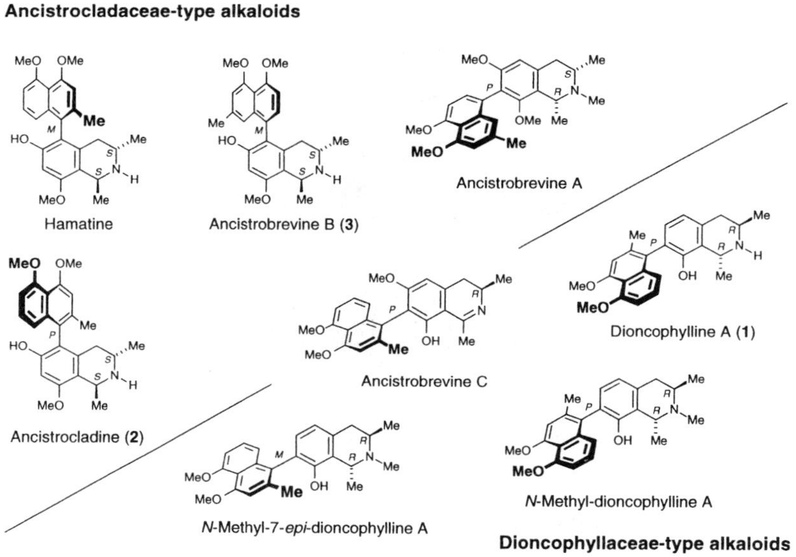

Fig. 1.3. Alkaloids from *Ancistrocladus abbreviatus* (West Africa) - a selection.

For the first total synthesis of these natural products, we have developed a novel procedure for the regio- and stereoselective construction of biaryl axes (Bringmann *et al.* 1990b, 1995) and have demonstrated its broad applicability in the preparation of dioncophylline A (**1**) (Bringmann *et al.* 1990a; Bringmann and Jansen 1991), ancistrocladine (**2**) (Bringmann *et al.* 1986; Bringmann and Jansen 1989), and a whole series of further alkaloids (Bringmann and Reuscher 1989a, 1989b; Bringmann *et al.* 1990b, 1992a, 1995).

Naphthylisoquinoline producing plants and cell cultures

Finally we have achieved the first cultivation of *Ancistrocladus* plants, initially starting with the Indian species *A. heyneanus* (Bringmann *et al.* 1991c, 1993c), from its typical seeds, which are nuts with five wing-like extended sepals (Fig. 1.4). Whereas even in India, its country of origin, this plant had not been cultivated before, we are now capable of growing prosperous plants of this species in the green house, they form their typical hooks, from which the name *Ancistrocladus* (= hooked branch) is derived. We have even obtained blooms and fruits in Würzburg. Similar cultivation success was achieved (Bringmann and Pokorny 1995; Bringmann *et al.* 1993c, 1994a) with the related African species *A. abbreviatus* (for its chemical constituents, see Fig. 1.3), *A. barteri* (likewise from West Africa), *A. robertsoniorum* (from East Africa), and *Triphyophyllum peltatum,* an interesting plant with hooked leaves and sometimes carnivorous organs.

Fig. 1.4. A *ca.* 6 weeks old seadling of *A. heyneanus.*

Biosynthetic origin of the alkaloids

This first cultivation of *Ancistrocladus* and *Triphyophyllum* plants is an important precondition for biosynthetic investigations. The unusual structures of the alkaloids as well as model reactions (Bringmann 1985a, 1985b) and feeding experiments in our group (Bringmann and Pokorny 1995a; Bringmann *et al.* 1991d) suggest that these are the first tetrahydroisoquinoline alkaloids that are not formed from aromatic aminoacids, but from acetate units, *via* ß-polyketides of type **4**, which even act as precursors for *both* molecular halves - for the isoquinoline **5** (resp. **7**) *and* the naphthalene **6**!

Naphthylisoquinolines as biologically active compounds and lead structures

The biological activities of the alkaloids are remarkable, too: Thus, together with the BASF AG, we found distinct fungicidal activities with relevance to crop protection (Bringmann *et al.* 1993e). Already extracts of *Triphophyllum peltatum* are highly active against important plant-pathogenic fungi. Among the pure isolated compounds tested, dioncophylline B (**8**) proved to be the most efficient one. Other active alkaloids are dioncophylline A (**1**), dioncopeltine A (**9**), and dioncophylline C (**10**). This fungicidal activity is most likely to give the plant a specific advantage in the humid rain forest.

The same protective effect can be expected from a strong antifeedant and growth retarding activity against insects, tested on the polyphagous generalist *Spodoptora littoralis* - a well-established (largely toxin-resistant) herbivore model

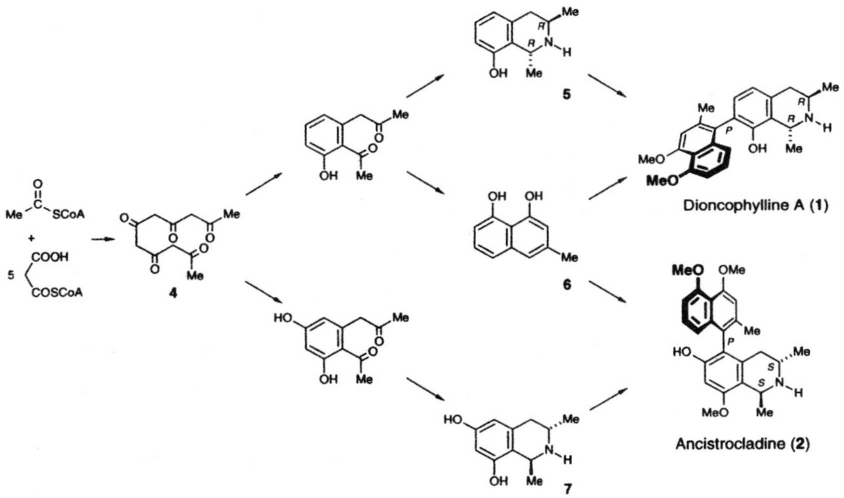

Fig. 1.5. Proposed biosynthetic origin of acetogenic isoquinoline alkaloids.

system (Bringmann *et al.* 1992e). Antifeedant and growth retarding activities towards the larvae were investigated by adding the test substance to the diet, preferentially at concentrations in which the alkaloids occur in the plants. The by far highest activity was thus found for dioncophylline A (**1**), but other alkaloids such as ancistrocladidine (**11**), ancistrocladinine (**12**) were also quite active.

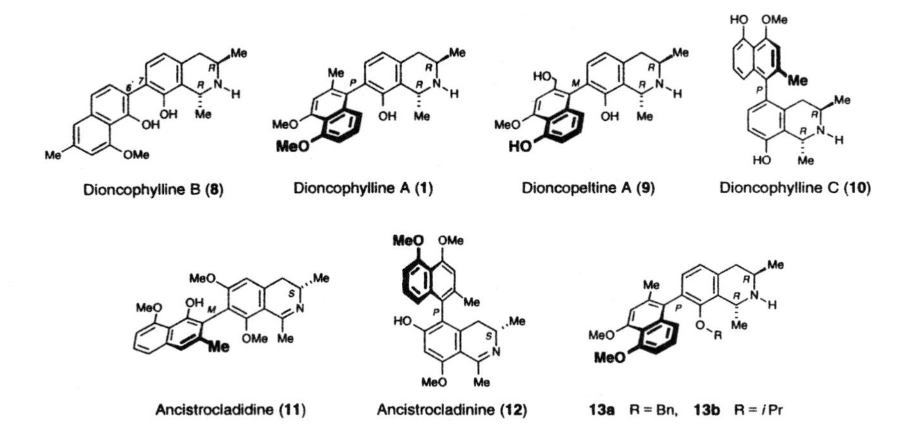

Fig. 1.6. Naphthylisoquinoline alkaloids with fungicidal, insect antifeedant and growth retarding, resp. molluscicidal activity.

By chemical modification of the dioncophylline A molecule, we got first in sight into the structural requirements for the growth retardation activity and even found derivatives like **13a** that are more active than the natural product dioncophylline A, by a factor of 4-5.

Of even higher importance is the molluscicidal activity of our alkaloids against the tropical snail *Biomphalaria glabrata*, the intermediate host of the tropical disease bilharzia (schistosomiasis), which is so widespread in Africa. Besides the direct chemotherapy of bilharzia in the patients, another strategy is to kill the snails in order to disrupt the parasitical cycle. Together with Professor Hostettmann's group in Lausanne we found that extracts of naphthylisoquinoline-containing plants and some of the alkaloids themselves show good anti-snail activities, in particular again dioncophylline A (**1**), with an $LD_{100(24\ h)}$ of 20 ppm (Bringmann *et al.*, *Planta med.*, in press). Again, by systematic chemical modification of the molecule, more active derivatives like **13b** were found. This work is still in progress.

One of the presently most exciting biological properties of our alkaloids is their antimalarial activity. Approximately a third of the world's population lives in malaria-endemic areas, and more than 2-3 million people die of malaria each year. Many *Plasmodium* strains, especially of *P. falciparum*, have become resistant to classical antimalarial drugs such as chloroquine. Therefore and due to the rapid spread of the disease, the search for *new* antimalarial agents is an urgent task. Some *Ancistrocladus* species are used in traditional medicine for the treatment of malaria. Their extracts indeed exhibit high *in vitro* activities against *P. falciparum,* also against chloroquine-resistant strains (François *et al.* 1994). Among the pure isolated alkaloids, the by far most active ones were dioncophylline C (**10**) and dioncopeltine A (**9**), with excellent IC_{50} values (Fig. 1.7). In this case, none of the more than 100 further natural or modified analogs that we have prepared and tested, was more active than dioncophylline C (**10**), itself (François *et al.* 1996).

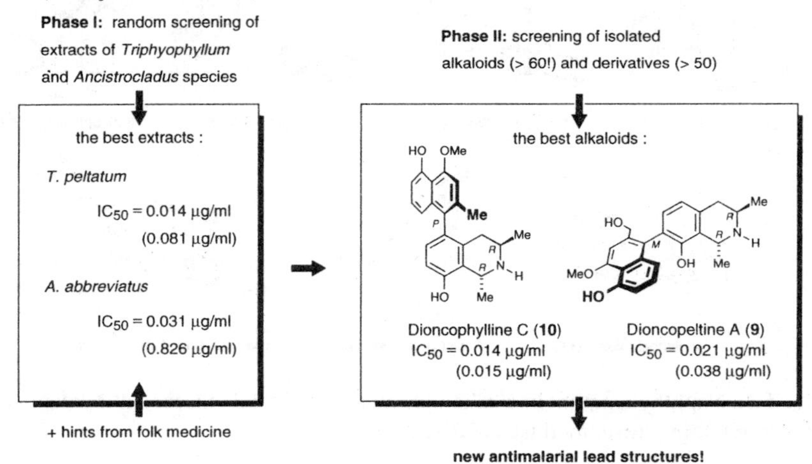

Fig. 1.7. Naphthylisoquinoline alkaloids are active against *Plasmodium falciparum* (and *P. berghei*) *in vitro.*

Similar results were obtained for the related parasite *P. berghei* (François *et al.* 1995; François *et al., Intern. J. of Pharmacogn.*, in press) (see the values in parantheses). The availability of this rodent malarial system furthermore allowed first investigations on the *in vivo* antimalarial activity of our alkaloids (Bringmann and Pokorny 1995) (see Fig. 1.8). Thus, OF1 mice were inoculated with *P. berghei* erythrocytic forms and then treated with dioncophylline C (**10**). Their parasitaemia was found to be reduced to 0% from day 4 on. Even after 3 months, the treated animals were still alive and looked absolutely normal, whereas all of the control animals had died soon after the infection. Similar, but weaker, *in vivo* activities were found for dioncopeltine A (**9**) and dioncophylline B (**8**).

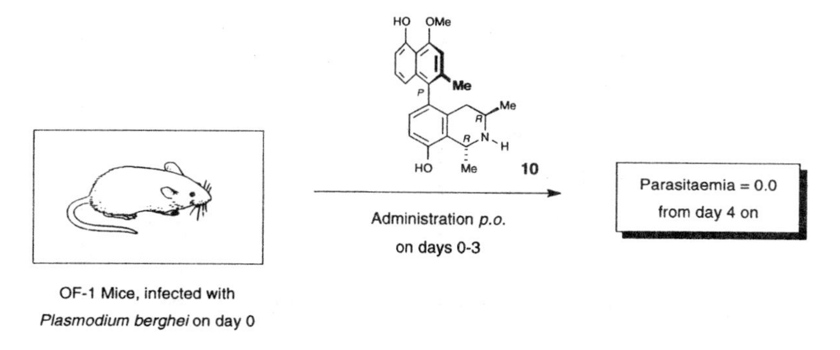

Fig. 1.8. Antimalarial activity also *in vivo*!

In particular these antimalarial and antibilharzia activities of the alkaloids are of greatest importance for our work, since such tropical diseases take a high toll of lives in those countries where the plants grow - a further motivation for us to focus on these fields of indication, predominantly.

Michellamines - natural dimers of naphthylisoquinoline alkaloids

Discovery, constitution, and anti-HIV activity

Very recently (see chapter 3), *Ancistrocladus* alkaloids have even become candidates as possible drugs against the immunodeficiency disease AIDS - one of the great medicinal and social challenges of our time. Within the broad NCI anti-HIV screening program, a particularly high activity was found for a Cameroonian *Ancistrocladus* species. The extract of this vine showed one of the highest activities out of more than 70,000 tested samples! Meanwhile it turned out that the active plant was *not*, as initially assumed, the long-known and well-investigated (cf. Fig. 1.3) species *A. abbreviatus*, but a 'new' *Ancistrocladus* species that had not yet been described in science - it was even unknown to folk medicine! It has meanwhile been named *Ancistrocladus korupensis* (Thomas and Gereau 1993), after the Parc National de Korup in Cameroon, where it was detected.

The antiviral constituents (*i.e.* the active principle) represent an entirely novel type of quateraryl alkaloids (see Fig. 1.9), called michellamines (Manfredi *et al.* 1991; Boyd *et al.* 1994), with 4 aromatic systems and 3 biaryl axes, they are highly polar with 6 free phenolic functions and 2 amino groups. They are constitutionally symmetric dimers **14** of a 5,8'-coupled naphthylisoquinoline. The formal monomer resembles ancistrobrevine B (**3**), which was among those alkaloids (see Fig. 1.3) that had previously been isolated from the 'true' species *Ancistrocladus abbreviatus* (Bringmann *et al.* 1991e, 1992b).

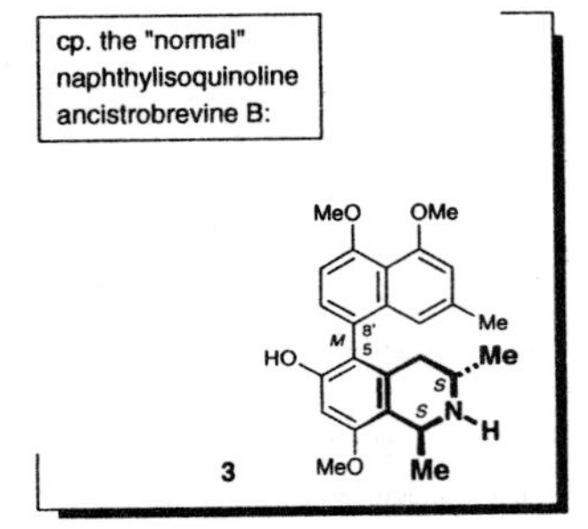

Fig. 1.9. Constitution of michellamines A-C from *A. korupensis,* and structure of the related monomeric alkaloid ancistrobrevine B (**3**) from *A. abbreviatus.*

Michellamines, in particular michellamine B, are active against various HIV strains in different host cells, against HIV-1 *and* 2, even against drug-resistant strains, and they act against both early and late stages of the viral life cycle, by inhibiting the reverse transcriptase and the cell-cell fusion process (McMahon *et al.* 1995).

The high antiviral activity of these novel compounds and their intriguing structures led us to contact Dr. Boyd and his NCI group in order to build up a collaboration on dimeric naphthylisoquinoline alkaloids, hopefully within a triangle 'NCI-Yaoundé-Würzburg', thus combining the excellent screening know-how of the NCI and the chemical and botanical expertise of the Cameroonian group with our large experience with *Ancistrocladus* plants and their *monomeric* alkaloids (see above) - a unique chance of rapidly exploring the potential of this novel type of compounds!

Besides the constitution **14** of these natural products, which was elucidated by Dr. Boyd's NCI group (Manfredi *et al.* 1991), it is particularly the stereochemical features of the michellamines that are interesting and challenging at the same time: There are not only two, but four stereocenters, and even three biaryl axes (Fig. 1.9), two of which are stereogenic (*), whereas the central axis between the two constitutionally identical molecular halves is configuratively unstable (°).

While the relative configuration at centers and axes was elucidated by the NCI group, the absolute configuration of michellamines, *e.g.* for michellamine A (see Fig. 1.10), was established in Würzburg (Bringmann *et al.* 1993d; Boyd *et al.* 1994). For the centers, this was done by our oxidative degradation reaction, which neatly cuts out C-1 and C-3 by transforming them into alanine and 2-aminobutyric acid, hence simple and easy-to-analyze amino acids. From the *R*-2-aminobutyric acid [(*R*)-**15**] and the D-alanine [(*R*)-**16**] formed, all the stereocenters of michellamine A clearly must be *R*-configured; given the known relative configuration at centers *vs.* axes, it comes out that **14a** must be the full stereostructure of michellamine A, with *P*-configuration at both axes.

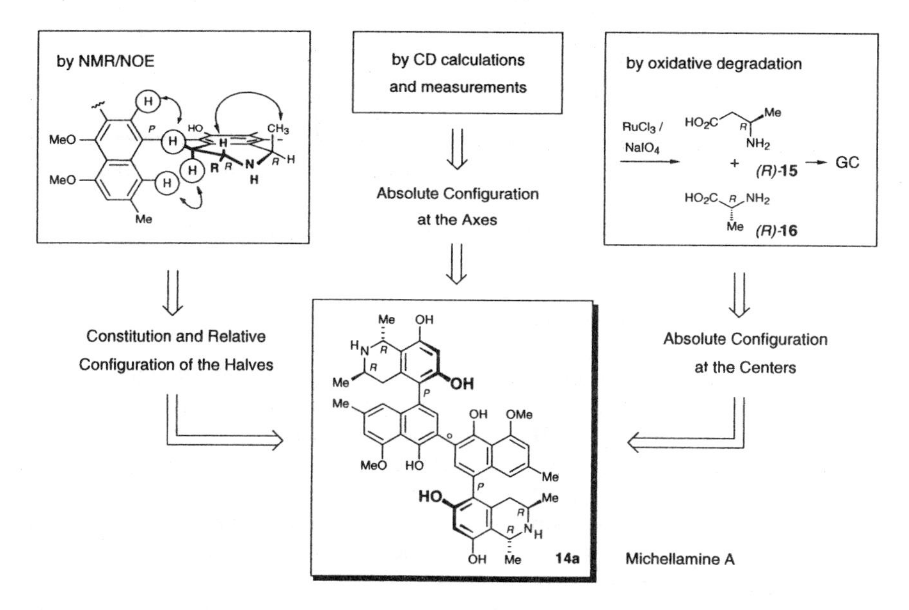

Fig. 1.10. The 3-dimensional stereostructure **14a** of michellamine A.

This axial *P*-configuration was independently confirmed by a specific method established and further developed in our group, the fruitful interplay between experimental and theoretical CD spectroscopy (Bringmann *et al.* 1994b): we calculate the CD spectrum for a particular stereostructure (*e.g.* for **14a**) and compare it with the experimental one - a most efficient novel procedure in particular for entirely novel structural types, which here fully confirms our attribution.

Similarly, again together with the NCI group, we have established the absolute stereostructures of michellamines B (**14b**) and C (**14c**). As can be seen, the michellamines are all identical concerning their constitution and their stereocenters: They are *R*-configured all over and thus differ only by their *axial* chirality, forming a complete set of atropo-diastereomers, with "*P,P*", "*M,P*", and

"*M,M*" configurations. Hence, only michellamines A and C are truly dimeric and thus C_2-symmetric molecules, whereas michellamine B with its heterochiral axes (and thus diastereomorphous 'halves') is a constitutionally symmetric coupling product of two atropodiastereomeric naphthylisoquinoline alkaloids.

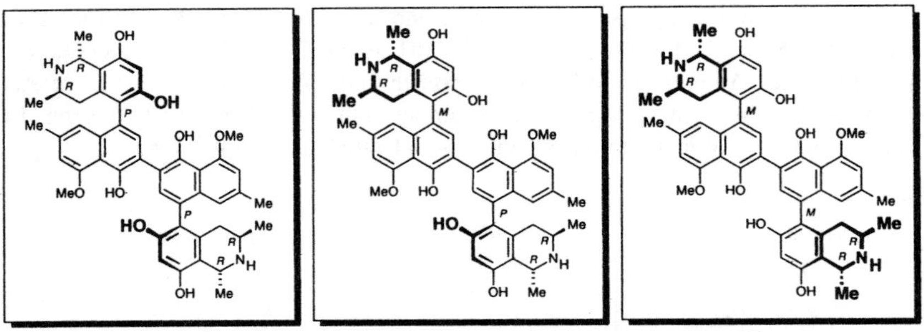

Michellamine A (**14a**) Michellamine B (**14b**) Michellamine C (**14c**)

Fig. 1.11. The absolute configurations of michellamines A, B, and C.

The total synthesis of michellamines

Having acquired the full information on the 3-dimensional structure of the antiviral michellamines, we could now start to develop a first total synthesis of these attractive synthetic goals. By retrosynthetically dissecting the molecule at the axes, it seemed suggesting to start the synthesis from the two aromatic building blocks **18** and **19** (Fig. 1.12), for which two principal synthetic strategies are imaginable: One might first couple the isoquinoline with the naphthalene part and thus build up the two outer axes and then the inner one, or, *vice versa,* first the inner axis and then the outer ones.

Both concepts have their advantages: For *Pathway 1*, one would first build up monomeric naphthylisoquinolines like **17a**, which are also natural products: they likewise occur in *A. korupensis* and were thus named korupensamines (Hallock *et al.* 1994). Like some other monomeric naphthylisoquinolines, they show high antimalarial activities *in vitro* and *in vivo* - but, as *all* the other monomeric naphthylisoquinolines, they are devoid of any anti-HIV activity! This underlines the necessity of synthesizing entire *dimeric* naphthylisoquinoline alkaloids if one wants to get antiviral activity. Consequently, these korupensamines would then have to be dimerized - possibly biomimetically, according to the principle of oxidative phenolic coupling. As an important intermediate goal, one would thus attain a first synthesis of these korupensamines, and might, *vice versa*, profit from the additional availability of these precursors from the plant material. This *biomimetic* pathway was consequently realized first (Bringmann *et al.* 1994c, 1994d; Bringmann *et al., Liebigs Ann. Chem.*, in press).

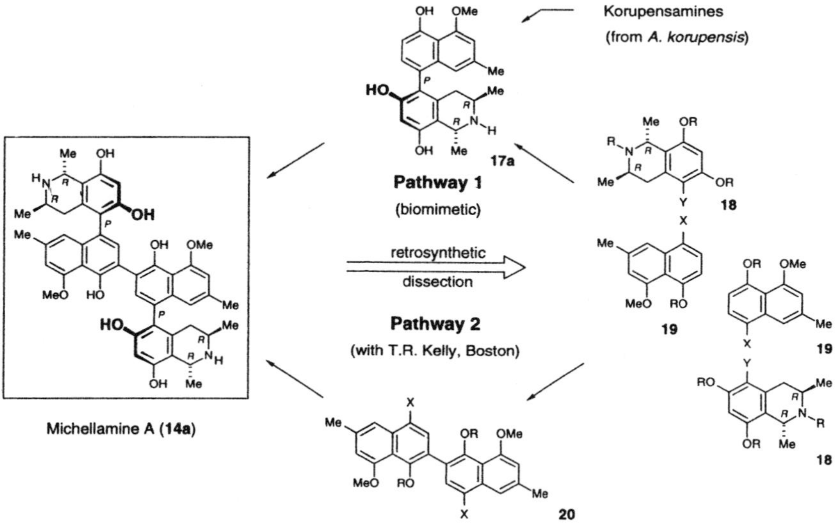

Fig. 1.12. Alternative strategies for the synthesis of michellamines.

In this paper, only the key steps of these total syntheses, the construction of the biaryl axes, will be presented. Linking the two molecular halves of **17** (see Fig. 1.13) was achieved by coupling the tin-activated naphthalene **21** with the brominated isoquinoline building block **22** with palladium catalysis, to give the 5,8'-coupling type. The separation of the two resulting atropisomers **23a** and **b** was done after setting free the authentic natural products, korupensamines A (**17a**) and B (**17b**), which proved to be identical with natural material in all spectroscopic, physical, chromatographic, and biological properties.

We had thus reached our first synthetic goal, the korupensamines, and could now start to build up the inner axis. From a direct oxidative dimerization of **17a** or **b**, however, we expected serious problems due to the large number of free phenolic oxygen and amino functions of such polar and sensitive compounds. It therefore seemed suggesting to predetermine selectivity in the coupling step on an earlier level, by specifically 'sealing' all these functionalities except for the crucial oxygen function adjacent to the scheduled coupling site (see Fig. 1.14). This strategy worked extremely well, by *N*-formylation and subsequent specific *O*-acetylation of the non-chelated OH groups, exclusively. The specifically monophenolic korupensamine A derivative **24a** was then submitted to oxidative coupling conditions. The by far best results were achieved with silver oxide (Bringmann *et al.* 1994d; Bringmann *et al.*, *Liebigs Ann. Chem.*, in press), which gave the nicely violet-colored dione **25a**. From here on, the rest was simple and straightforward: Reduction and cleavage of all the six protective groups in a single step gave michellamine A (**14a**), which, in all its properties, was found to be fully

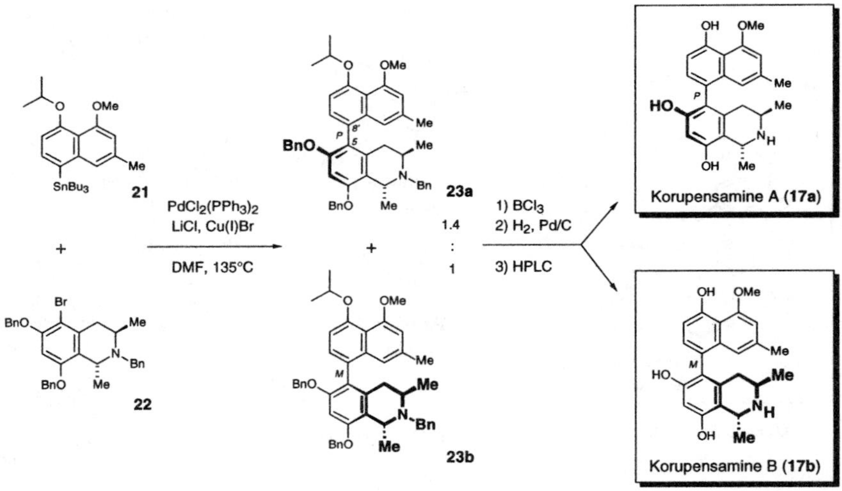

Fig. 1.13. Construction of the "outer" biaryl axis: first total synthesis of korupensamines A and B.

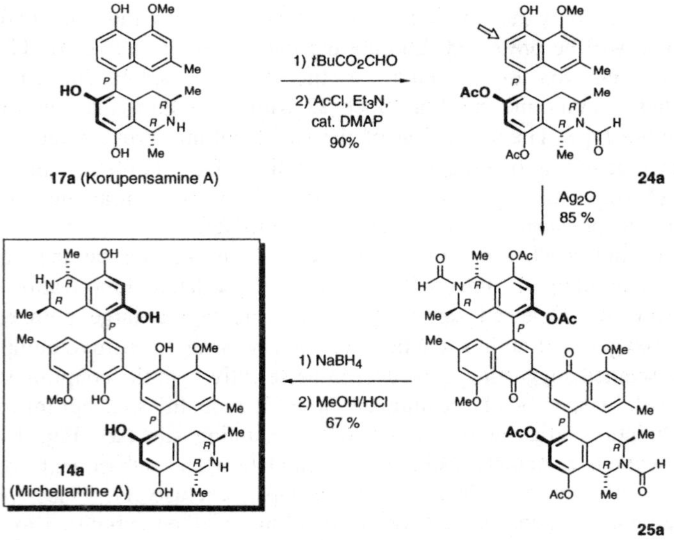

Fig. 1.14. Biomimetic oxidative dimerization of korupensamine A: first partial (and total) synthesis of michellamine A (**14a**).

identical with natural michellamine A from *A. korupensis*. Similarly, the other michellamines (**14b**, **14c**) have been prepared, including even more efficient protective strategies (*via O,N*-benzylated intermediates) and enzymic methods, using an enzyme from *Ancistrocladus* plants for the oxidative coupling of authentic, non-protected korupensamines (Bringmann *et al.*, submitted for publication).

This biomimetic "dimerization" of a naphthylisoquinoline alkaloid completes the very first total synthesis of michellamines - against strong international competition. Simultaneously it fully confirms our proposed stereostructure for the michellamine, in all details, and demonstrates how efficient and simple such biomimetic syntheses of natural products can be, as realized in this *pathway no. 1*.

But also *pathway no. 2*, which we have realized together with T.R. Kelly's group in Boston (Kelly *et al.* 1994), worked tremendously well, too (see Fig. 1.15): By coupling the central binaphthalene fragment **20** in the form of the bis-*O*-triflate **20a**, *i.a.* with the inner biaryl axis already present, the michellamine framework is very rationally built up in a single step, by reaction with two equivalents of the boron-activated isoquinoline building block **26**, with simultaneous formation of the two outer axes. In a high 74 % yield, we now directly obtain a mixture of the still protected atropo-diastereomeric quateraryls **27**, from which the michellamines can easily be set free by deprotection.

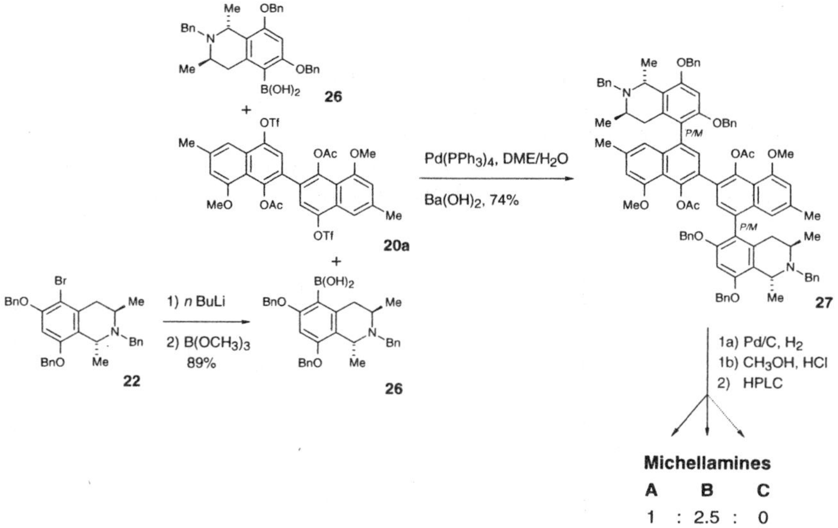

Fig. 1.15. Successful double coupling: completion of the second michellamine synthesis.

To our great surprise, we do not find the three imaginable michellamines A-C that we statistically would have expected in a 1:2:1 ratio, but only michellamines A and B, whereas michellamine C is not present in the mixture. Apparently, one of the two coupling steps is extremely stereoselective. This is all the more remarkable as also in the plants, one finds a ratio of about 1:2 for michellamines A (**14a**) and B (**14b**), whereas it seems that michellamine C (**14c**) is *not* a natural product, but just an artifact that is formed under too harsh isolation conditions.

Thus, a second most efficient synthetic access to this structural type of michellamines - which appears so complex at first sight - had been found. Meanwhile, further michellamine syntheses have been published (Hoye *et al.* 1994; Hobbs *et al.* 1996).

By these first total syntheses of michellamines we do not intend to compete with the biological production of michellamines, which has already begun by cultivating *A. korupensis* in Cameroon. The value of our syntheses is rather the confirmation of the absolute stereostructure as established in our lab, and furthermore, our synthetic work allows the possibility of now preparing structural analogs of michellamines. This indeed seems necessary, because michellamine B (**14b**) does not only show antiviral activity, but also a certain toxicity that possibly may jeopardize the development of this potential drug. Consequently, it *must* be our aim to look for less toxic and (if possible) more active analogs of this new lead - by isolation of further representatives from the plants or by synthetic work in the chemical lab.

Isolation of further natural michellamines

Indeed, more recent isolation work by the NCI group shows that *A. korupensis* contains further, now constitutionally *unsymmetric* (see Fig. 1.16) michellamines D, E, and F (**14d-f**), although only in traces. Again, using our CD methodology and the degradation procedure, we succeeded in unequivocally establishing their stereostructures (Boyd *et al.*, unpublished results; Bringmann and Pokorny 1995).

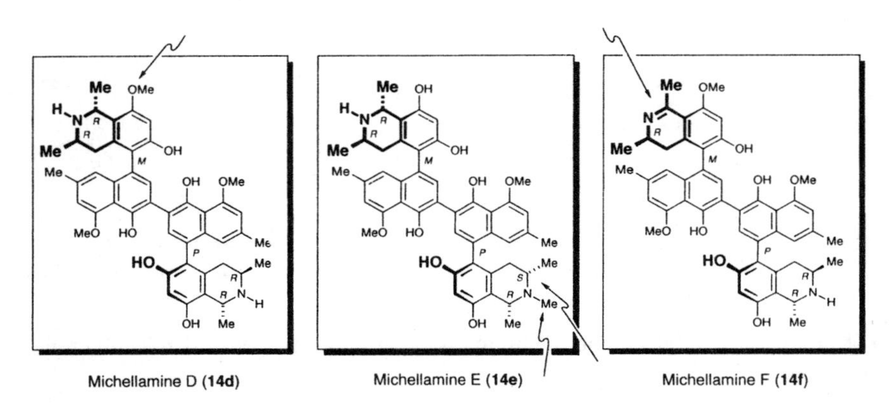

Fig. 1.16. Further new anti-HIV michellamines from *A. korupensis* (structural differences between the two 'halves' marked by arrows).

These three new michellamines show similar anti-HIV activities as the other michellamines A-C, yet, they exhibit a comparable toxicity. Given the fact that all the numerous other *Ancistrocladus* plants that we have so far investigated, were found to produce *monomeric* naphthylisoquinolines exclusively, and since all such monomeric halves are inactive towards HIV, it becomes evident that only synthetic procedures can give rise to rationally modified michellamine analogs. Such structural analogs might be prepared either

- by the modification of natural michellamines by derivatization,
- or by the chemical dimerization of isolated natural monomeric naphthylisoquinolines,
- or by stereoselective total synthesis.

Synthetic preparation of new, unnatural michellamines

Besides the chemical modification of natural michellamines as isolated from plant material, which is quite limited with respect to larger changes of the carbon skeleton, the probably most rational option to prepare a broad series of modified michellamines is the chemical dimerization of naturally occurring monomeric naphthylisoquinolines, since we have so many of them in hands from our extended isolation work over the past years. Our first step in this direction (see Fig. 1.17) is the successful dimerization of the easily available alkaloid dioncophylline A (**1**), which can be brominated, protected, coupled, and finally deprotected (Bringmann *et al., Tetrahedron,* submitted for publication). Compared with the michellamines, the dimeric dioncophylline A (**28**), which we have named jozimine A (for jozi = pair in *Kiswahili*), shows a different constitution, since now the two halves are joined together not *via* the naphthalene part, but *via* the isoquinoline moiety, with a connectivity: naphthalene-isoquinoline-isoquinoline-naphthalene. This has also stereochemical consequences: different from the michellamines, now even the central axis is configuratively stable to a certain degree, leading to the two C_2-symmetric atropoisomers **28a** and **b**, which can be separated and fully structurally attributed. These atropisomers very slowly interconvert at room temperature.

Jozimine A (**28**) is a prototype - it is the first artificial dimer of a natural monomeric naphthylisoquinoline alkaloid. Regrettably, it does not exhibit improved anti-HIV activity, but, most unexpectedly, it shows a high antimalarial activity - which is astonishing because the other dimers, in particular the michellamines, are entirely inactive against *Plasmodium falciparum*, and now this dimer **18** is even far more active than its monomeric precursor **1**, dioncophylline A.

For this reason, we have prepared further unnatural dimers of natural monomeric alkaloids (Fig. 1.18), like the dimers **30** and **31** of ancistrocladine (**2**) and dioncophylline C (**10**), and have even prepared a dimer **32**, pindikamine (pindika = skew in *Kiswahili*), of a 6,8'-coupled (and thus 'skew') *unnatural,* fully

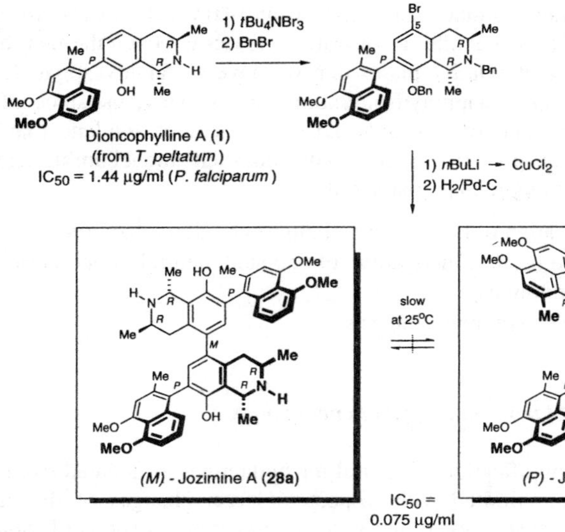

1) tBu4NBr3
2) BnBr
Dioncophylline A (1)
(from T. peltatum)
IC50 = 1.44 µg/ml (P. falciparum)
1) nBuLi → CuCl2
2) H2/Pd-C
slow
at 25°C
(M) - Jozimine A (28a)
(P) - Jozimine A (28b)
IC50 =
0.075 µg/ml

Fig. 1.17. Jozimine A (dimeric dioncophylline A): the first *non-natural* 'dimer' of a natural 'monomeric' naphthylisoquinoline alkaloid.

2 (Ancistrocladine,
from Ancistrocladus spec.)
10 (Dioncophylline C,
from T. peltatum)
29 (non-natural)
30 ('Jozimine B')
31 ('Jozimine C')
32 ('Pindikamine A')

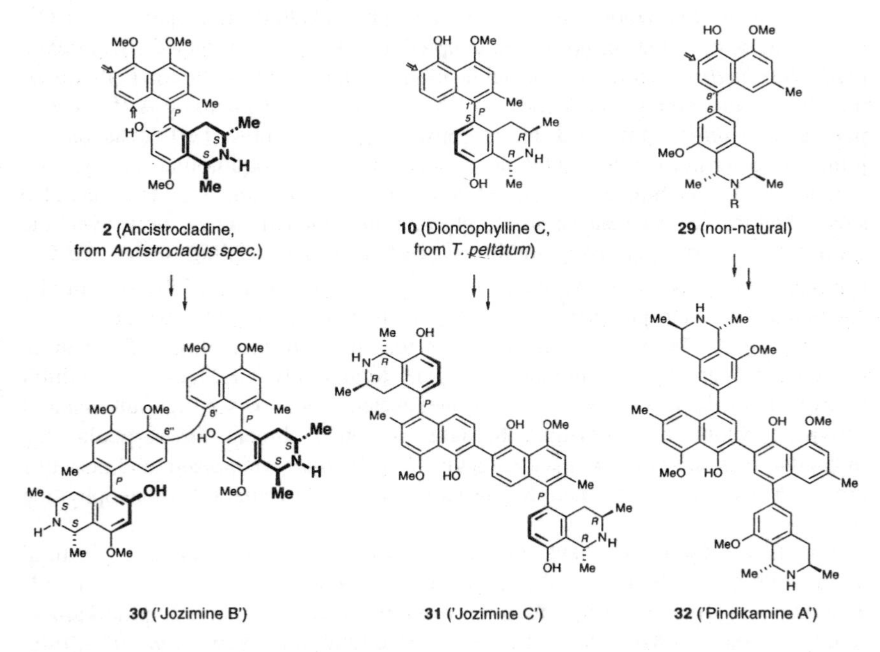

Fig. 1.18. Novel artificial dimers of natural or non-natural naphthylisoquinolines.

synthetic monomer **29** (Bringmann *et al.*, unpublished data). Their testing, as well as the designing and preparation of a whole series of further artificial dimers of natural or unnatural monomeric naphthylisoquinolines, is underway.

Conclusion

This intriguing novel class of compounds, the antimalarial monomeric naphthylisoquinoline alkaloids and their dimers, the anti-HIV michellamines (as well as their unnatural analogs), will be subject to further intensive research in the future, because of their unprecedented structures, their interesting biogenetic origin (as far as *natural* products are concerned), and their promising perspectives as potential anti-HIV or antimalarial drugs.

Acknowledgements

For generous financial support of the presented work, we thank the Deutsche Forschungsgemeinschaft (SFB 251 "Ökologie, Physiologie und Biochemie pflanzlicher und tierischer Leistungen unter Streß", Graduiertenkolleg "Magnetische Kernresonanz *in vivo* und *in vitro* für die biologische und medizinische Grundlagenforschung", and Normalverfahren), the UNDP / World Bank / WHO Special Program for Research and Training in Tropical Diseases (TDR), the Fonds der Chemischen Industrie, the BASF AG, and the Max-Buchner-Stiftung.

References

Boyd, M.R., Hallock, Y.F., Cardellina II, J.H., Manfredi, K.P., Blunt, J.W., McMahon, J.B., Buckheit Jr., R.W., Bringmann, G., Schäffer, M., Cragg, G.M., Thomas, D.W., and Jato, J.G. (1994). Anti-HIV Michellamines from *Ancistrocladus korupensis*. *Journal of Medicinal Chemistry* **37**, 1740-1745.

Bringmann, G. (1985a). Aufbau und Cyclisierung zentral modifizierter ß-Pentaketone: Synthese monocyclischer Isochinolinalkaloid-Vorstufen. *Liebigs Annalen der Chemie*, 2105-2115.

Bringmann, G. (1985b). Biomimetische Synthesen beider Molekülhälften der *Ancistrocladus*- und der *Triphyophyllum*-Alkaloide aus gemeinsamen Vorstufen. *Liebigs Annalen der Chemie*, 2126-2134.

Bringmann, G. and Jansen, J.R. (1989). Chiral economy with respect to rotational isomerism: Rational synthesis of hamatine and (optionally) ancistrocladine from joint helical precursors. *Heterocycles* **28**, 137-142.

Bringmann, G. and Jansen, J.R. (1991). Stereocontrolled ring opening of *axially prostereogenic* biaryl lactones with hydrogen nucleophiles: Directed synthesis of a dioncophylline A precursor and (optionally) its atropdiastereomer. *Synthesis*, 825-827.

Bringmann, G. and Pokorny, F. (1995). The naphthylisoquinoline alkaloids. In *The Alkaloids*, vol . 46 (ed. G. Cordell), pp. 127-271. Academic Press, New York.

Bringmann, G. and Reuscher, H. (1989a). Aryl-coupling via "axially prostereogenic" lactones: First total synthesis of (+)-ancistrocladisine and (optionally) its atropisomer. *Tetrahedron Letters* **30**, 5249-5252.

Bringmann, G. and Reuscher, H. (1989b). Atropdiastereoselective ring opening of bridged, "axial-prostereogenic" biaryls: Directed synthesis of (+)-ancistrocladisine. *Angewandte Chemie* **101**, 1725-1726; *Angewandte Chemie International Edition in English* **28**, 1672-1673.

Bringmann, G., Jansen, J.R., and Rink, H.-P (1986). Regioselective and atropisomeric-selective aryl coupling to give naphthyl isoquinoline alkaloids: The first total synthesis of (-)-ancistrocladine. *Angewandte Chemie* **98**, 917-919; *Angewandte Chemie International Edition in English* **25**, 913-915.

Bringmann, G., Jansen, J.R., Reuscher, H., Rübenacker, M., Peters, K., and von Schnering, H.G. (1990a). First total synthesis of (-)-dioncophylline A ("triphyophylline") and of selected stereoisomers: Complete (revised) stereostructure. *Tetrahedron Letters* **31**, 643-646.

Bringmann, G., Walter, R., and Weirich, R. (1990b). The directed synthesis of biaryl compounds: Modern concepts and strategies. *Angewandte Chemie* **102**, 1006-1019; *Angewandte Chemie International Edition in English* **29**, 977-991.

Bringmann, G., Geuder, T., Rübenacker, M., and Zagst, R. (1991a). A facile degradation procedure for the determination of absolute configuration in 1,3-dimethyltetra- and dihydroisoquinolines. *Phytochemistry* **30**, 2067-2070.

Bringmann, G., Lisch, D., Reuscher, H., Aké Assi, L., and Günther, K. (1991b). Atrop-diastereomer separation by racemate resolution techniques: *N*-Methyl-dioncophylline A and its 7-epimer from *Ancistrocladus abbreviatus*. *Phytochemistry* **30**, 1307-1310.

Bringmann, G., Pokorny, F., and Zinsmeister, H.-D. (1991c). *Ancistrocladus*, eine botanisch und chemisch bemerkenswerte Gattung. *Der Palmengarten* **55/3**, 13-18.

Bringmann, G., Pokorny, F., Stäblein, M., Govindachari, T.R., Almeida, M.R., and Ketkar, S.M. (1991d). On the biosynthesis of acetogenic tetrahydroisoquinoline alkaloids: First *in vivo* feeding experiments. *Planta Medica* **57 (Suppl. 2)**, 98.

Bringmann, G., Zagst, R., and Aké Assi, L. (1991e). Ancistrobrevine B: A naphthylisoquinoline alkaloid with a novel coupling type from *Ancistrocladus abbreviatus*. *Planta Medica* **57 (Suppl. 1)**, 96-97.

Bringmann, G., Kinzinger, L., Busse, H., and Zhao, C. (1992a). Isolation, structure elucidation, and total synthesis of ancistrocline, an alkaloid of *Ancistrocladus tectorius*. *Planta Medica* **58 (Suppl. 1)**, 704.

Bringmann, G., Zagst, R., Reuscher, H., and Aké Assi, L. (1992b). Ancistrobrevine B, the first naphthylisoquinoline alkaloid with a 5,8'-coupling site, and related compounds from *Ancistrocladus abbreviatus*. *Phytochemistry* **31**, 4011-4014.

Bringmann, G., Zagst, R., Lisch, D., and Aké Assi, L. (1992c). Dioncoline A and its atropisomer: "inverse hybrid type" Ancistrocladaceae/Dioncophyllaceae alkaloids from *Ancistrocladus abbreviatus*. *Planta Medica* **58 (Suppl. 1)**, 702-703.

Bringmann, G., Weirich, R., Lisch, D., and Aké Assi, L. (1992d). Ancistrobrevine D: An unusual Alkaloid from *Ancistrocladus abbreviatus*. *Planta Medica* **58 (Suppl. 1)**, 703-704.

Bringmann, G., Gramatzki, S., Grimm, C., and Proksch, P. (1992e). Feeding deterrency and growth retarding activity of the naphthylisoquinoline alkaloid dioncophylline A against *Spodoptera littoralis*. *Phytochemistry* **31**, 3821-3825.

Bringmann, G., Gulden, K.P., Busse, H., Fleischhauer, J., Kramer, B., and Zobel, E. (1993a). Circular dichroism of naphthyltetrahydroisoquinoline alkaloids: Calculation of CD Spectra by semiempirical methods. *Tetrahedron* **49**, 3305-3312.

Bringmann, G., Pokorny, F., Stäblein, M., Schäffer, M., and Aké Assi, L. (1993b). Ancistrobrevine C from *Ancistrocladus abbreviatus:* the first mixed "Ancistrocladaceae/ Dioncophyllaceae-type" naphthylisoquinoline alkaloid. *Phytochemistry* **33**, 1511-1515, 1663.

Bringmann, G., Schneider, Ch., Pokorny, F., Lorenz, H., Fleischmann, H., Sankaranarayanan, A.S., Almeida, M.R., Govindachari, T.R., and Aké Assi, L. (1993c). The cultivation of tropical lianas of genus *Ancistrocladus*. *Planta Medica* **59 (Suppl. 1)**, 623-624.

Bringmann, G., Zagst, R., Schäffer, M., Hallock, Y.F., Cardellina II, J.H., Boyd, M.R. (1993d). The absolute configuration of michellamine B, a "dimeric", anti-HIV-active naphthylisoquinoline alkaloid. *Angewandte Chemie* **105**, 1242-1243; *Angewandte Chemie International Edition English* **32**, 1190-1191.

Bringmann, G., Haller, R.D., Bär, S., Isahakia, M.A., and Robertson, S.A. (1994a). *Ancistrocladus robertsoniorum* J. Léonard: eine erst spät entdeckte *Ancistrocladus*-Art. *Der Palmengarten* **58**, 148-153.

Bringmann, G., Gulden, K.-P. Hallock, Y.F., Manfredi, K.P., Cardellina II, J.H., Boyd, M.R., Kramer, B., and Fleischhauer, J. (1994b). Circular dichroism of michellamines: Independent assignment of axial chirality by calculated and experimental CD spectra. *Tetrahedron* **50**, 7807-7814.

Bringmann, G., Götz, R., Keller, P.A., Walter, R., Henschel, P., Schäffer, M., Stäblein, M., Kelly, T.R., and Boyd, M.R. (1994c). First total synthesis of korupensamines A and B. *Heterocycles* **39**, 503-512.

Bringmann, G., Harmsen, S., Holenz, J., Geuder, T., Götz, R., Keller, P.A., Walter, R., Hallock, Y.F., Cardellina II, J.H., and Boyd, M.R. (1994d). 'Biomimetic' oxidative dimerization of korupensamine A: Completion of the first total synthesis of michellamines A, B and C. *Tetrahedron* **50**, 9643-9648.

Bringmann, G., Walter, R., and Weirich, R. (1995). Axially chiral compounds - Biaryls. In *Houben-Weyl, Methods of Organic Chemistry*, vol E 21a (eds. G. Helmchen, R. W. Hoffmann, J. Mulzer, and E. Schaumann), pp. 567-587. Thieme Stuttgart, New York.

François, G., Bringmann, G., Phillipson, J.D., Aké Assi, L., Dochez, C., Rübenacker, M., Schneider, Ch., Wéry, M., Warhurst, D.C., and Kirby, G.C. (1994). Activity of extracts and naphthylisoquinoline alkaloids from *Triphyophyllum peltatum, Ancistrocladus abbreviatus* and *A. barteri* against *Plasmodium falciparum in vitro. Phytochemistry* **35**, 1461-1464.

François, G., Bringmann, L., Dochez, C., Schneider, C., Timperman, G., and Aké Assi, L. (1995). Activities of extracts and naphthylisoquinoline from *Triphyophyllum peltatum, Ancistrocladus abbreviatus* and *Ancistrocladus barteri* against *Plasmodium berghei* (Anka strain) *in vitro. Journal of Ethnopharmacol.* **46**, 115-120.

François, G., Timperman, G., Holenz, J., Aké Assi, L., Geuder, T., Maes, L., Dubois, J., Banocq, M., and Bringmann, G. (1996). Naphthylisoquinoline alkaloids exhibit strong growth-inhibiting activities against *Plasmodium falciparum* and *P. berghei in vitro* - Structure-activity relationship of dioncophylline C. *Annals of Tropical Medicine and Parasitology* **90**, 115-123.

Hallock, Y.F., Manfredi, K.P., Blunt, J.W., Cardellina II, J.H., Schäffer, M., Gulden, K.-P., Bringmann, G., Lee, A.Y., Clardy, J., François, G., and Boyd, M.R. (1994). Korupensamines A-D, novel antimalarial alkaloids from *Ancistrocladus korupensis. Journal of Organic Chemistry* **59**, 6349-6355.

Hobbs, P.D., Upender, V., Liu, J., Pollart, D.J., Thomas, D.W., and Dawson, M.I. (1996). The first stereospecific synthesis of michellamine B. *Journal of the Chemical Society, Chemical Communications,* 923-924.

Hoye, T.R., Cheng, M., Mi, L., and Priest, O.P. (1994). Total synthesis of michellamines A-C: Important anti-HIV agents. *Tetrahedron Letters* **47**, 8747-8750.

Kelly, T.R., Garcia, A., Lang, F., Walsh, J.J., Bhaskar, K.V., Boyd, M.R., Götz, R., Keller, P.A., Walter, R., and Bringmann, G. (1994). Convergent total synthesis of the michellamines. *Tetrahedron Letters* **35**, 7621-7624.

Manfredi, K.P., Blunt, J.W., Cardellina II, J.H., McMahon, J.B., Pannell, L.L., Cragg, G.M., and Boyd, M.R. (1991). Novel alkaloids from the tropical plant *Ancistrocladus abbreviatus* inhibit cell killing by HIV-1 and HIV-2. *Journal of Medicinal Chemistry* **34**, 3402-3405.

McMahon, J.B., Currens, M.J., Gulakowski, R.J., Buckheit Jr., R.W., Lackman-Smith, C., Hallock, Y.F., and Boyd, M.R. (1995). Michellamine B, a novel plant alkaloid, inhibits human immunodeficiency virus-induced cell killing by at least two distinct mechanisms. *Antimicrobial Agents and Chemotherapy* **39**, 484-488.

Thomas, D.W. and Gereau, R.E. (1993). *Ancistrocladus korupensis* (Ancistrocladaceae): A new species of liana from Cameroon. *Novon* **3**, 494-498

2. Strategy in the search for bioactive plant constituents

K. HOSTETTMANN, J.-L. WOLFENDER, S. RODRIGUEZ AND A. MARSTON

Institut de Pharmacognosie et Phytochimie, Université de Lausanne, BEP, CH-1015 Lausanne, Switzerland

---

Introduction

The potential of higher plants as sources for new drugs is still largely unexplored. Among the estimated 250000-500000 plant species, only a small percentage has been investigated phytochemically and the fraction submitted to biological or pharmacological screening is even smaller. Although, for example, the National Cancer Institute (NCI) of the United States screened some 35'000 plant species for antitumour activity from 1957 to 1981 (Suffness and Douros 1982) and is currently in the process of acquiring some 20000 tropical species from Latin America, Africa and Southeast Asia (Cassady *et al.* 1990), these plants will still have to be considered as 'uninvestigated' with respect to any other pharmacological activity. Plants contain hundreds or even thousands of metabolites. Thus, any phytochemical investigation of a given plant will reveal only a very narrow spectrum of its constituents.

When screening for biologically active plant constituents, the selection of the plant species to be studied is obviously a crucial factor for the ultimate success of the investigation. Besides random collection of plant material, targeted collection based on consideration of chemotaxonomic relationships and exploitation of ethnomedical information is currently performed. With the wealth of information from African traditional healers, for example, a large area of study is possible. Plants used in traditional medicine are more likely to yield pharmacologically active compounds. In the field of anticancer activity, a correlation between biological activity and use in traditional medicine has been demonstrated (Farnsworth and Kass 1981; Spjut and Perdue 1976). Field observation may be useful in screening programs aimed at antimicrobial or insect deterrent/insecticidal activities.

The process that leads from the plant to a pharmacologically active pure constituent is very long and tedious, and requires a multidisciplinary collaboration of botanists, pharmacognosists, chemists, pharmacologists and toxicologists. This approach involves the following steps (see Fig. 2.1.) (Hostettmann *et al.* 1995):

- collection, proper botanical identification and drying of the plant material;
- preparation of appropriate extracts;
- **biological and pharmacological screening of crude extracts;**
- **chemical screening of crude extracts by combined LC/UV, LC/MS and LC/NMR;**
- several consecutive steps of chromatographic separation, where each fraction obtained has to be submitted to bioassays in order to follow the activity (activity-guided fractionation);
- verification of the purity of the isolated compounds;
- structure elucidation by chemical and physicochemical methods;
- partial or total synthesis;
- preparation of derivatives/analogues for the investigation of structure-activity relationships;
- large-scale isolation for further pharmacological and toxicological tests.

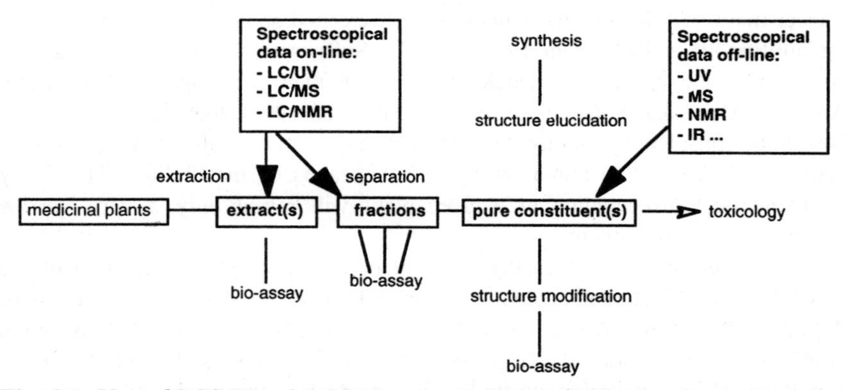

Fig. 2.1. Use of LC/UV and LC/MS as strategic analytical screening tools during the isolation process of a plant extract. These hyphenated techniques allow the early recognition of common compounds in the extract and at the same time, the localisation of interesting new ones.

In this paper the role of biological and chemical screening of crude extracts will be outlined, with emphasis on plants from Africa.

Biological and chemical screening of crude extracts

Crucial to any investigation of plants with biological activities is the availability of suitable bioassays for monitoring the required effects (Hostettmann *et al.* 1995). In order to cope with the number of extracts and fractions from bioactivity-guided fractionation steps, the capacity for high sample throughput is necessary. The test systems should ideally be simple, rapid, reproducible and inexpensive. If active principles are only present at low concentrations in the crude extracts, the bioassays have to be sensitive enough for their detection. At the same time, the number of false positives should be reduced to a minimum. Another factor of special relevance to plant extracts is the solubility of the sample. Finding a suitable solvent can pose problems.

When deciding which bioassays to employ in research on plant constituents, the first step is to choose suitable target organisms. These can be lower organisms (microorganisms, insects, crustaceans, molluscs), isolated subcellular systems (enzymes, receptors, organelles), cultured cells of human or animal origin, isolated organs of vertebrates or whole animals. However, the right target has to be found for the disease in question.

Table 2.1. Simple bioassays for the screening of crude plant extracts

Activity	Target
Antibacterial activity	Human/plant pathogenic bacteria (*e.g. Staphylococcus aureus*, *Escherichia coli*, *Erwinia* sp.)
Antifungal activity	Human/plant pathogenic fungi and yeasts (*e.g. Candida albicans*, *Aspergillus* sp., *Cladosporium* sp.)
Brine shrimp toxicity	*Artemia salina*
Crown gall tumour inhibtion (potato disk assay)	Cells of potato tubers (*Solanum tuberosum*) transformed by *Agrobacterium tumefasciens*
Antimitotic activity	Sea urchin eggs (*Strongylocentrotus* sp. etc.)
Insect antifeedant activity	*Spodoptera* sp. (African army worms), *Epilachna varivestis* (Mexican bean beetle)
Larvicidal activity	*Aedes aegypti* (yellow fever mosquito)
Molluscicidal activity	Schistosomiasis-transmitting snails (*Biomphalaria glabrata*)

The complexity of the bioassay has to be designed as a function of the facilities, resources and personnel available. In most phytochemical laboratories engaged in the investigation of bioactive medicinal plants, neither specialised animal facilities nor properly qualified technicians are on hand. Consequently, efforts have been made to introduce simple, inexpensive "bench-top" bioassays for the rapid screening of plant extracts and fractions. Care must sometimes be taken in the interpretation and predictive ability of these tests, but in general, they provide very important preliminary information for the evaluation of vegetable (or other) material under study. A selection of these bioassays is shown in Table 2.1. The list covers a variety of targets, ranging from a test for general toxicity with brine shrimp to snail killing activity with *Biomphalaria glabrata.* In this way, different properties and types of ailment, including microbial afflictions and parasitic diseases, can be investigated.

Chemical screening of crude extracts by combined LC/UV, LC/MS and LC/NMR

As the number of targets for biological screening is limited, an efficient system for the chemical screening of the extracts is also needed in order to detect new leads which can be potentially interesting from a chemical viewpoint. Early recognition of plant metabolites, at the earliest stage of separation as possible, is also essential in order to avoid a time-consuming isolation of common constituents.
Such a system requires hyphenated techniques which are able to provide efficient separation of the metabolites as well as good selectivity, sensitivity of detection and capacity to provide important structural information on-line. For this reason, screening of the extracts has been performed with high performance liquid chromatography (HPLC) coupled to a UV photodiode array detector (LC/UV) and a mass spectrometer detector (LC/MS). These hyphenated techniques have been fully integrated into the isolation process (Fig. 2.1.). They play an important role as an analytical support in the work of phytochemists for the efficient detection and rapid characterisation of natural products .

LC/UV has already been widely used for the analysis of crude plant extracts, while LC/MS is a more recent addition and is still not widely spread in the phytochemical community (Markham 1965). The main advantage of this latter technique is that it provides a type of 'universal' LC detection and a source of important structural information on-line (molecular weight and fragments). Nevertheless, the coupling of LC with MS is not as simple and straightforward as with UV. Furthermore, the MS response is dependent on the type of interface choosen for ionisation. Thermospray (TSP) (Blakley and Vestal 1983) and continuous-flow fast atom bombardment (CF-FAB) (Caprioli *et al.* 1986) interfaces are used routinely in our laboratory to screen crude plant extracts and other interfaces such as electrospray (ES) (Whitehouse *et al.* 1985) or atmospheric pressure chemical ionisation (APCI) have also been tested (Wolfender *et al.* 1995).

In order to obtain more precise structural information of the metabolites of interest, complementary techniques to LC/UV/MS such as LC/UV post-column reactions or LC/MS/MS experiments have been performed. In the screening for polyphenols, for example, post-column addition of UV-shift reagents prior to photodiode array detection has allowed the measurement of UV shifted spectra significant for the assignment of the positions of free hydroxyl groups on this type of nucleus. The combination of LC/UV, LC/MS and LC/UV shifted spectra permits the on-line identification of numerous polyphenols.

As this approach is not always sufficient for a full structural identification of flavonoids, for example, LC/MS/MS has also been investigated, providing precise information on the location of the substituents.

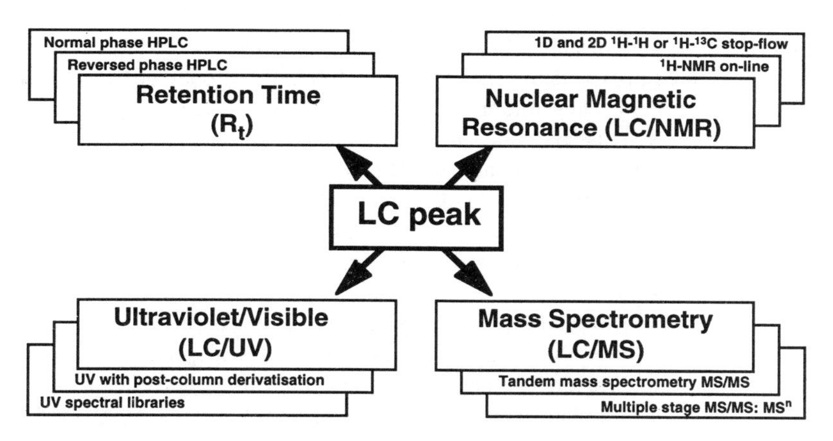

Fig. 2.2. Use of different complementary LC hyphenated techniques for a precise 4-D LC peak assignment.

Very recently, LC/NMR has been added to the LC/UV/MS setup. This technique, despite being known for over fifteen years, is not yet a widely accepted technique, mainly because of its lack of sensitivity. However, the recent progress in pulse field gradients, solvent suppression, the improvement in probe technology and the construction of high field magnets have given a new impulse to this technique. LC/NMR represents an important complementary hyphenated technique to LC/UV and LC/MS. NMR spectroscopy is indeed by far the most powerful spectroscopic technique for obtaining detailed structural information about organic compounds in solution. The use of LC/UV/MS and LC/NMR allows a 4-D peak assignment (Fig. 2.2.). With such techniques, the efficiency of the chemical screening and the potential for finding new leads is thus considerably enhanced.

Search for antidepressive compounds from Gentianaceae

In a general manner, depression is characterised by a weak neurotransmitter concentration in the synapses. From a therapeutical viewpoint, there are two ways to increase the neurotransmitter concentration: it is possible to slow down their recapture or to inhibit their deamination. With respect to the latter mechanism, the search for selective and reversible inhibitors of monoamine oxidase (MAO) has become of major importance (Kielholtz 1982; Delini-Stula 1983).

Disturbances in monoamine oxidase levels have been reported in a series of disorders (Parkinson's disease, Huntington's chorea, depression, anxiety and senile dementias; (Stroelin-Benedetti and Dostert 1992). The enzyme monoamine oxidase (MAO) plays a key role in the regulation of certain physiological amines in the human body. It causes a deactivation by deamination of neurotransmitters such as catecholamines and serotonin, and of endogenous amines such as tyramine. MAO exists as two isoenzymes, MAO-A and MAO-B, which exhibit different substrate specificities. Inhibitors of MAO, in particular of the A type isoenzyme, have a potential as antidepressive drugs since they increase both noradrenaline and serotonin levels in the brain. There is a current interest in the discovery of new reversible inhibitors (Stroelin-Benedetti and Dostert 1992) because existing drugs such as clorgyline (selectively inhibits MAO-A) and (-)deprenyl (selectively inhibits MAO-B) are irreversible inhibitors and exhibit serious side effects.

It is well known that plants contain inhibitors of MAO. For example, the bark of the tropical liana *Banisteriopsis caapi* (Malpighiaceae) contains ß-carboline alkaloids such as harmaline, a potent reversible inhibitor of MAO-A (Deulofeu 1967). Infusions of St. John's Wort, *Hypericum perforatum* (Guttiferae), have been used for the treatment of melancholia in European folk medicine and numerous preparations of this plant have now been commercialised for the management of different depressive states. The antidepressive activity of *H. perforatum* has been demonstrated *in vivo* (Okpanyi and Weischer 1987) and it was originally thought that hypericin, which has shown MAO-A and MAO-B inhibitory activity *in vitro* (Suzuki *et al.* 1984), was responsible for this activity. However, more recent investigations point towards xanthones and flavonoids as the major source of the antidepressive activity since these have potent MAO inhibitory properties (Sparenberg 1993).

Investigations with extracts of *Canscora decussata* (Gentianaceae) (Bhattacharya *et al.* 1972) *in vivo* showed that the fractions containing xanthones had a stimulating effect on the central nervous system in mice, rats and dogs (Ghosal *et al.* 1985). A strong inhibition of monoamine oxidase by trisubstituted xanthones was observed by Suzuki *et al.* (Suzuki *et al.* 1978). In fact, a number of xanthones which inhibit MAO have been identified from plant sources (Suzuki *et al.* 1981; Suzuki *et al.* 1980; Schaufelberger and Hostettmann 1988). Of those which have been tested, the aglycones are, in general, more active than the corresponding glycosides. Xanthones with the substitution patterns 1-hydroxy-

3,8-dimethoxy- and 1,3-dihydroxy-7,8-dimethoxy- show pronounced activity. They are also selective inhibitors, being more effective against MAO-A than MAO-B. Xanthones isolated from the aerial parts of *Gentiana lactea* (Gentianaceae), all 1,3,5,8-tetrasubstituted, have been tested against monoamine oxidases from rat brain mitochondria. Two glycosides and the aglycone swerchirin (methoxylated at C-3 and C-5) were inactive. Bellidifolin (1,5,8-trihydroxy-3-methoxyxanthone), on the other hand, inhibited MAO-A to the same extent as the reference compound pargyline. It was also selective, with a marked reduction in activity against MAO-B (IC_{50} MAO-A/IC_{50} MAO-B = 0.001). Desmethylbellidifolin (1,3,5,8-tetrahydroxyxanthone) gave an IC_{50} against MAO-A which was approximately ten times higher than that obtained with bellidifolin and, therefore, was a correspondingly weaker inhibitor (Schaufelberger and Hostettmann 1988).

In an effort to find further active xanthones, other plants (especially of the families Gentianaceae, Guttiferae and Polygalaceae) have been investigated. Several examples of this research with reference to the Gentianaceae are described below.

LC/UV/MS and LC/UV analyses of Chironia krebsii extracts with post-column addition of UV shift reagents

The extracts of the roots of *Chironia krebsii,* a member of the Gentianaceae from Malawi, showed an important inhibition of MAO-A *in vitro* (CH_2Cl_2 ext: 100 %; MeOH ext: 75 % IMAO-A at 15 μg/ml). These activities were of several orders of magnitude stronger than those measured for *H. perforatum* extracts (CH_2Cl_2 ext: 18 %; MeOH ext: 14 % IMAO-A at 15 μg/ml) (Thull 1995). The LC/UV analysis of *C. krebsii* showed that this plant was very rich in xanthones (see structures in Fig. 2.3.)

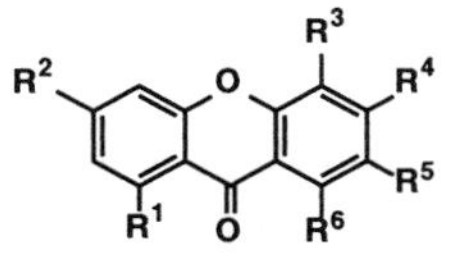

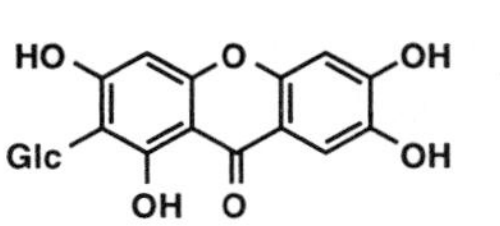

	R¹	R²	R³	R⁴	R⁵	R⁶
3	OPrim	OMe	OH	H	H	H
4	OGlc	OMe	OH	H	H	H
5	OPrim	OMe	H	H	OH	OH
6	OPrim	OMe	OMe	H	H	H
7	OH	OMe	OPrim	H	H	H
8	OPrim	OMe	OMe	OMe	OMe	OMe
9	OH	OH	H	H	OH	H
10	OH	OH	H	H	OH	OH
11	OH	OMe	OH	H	H	H
12	OH	OMe	OMe	OH	OMe	OMe
13	OH	OMe	H	H	OH	OH
14	OH	OMe	OMe	OMe	OMe	OMe
18	OH	OMe	H	H	OMe	OMe

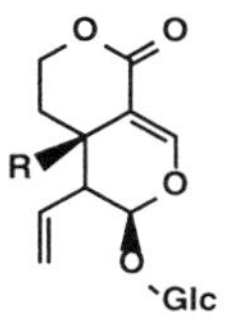

Mangiferin (15)

	R
1	OH
2	H

Figure 2.3. Structures of the xanthones and secoiridoids found in *Chironia krebsii* and *Gentiana rhodantha.*

The number of bands and the general aspects of the UV spectra of xanthones (UV spectra **3-14** in Fig. 2.4.) allowed a first attribution of the type of oxygenation patterns encountered (Hostettmann and Hostettmann 1989; Lins Mesquita *et al.* 1968). The constituents having a weaker chromophore (UV max *ca.* 250 nm) were attributable to the secoiridoids swertiamarin (**1**) and sweroside (**2**), widespread bitter principles of the Gentianaceae.

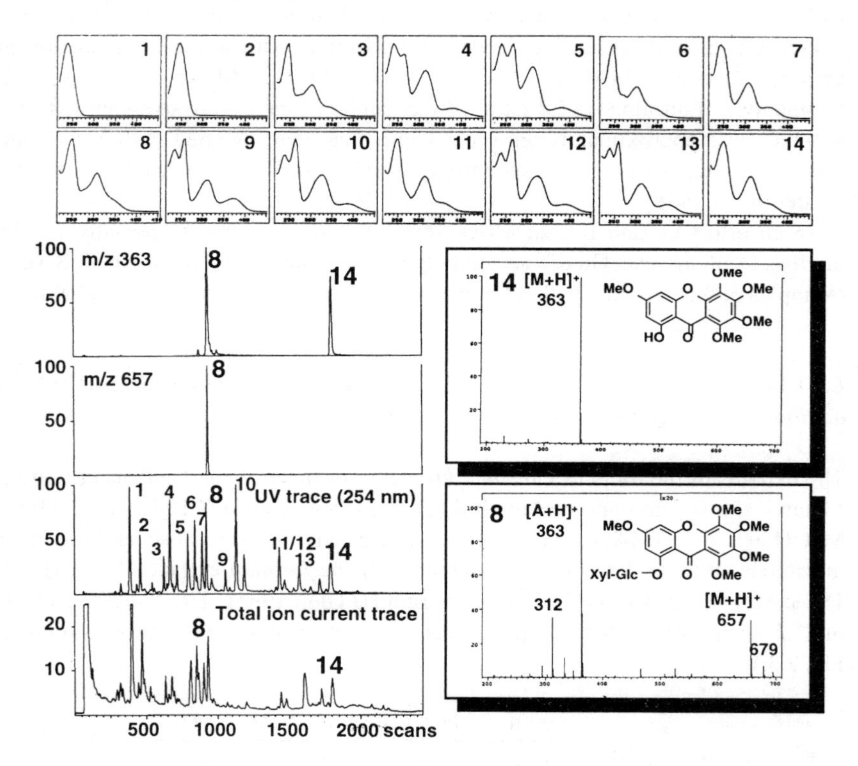

Fig. 2.4. LC/UV and LC/TSP-MS analysis of the root methanolic extract of *Chironia krebsii.* (Gentianaceae). UV spectra **1-2** are characteristic for secoiridoids. Peaks **1-10** are xanthone-O-glycosides, **11-14** are xanthone aglycones. UV traces were recorded at 254 nm and UV spectra from 200-500 nm. HPLC: Column; RP-18 NovaPak (4μm, 150 x 3.9 mm i.d.); gradient, CH_3CN-H_2O (0.1% TFA) 5:95 -> 65:35 in 50 min (1 ml/min).

In order to get more information on the molecular weight, the molecular formula and the sugar sequence (glycosides) of the xanthones, LC/TSP-MS analysis of the extract was carried out. LC/TSP-MS was performed in the positive ion mode with the filament off mode and by using ammonium acetate as buffer. The source was set at 280°C and the vaporiser at 100°C (Wolfender and Hostettmann 1993). Under these conditions, the total ion current trace recorded with LC/TSP-MS showed ionisation of all the peaks recorded in the UV trace of the extract (254 nm) (Fig. 2.4.). Nevertheless, the total ion current response for highly hydroxylated xanthones with high melting point temperature (*i.e.* **10** Fig. 2.4.) was found to be very weak in comparison to the UV trace of the corresponding peak, leading to a difficult MS detection (Wolfender *et al.* 1994).

The TSP mass spectra of the xanthone aglycones recorded on-line after HPLC separation exhibited only $[M+H]^+$ ions as the main peak (see TSP spectra of peak **14** in Fig. 2.4.). Corresponding mass spectra of the xanthone glycosides usually showed two weak ions due to $[M+H]^+$ and $[M+Na]^+$ adducts, and a main peak for the protonated aglycone moiety $[A+H]^+$ (see TSP spectra of peak **8** in Fig. 2.4.). In the case of diglycosides, the successive losses of the monosaccharide units were marked by the corresponding peaks in the spectrum. Thus, for the diglycosides of *Chironia* species, a loss of 132 u was first observed, corresponding to a pentosyl residue, followed by a loss of 162 u corresponding to a hexosyl moiety, leading to the aglycone ion $[A+H]^+$ (see TSP spectra of peak **8** in Fig. 2.4.). After subsequent isolation of these glycosides, the disaccharide was shown to be primeverose, a β-D-xylopyranosyl-(1->6)-β-D-glucopyranoside disaccharide unit, often encountered in the Gentianaceae family (Hostettmann and Wagner 1977).

The MS detection allows the selective recording of the trace of each ion. By this means, it was possible for example to reconstruct the chromatogram of the ion *m/z* 657 and to obtain a precise assignment of the peak corresponding to compound **8** in the extract. Similarly, the specific ion trace at *m/z* 363 gave responses for the HPLC peaks of compound **8** (this ion is the main fragment of **8**) and also of compound **14**, a slower-running component of the extract. According to the UV and TSP mass spectra of **14**, this compound was readily identified as the corresponding free aglycone of **8**. This information allowed the rapid identification of both aglycones and their corresponding glycosides simultaneously in a crude plant extract.

In order to obtain more structural information on the position of the free hydroxyl groups on the xanthone nucleus, LC/UV with post-column addition of shift reagents was performed. The structural information obtained on-line by the combination of LC/UV, LC/TSP-MS and LC/UV with post-column addition of shift reagents is illustrated for the xanthone **10** (Fig. 2.5.).

According to Kaldas (1977), the UV spectra of **10** with four absorption maxima and a higher intensity for band II were characteristic for 1,3,7,8-tetraoxygenated xanthones with a free hydroxyl at position C-1. The TSP-MS spectra recorded on-line permitted the molecular weight determination and the assignment of the number and of the type of substituents of **10** (MW 260: four OH). The shifted UV spectra recorded on-line for **10** confirmed this compound to be 1,3,7,8-

tetrahydroxyxanthone. Indeed the presence of a free hydroxyl at positions C-1 and C-8 was characterised by the important shift recorded with $AlCl_3$ and the *ortho*-dihydroxyl group at C-7 and C-8 was confirmed by the shift due to the complexation of boric acid. The position of the fourth hydroxyl at C-3 was confirmed by a shift registered with the weak base NaOAc. The structures of all compounds (**3-15**) were deduced following the same procedure (Fig. 2.5.).

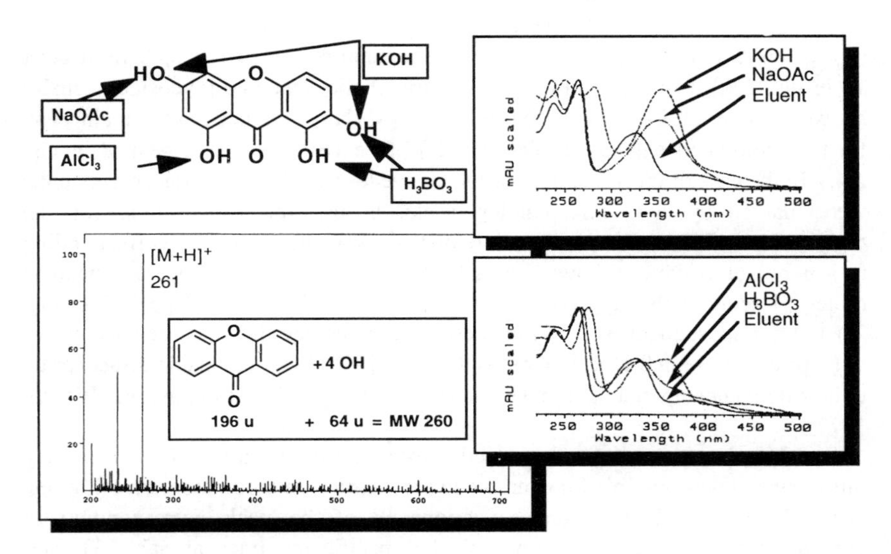

Fig. 2.5. Summary of all the on-line spectral data obtained by LC/TSP-MS, LC/UV and LC/UV with post column addition of shift reagents for the xanthone **10** from the root methanolic extract of *Chironia krebsii* (Gentianaceae). LC/UV/MS: same conditions as for Fig. 2.4.

Xanthones **3-15** were isolated (Wolfender *et al.* 1991) in order to test their IMAO activity. Several of them showed important selective and reversible inhibition of MAO-A. The most active xanthone was 1,5-dihydroxy-3-methoxyxanthone (**11**) with an inhibitory activity of 0.04 µM (IC_{50} MAO-A) (Thull 1995). The structures of the isolated compounds were in good agreement with those obtained from on-line data. The UV spectra of the fully identified pure products were registered in a UV data base in order to permit a computer-fitting search with xanthones from other extracts and thus accelerate the on-line identification procedure of these compounds in other Gentianaceae species (Rodriguez *et al.* 1995b).

LC/TSP-MS and LC/CF-FAB-MS of Gentiana rhodantha

Among the other Gentianaceae species screened by LC/UV/MS, *Gentiana rhodantha* from China presented a completely different extract composition from that found for *C. krebsii*. The LC/UV analysis of *G. rhodantha* showed the presence of only one predominant xanthone **15** and different secoiridoids. Compound **15** was rapidly identified as the widespread xanthone C-glycoside mangiferin (see structure in Fig. 2.3.). The TSP-MS spectrum of **15** exhibited a pseudomolecular ion at m/z 423, characteristic fragments for C-glycosides at $[M+H-90]^+$ and $[M+H-120]^+$ and a weak aglycone ion at m/z 261 (xanthone with four OH). The computer fitting of the UV spectrum of **15** with our in-house UV spectral library allowed the definitive characterisation of **15**. With the help of the LC/TSP-MS spectra recorded on-line, chemotaxonomical considerations and comparison with pure standards, the secoiridoids with retention times less than 10 min were easily identified (Ma *et al.* 1994). Among them a very minor secoiridoid, sweroside (**2**) (MW:358), was found to be present. The slower running peak, **16**, also exhibited the same characteristic UV spectra of secoiridoids (one band at around 240 nm) (Fig. 2.6.). This compound, which was less polar than the common secoiridoids, was studied in more detail. The LC/TSP-MS analysis of **16** gave in each case a spectrum identical to that obtained for sweroside **2**; it exhibited an intense ion at 359 u and no ion at higher masses as shown by the display of the ion trace m/z 359 of the extract (Fig. 2.6.). However, the chromatographic behaviour of **16** and sweroside (**2**) were quite different. In order to obtain complementary information on these constituents, a second LC/MS analysis with CF-FAB was performed, using the same HPLC conditions. The total ion current recorded for the whole chromatogram showed a very important MS response for compound **16,** while the more polar metabolites were only weakly ionised (Fig. 2.6B.). The CF-FAB spectrum of **16** recorded on-line exhibited a very intense pseudomolecular ion $[M-H]^-$ at m/z 913 together with a weak ion at m/z 555 corresponding to the loss of a "sweroside like" unit $[M-H-358]^-$ (CF-FAB spectrum of **16**, Fig. 2.6B.). This complementary information indicated clearly that the molecular weight of **16** was 914 u. According to the different results obtained on-line for **16** in the HPLC screening of the extract of *G. rhodantha,* it was concluded that **16** was probably a type of moderately polar large secoiridoid containing at least one unit very similar to sweroside (**2**).

Following the LC/MS screening results, a targeted isolation of **16** was undertaken. A full structure determination of **16** with the help of 1D and 2D NMR experiments as well as with different chemical reactions showed that **16** consisted of two secoiridoid units linked together with a monoterpene unit through two ester groups (Ma *et al.* 1994). This compound was found to be a natural product of a new type. The structure determination of other closely related compounds from this extract is still in progress.

This example shows the use of both LC/MS ionisation techniques for targeting unknowns. LC/TSP-MS indicated that compounds **2** and **16** had common sub-units (identical fragments (MW 358)), while CF-FAB allowed the on-line molecular weight determination of all of the oligomeric compounds. The combination of both types of information thus allowed the early recognition of this type of large secoiridoid glycoside.

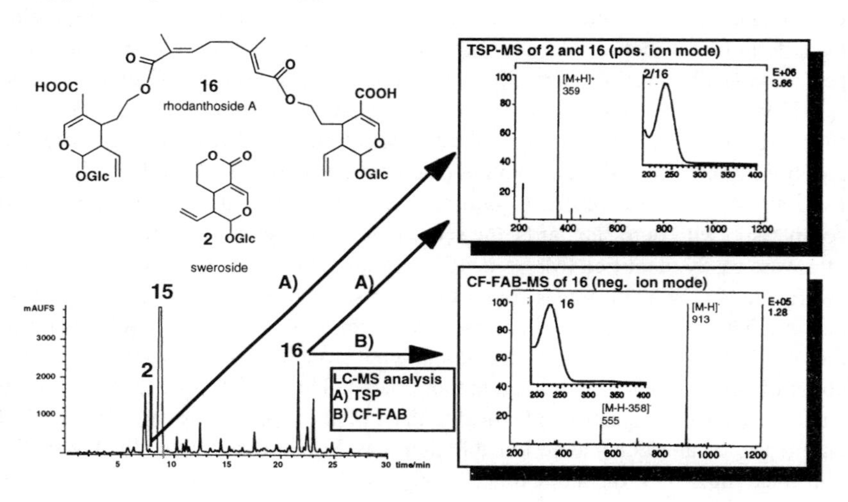

Fig. 2.6. Combined TSP (A) and CF-FAB (B) LC/MS of the enriched BuOH fraction of the methanolic extract of *Gentiana rhodantha* (Gentianaceae). HPLC: Column, RP-18 Novapak (4µm, 150 x 3.9 mm i.d.); gradient, CH_3CN-H_2O (0.05% TFA) 5:95 -> 50:50 in 30 min (0.9 ml/min).

LC/UV/MS and LC/NMR analysis of the dichloromethane extract of Swertia calycina

The dichloromethane extract of *Swertia calycina* (Gentianaceae), from Rwanda, presented a strong antifungal activity against *Cladosporium cucumerinum* and *Candida albicans*. According to the TLC assay, this activity was linked to a strong UV visible spot (Rf = 0.38; petroleum ether - ethylacetate 1:1). As some xanthones are known to be antifungal agents (Marston *et al.* 1993), one of these could be the active principle.

The LC/UV chromatogram of dichloromethane extract of *S. calycina* was simpler than the methanolic extract and three main peaks (**2**, **17** and **18**) were detected (Fig. 2.7.). Compound **18** presented a UV spectrum with four absorption bands characteristic of a xanthone. Its TSP-MS spectrum exhibited a strong protonated ion at 303 u, indicating a xanthone with a molecular weight of 302, thus substituted by one hydroxyl and three methoxyl groups (Fig. 2.7.). This information, together with the use of a home-made UV spectral library, permitted the identification of **18** as decussatin, a widespread xanthone in the Gentianaceae family, also previously found in *C. krebsii*.

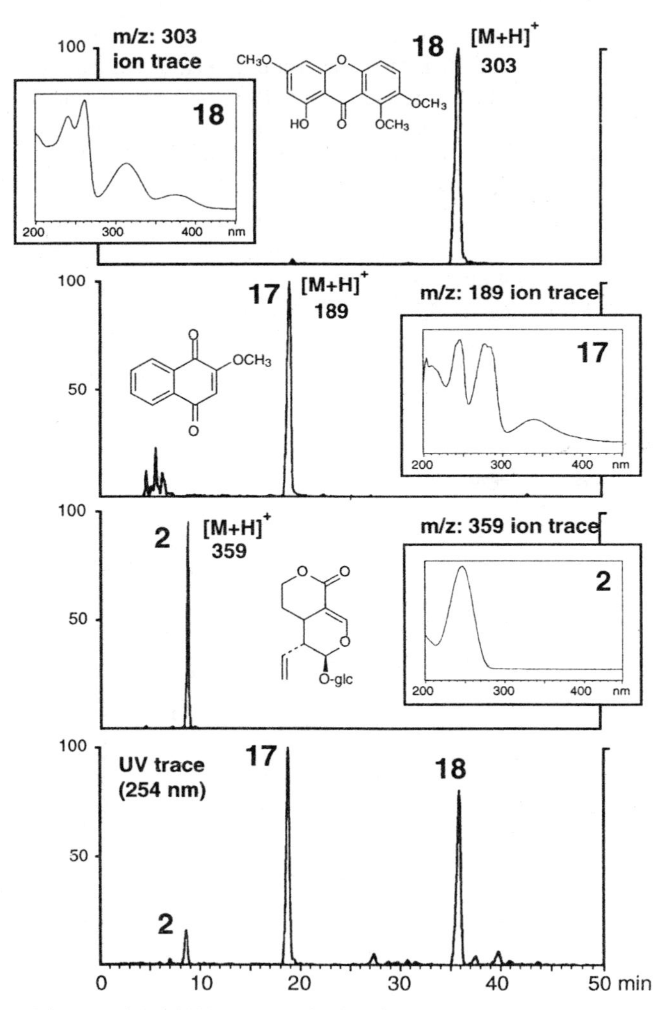

Fig. 2.7. LC/UV and LC/TSP-MS analysis of the crude CH_2Cl_2 extract of *Swertia calycina*. For each major peak, the single ion LC/MS traces of the protonated molecular ions $[M+H]^+$ are displayed, together with the UV spectra obtained on-line.

The on-line data obtained for compound **2** indicated the presence of a secoiridoid-type molecule with a molecular weight of 358 u. The loss of 162 u observable in the TSP spectrum was characteristic for the presence of a hexosyl moiety. These data suggested strongly that **2** was most probably sweroside. The UV spectrum of **17** was not attributable to a common polyphenol of the Gentianaceae such as flavones or xanthones. It was very weakly ionised in TSP (see the background ions of the TSP spectrum in Fig. 2.9.), but a protonated molecular ion was nevertheless found at m/z 189. This small molecular weight (188 u) and the UV spectrum suggested that **2** could be a quinonic compound, but as no metabolite of this type had previously been found in Gentianaceae, it was not possible to identify it on-line.

In order to confirm the attributions and to obtain more structure information on-line, the same extract of *S. calycina* was submitted to LC/^{1}H-NMR analysis on a 500 MHz instrument (Wolfender *et al.* 1996, *Phytochemical Analysis*, in press). The same LC conditions as for the LC/UV/MS analysis were used except that the water of the LC gradient system was replaced by D_2O. However the quantity of extract injected onto the column was increased to 1 mg to obtain at least 20 μg for each peak of interest.

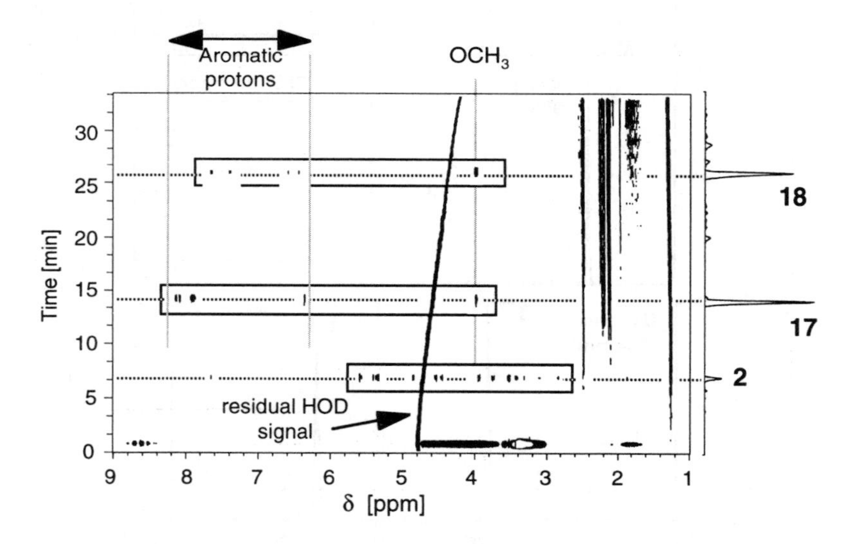

Fig. 2.8. Bidimensional LC/^{1}H-NMR chromatogram of the crude CH_2Cl_2 extract of *Swertia calycina*. Methoxyl groups and aromatic proton signals of **17** and **18** are clearly visible together with all the resonances of the monoterpene glycoside **2**. The signal of HOD is negative and was continually shifted during the LC gradient.

For the suppression of the solvent signal of MeCN and its two ^{13}C satellites, as well as the residual HOD peak, a fast sequence called WET (Smallcombe *et al.* 1995) was run before each acquisition. In the gradient LC run, the solvent peaks change frequency during the course of the experiment. As a result, the solvent suppression must be continuously adjusted for optimal performance. To do this, a one-transient-one-pulse experiment is used to find the solvent peaks prior WET suppression (Scout Scan) (Smallcombe *et al.* 1995). Thank to this sequence, the transmitter is automatically adjusted to keep the biggest solvent peak at a constant frequency (MeCN), while the spectrometer is locked on D_2O (Wolfender *et al.* 1996).

The on-line LC/NMR analysis of *S. calycina* provided ^1H-NMR spectra for all the major constituents. A plot of the retention time (y axis) versus the NMR shifts (x axis) permitted the localisation of the resonances of compounds **2**, **17** and **18** (Fig. 2.8.). On this unusual two-dimensional chromatogram, strong signals of aromatic methoxyl groups were observed around 4 ppm for **17** and **18**. Xanthone **18** exhibited two pairs of aromatic protons, while the quinonic compound **17** presented five other low field protons. The more polar secoiridoid **2** showed different signals between 3 and 6 ppm. The important trace starting from 4.8 ppm (at 0 min) and ending to 4 ppm (at 30 min) was due to the change of the chemical shift of the residual negative water (HOD) signal during the LC gradient. The traces between 1 and 2.6 ppm were due to residual MeCN signal and solvent impurities.

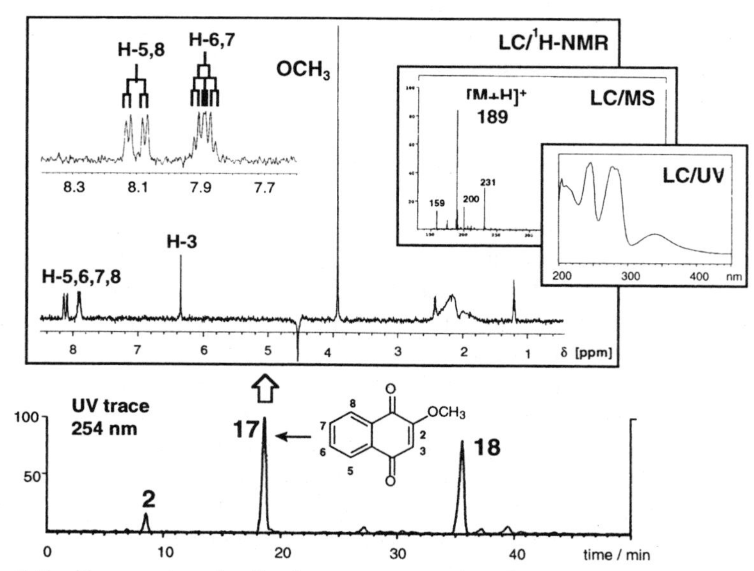

Fig. 2.9. Summary of all the spectroscopic data obtained on-line for naphthoquinone **17** in the CH_2Cl_2 extract of *Swertia calycina*.

A slicing of this bidimensional plot in single on-line LC/^{1}H-NMR spectra for each constituent allowed a precise assignment of their specific resonances. Xanthone **18** exhibited three methoxyl groups at 3.92, 3.93 and 3.95 ppm. A pair of *meta* coupled aromatic protons (1H, δ 6.55, *d*, *J*=2.4, H-4) and (1H, δ 6.41, *d*, *J*=2.4, H-2) was indicative of a 1,3-disubstituted A-ring. The B-ring protons exhibited a pair of *ortho* coupled protons (1H, δ 7.36, *d*, *J*=9.2, H-5) and (1H, δ 7.62, *d*, *J*=9.2, H-6), suggesting a 1,3,5,6 or 1,3,7,8 substitution pattern for **18**. A comparison of these NMR data together with UV and MS information allowed **18** to be identified as 1-hydroxy-3,7,8-trimethoxyxanthone (decussatin), as already suggested by the LC/UV/MS data.

On the LC/^{1}H NMR spectrum of **17** (Fig. 2.9.), two signals (2H, δ 8.11, *m*, H-5,8 and 2H, δ 7.89, *m*, H-6,7) were characteristic of four adjacent protons of an aromatic ring with two equivalent substituents. The low field shift of the H-5,8 signal indicated that these two protons were in a *peri* position to the carbonyl functions, suggesting most probably the presence of a naphthoquinone nucleus (Thomson 1987). The strong bands recorded in UV at 243, 248, 277, 330 nm confirmed this deduction. The singlet at 4.35 ppm was attributed to H-3 and the remaining methoxyl group was thus at position C-2. With this on-line data and the molecular weight deduced from the LC/TSP-MS spectrum (MW 188), **17** was finally identified as 2-methoxy-1,4-naphthoquinone (Fig. 2.9.). As this was the first naphthoquinone to be reported in Gentianaceae, it was isolated and was found to be the compound responsible for the strong antifungal activity of the extract of *S. calycina* (Rodriguez *et al.* 1995a).

This example showed that the LC/UV/MS and LC/NMR information obtained for *S. calycina* permitted a full structure identification of its main constituents.

Search for molluscicidal compounds from plants

Schistosomiasis is a parasitic disease affecting more than 200 million people in 70 countries throughout South America, Africa and the Far East (D'Arcy and Harron 1983). Certain species of aquatic snail (of the genera *Biomphalaria*, *Bulinus* or *Oncomelania*) act as intermediate hosts in the life cycle of the parasitic schistosomes, trematodes which are responsible for the symptoms of the disease in humans. Destroying the snails that harbour the developing schistosome larvae is one way of interrupting the parasite's life cycle and preventing human infection. Although chemotherapy with orally administered anti-schistosomal drugs is very successful, the best method of controlling schistosomiasis is to use a combination of chemotherapy, together with molluscicidal (snail-killing) treatment of snail habitats. As plants provide some very powerful molluscicidal agents, a large amount of effort has been spent over the last few years in the discovery of new plant molluscicides (Mott 1987).

Studies on the constituents of Swartzia madagascariensis

One of the most promising plants for the potential control of schistosomiasis-transmitting snails in Africa, besides *Phytolacca dodecandra* (phytolaccaceae) (see chapter 19), is *Swartia madagascariensis* (Leguminosae). Aqueous extracts of the seed pods contain large amounts of saponins with high molluscidal activity. These saponins have been characterised and field trials with extracts of the fruits have been performed in Tanzania (Hostettmann 1989). The most active saponins are glycosides of oleanolic acid.

LC/TSP-MS, LC/ES-MS and LC/CF-FAB-MS of Swartzia madagascariensis

LC/ES-MS was compared to LC/TSP-MS and LC/CF-FAB-MS for the analysis of saponins in the crude MeOH extracts of the fruits of *Swartzia madagascariensis* (Leguminosae) (Borel *et al.* 1987). For the LC/MS analysis of *S. madagascariensis,* three oleanolic acid saponins **19**, **20** and **21**, bearing respectively 4, 3 and 2 sugar units were selected (Fig. 2.10) (Wolfender *et al.* 1995).

The LC/TSP-MS (P.I., NH_4OAc buffer) analysis of the extract exhibited the presence of triterpene glycosides derived from oleanolic acid (MW: 456). Indeed, the TSP trace recorded at *m/z* 439 was characteristic for dehydrated oleanolic acid moieties $[A+H-H_2O]^+$ (Fig. 2.11a.). For **19**, a distinctive ion at *m/z* 796 and a fragment ion at *m/z* 650 were characteristic for a saponin bearing a diglycosidic moiety consisting of a terminal deoxyhexose unit (-146 u) and a glucuronic acid (-176 u) moiety. As rhamnose is the most frequent deoxyhexose occurring in saponins, it can be assumed from these on-line MS data that **19** was a saponin of oleanolic acid, substituted by a glucuronic acid unit and a rhamnose in the terminal position. The TSP spectra of saponins **20** and **21** were less clear than those of **19**. In both cases, characteristic signals for the oleanolic acid moiety were present, and fragment ions at *m/z* 795 were indicative of the presence of at least a glucuronic acid with a hexose unit. On the traces at *m/z* 941 and *m/z* 1103, no clear molecular ions for tri- or higher glycosylation were visible. For these two metabolites, the LC/TSP-MS analysis alone could not give enough structural information on-line.

In the LC/CF-FAB-MS (N.I., glycerol matrix) analysis of the same extract (Fig. 2.11b.), in contrast to the TSP results, all the saponins found in the extract exhibited intense deprotonated molecular ions $[M-H]^-$ and very weak ions characteristic for the aglycone moiety $[A-H]^-$ (*m/z* 455) and $[A-H-H_2O]^-$ (*m/z* 437). Furthermore, different characteristic cleavages were distinguishable. For **19**, ions at *m/z* 777 $[M-H]^-$, *m/z* 631 $[M-H-146]^-$ and *m/z* 455 $[A-H]^-$ confirmed the results obtained with TSP. For **20**, an intense $[M-H]^-$ ion at *m/z* 939 was observed in the CF-FAB spectrum, showing that it was a triglycosylated saponin (see trace *m/z* 939, Fig. 2.11b.). The different fragment ions recorded in the CF-FAB spectrum of **20** (*m/z* 777 $[M\text{-}H\text{-}162]^-$ and *m/z* 793 $[M\text{-}H\text{-}146]^-$) confirmed that **20** was probably similar to **19** with one more hexose unit in position C-28 or branched at the diglycoside moiety. The CF-FAB spectra of **21** exhibited an intense molecular ion at *m/z* 1101 $[M\text{-}H]^-$ (see trace *m/z* 1101, Fig. 2.11b.). This indicated that **21** has one hexose unit (164 u) more than **20**. Saponin **21** was thus a tetraglycosylated triterpene.

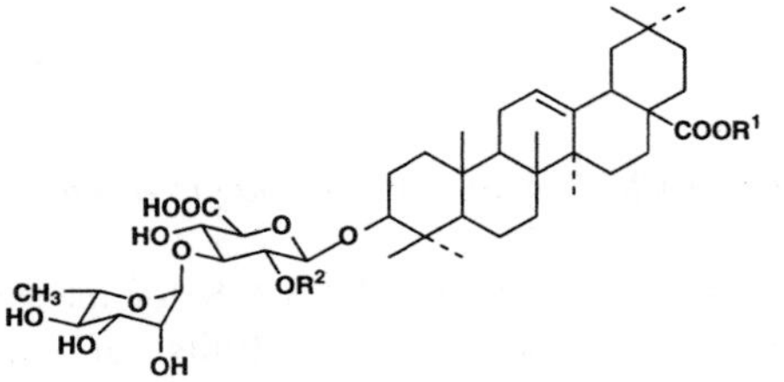

	R^1	R^2	MW	Sug.
21	H	H	778	2
20	H	Glc	940	3
19	Glc	Glc	1102	4

Fig. 2.10. Structures of selected saponins isolated from *Swartzia madagascariensis* (Leguminosae).

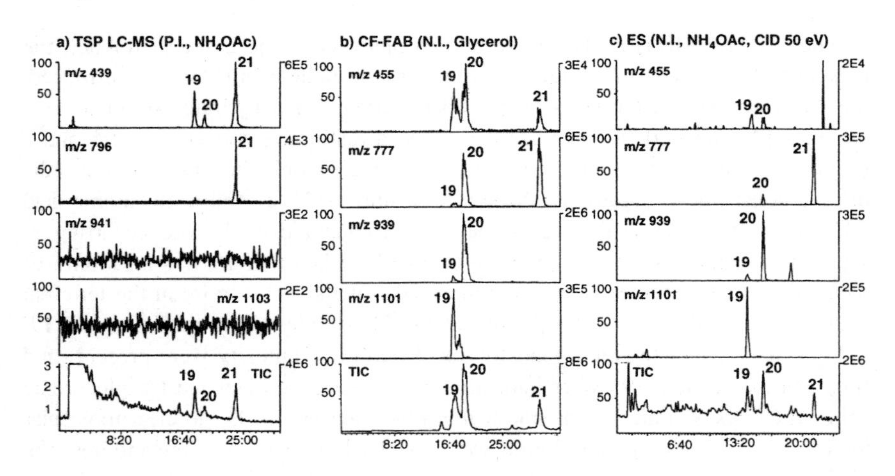

Fig. 2.11. Combined LC/TSP-MS, LC/CF-FAB-MS and LC/ES-MS of the MeOH extract of *Swartzia madagascariensis* (Leguminosae). HPLC: C18 Nova-Pak (4μm, 150 x 3.9 mm i.d.); gradient, $CH_3CN\text{-}H_2O$ (0.05% TFA) 30:70 -> 50:50 in 30 min (0.9 ml/min).

The LC/ES-MS analysis of the same extract (N.I., NH_4OAc buffer, CID 50V) gave intense and clearly discernible $[M\text{-}H]^-$ ions and weak acetate anion adducts $[M+CF_3COO^-]$ for saponins **19-21** (Fig. 2.11c.). With the aid of up front CID in the ES source, ions due to the loss of different sugar moieties were also observable.

Aglycone ions at *m/z* 455 appeared only for **19** and **20** but not for **21**. In ES, almost no peak broadening was observed producing clearly defined peaks in the selected ion traces (Fig. 2.11c.). In CF-FAB, on the contrary, large diminution of the chromatographic resolution was due to the important splitting and the post column addition of the glycerol matrix (Fig. 2.11b.).

This example clearly shows the importance of the choice of the right interface for the ionisation of a given type of molecule. Conditions have to be chosen carefully according to the type of compounds that have to be screened by LC/MS. LC/ES-MS gave mainly molecular ions but structural information can be efficiently obtained by the subsequent fragmentation of these ions in MS/MS, as shown for *Phytolacca dodecandra* (Phytolaccaceae) (See chapter 19).

Conclusion

Today the work of phytochemists is based mainly on the bioassay-guided fractionation of crude plant extracts. This type of approach has led to the isolation of numerous compounds with interesting activities. Plants contain thousands of constituents with a variety of different biological properties. Obviously for any successful investigation of biological material with such a wide range of properties, the future lies in the ability to have as large a number of biological screens as possible. It is also important to take into account the valuable contribution provided by traditional medicine (Marston and Hostettmann 1987). There are, however, many obstacles which seriously hamper this type of investigation. First, it is possible that a broad range of structurally diverse compounds contributes to the overall pharmacological activity of a plant and synergistic effects between the active principles may exist. This is particularly the case for medicinal plants possessing less specific activities. Secondly, there is an urgent need for more appropriate pharmacological models (Hamburger and Hostettmann 1991). Existing assays are quite often not reliably predictive for clinical efficiency.

Chemical screening of crude plant extracts which allows the localization and efficient targeted isolation of new types of constituents with potential activities, can be considered as an efficient complementary approach to bioassay-guided fractionation. The addition of highly sensitive analytical techniques such as HPLC, GC and coupled instrumentation (LC/UV and LC/MS) enables rapid initial screening of crude plant extracts. Hyphenated techniques provide a great deal of preliminary information about the content and nature of constituents of these extracts. In certain cases, combination with a spectral library and pre- or post-column derivatization allows structure determination on-line. This is very useful when large numbers of samples have to be processed because unnecessary isolation of compounds is avoided. Once the novelty or utility of a given constituent is established, it is then important to process the plant extracts in the usual manner, to obtain samples for full structure elucidation and biological or pharmacological testing. In this chapter, it has been shown that in the search for xanthones with potential monoamine oxidase-inhibitory activity, LC/MS and LC/UV analysis have been of great value for the preliminary screening of numerous different species of Gentianaceae: this method has been used not only to determine the presence of xanthones but also to provide an indication of their structure. The use of LC/UV and LC/MS has also allowed the localization and targetted isolation of new types of constituents such as high molecular weight secoiridoids. Furthermore, by selective ion monitoring in LC/MS or even LC/MSMS, it is possible to achieve the detection of specific target molecules - those, for example, which have already been found to exhibit a particular activity. In this field, the recent introduction of other hyphenated techniques such as LC-NMR (Spraul *et al.* 1993) will render the on-line structural determination of metabolites more and more accurate and rapid. All these hyphenated techniques will provide a very efficient investigation tool for the phytochemist and allow a rapid chemical screening of a large number of crude plant extracts with a minute amount of material.

Both biological and chemical screening provide important information on the plant constituents but these results will not be sufficient enough for the discovery of new potent drugs if no suitable pharmacological model exists. Thus, an important condition for success in the discovery of new bioactive plant constituents is the establishment of effective collaborations between botanists, phytochemists and pharmacologists in order to realise all the steps involved, starting with plant material in the field and concluding with viable pharmacologically-active compounds. Only in this manner can medicinal plants be investigated efficiently and new leads rapidly discovered .

References

Bhattacharya, S. K., Ghosal, S., Chaudhuri, R. K. and Sanyal, A. K. (1972). *Canscora decussata* (Gentianaceae) xanthones. III. Pharmacological studies. *Journal of Pharmaceutical Sciences* **61**, 1838-1840.

Blakley, C. R. and Vestal, M. L. (1983). Thermospray interface for liquid chromatography/mass spectrometry. *Analytical Chemistry* **55**, 750-754.

Borel, C. and Hostettmann, K. (1987). Molluscicidal saponins from *Swartzia madagascariensis* Desvaux. *Helv. Chim. Acta* **70**, 570-576.

Caprioli, R. M., Tan, F. and Cotrell, J. S. (1986). Continuous-flow sample probe for fast atom bombardment mass spectrometry. *Analytical Chemistry* **58**, 2949-2954.

Cassady, J. M., Baird, W. M. and Chang, C. J. (1990). Natural products as a source of potential cancer chemotherapeutic and chemopreventive agents. *Journal of Natural Products* **53**, 23-41.

D'Arcy, P. F. and Harron, D. W. G. (1983). Schistosomiasis. *Pharmacy International* **4**, 16-20.

Delini-Stula, M. (1983). *The Origin of Depression: Current Concept and Approaches,* pp. 351-365. Dahlem Konferenzen, Springer, Heidelberg.

Deulofeu, V. (1967). Chemical compounds isolated from *Banisteriopsis* and related species. In *Ethnopharmacologic Search for Psychoactive Drugs* (eds. D.H. Efron, B. Holmstedt and N.S. Kline), pp. 393-402. US Public Health Service publ. No 1645, Washington.

Farnsworth, N. R. and Kass, C. J. (1981). An approach utilizing information from traditional medicine to identify tumor-inhibiting plants. *Journal of Ethnopharmacology* **5**, 85-99.

Ghosal, S., Jaiswal, D. K., Singh, S. K. and Srivastava, R. (1985). Chemical constituents of Gentianaceae. Part 32. Dichotosin and dichotosinin, two adaptogenic glucosyloxy flavans from *Hoppea dichotoma*. *Phytochemistry* **24**, 831-833.

Hamburger, M. and Hostettmann, K. (1991). Bioactivity in plants: the link between phytochemistry and medicine. *Phytochemistry* **30**, 3864-3874.

Hostettmann, K. and Wagner, H. (1977). Xanthone glycosides. *Phytochemistry* **16**, 821-829.

Hostettmann, K. (1989). Plant-derived molluscicides of current importance. In *Economic and Medicinal Plant Research,* Vol. 3 (eds. H. Wagner, H. Hikino and N.R. Farnsworth), pp. 73-102. Academic Press, London.

Hostettmann, K. and Hostettmann, M. (1989). Xanthones. In *Methods in Plant Biochemistry* (ed. J.B. Harborne), pp. 493-508. Academic Press, London.

Hostettmann, K., Marston, A. and Wolfender, J.-L. (1995). Strategy in the search for new biologically active plant constituents. In *Phytochemistry of Plants Used in Traditional Medicine* (eds. K. Hostettmann, A. Marston, M. Maillard and M. Hamburger), pp. 17-45. Oxford Science Publications, Oxford.

Kaldas, M. (1977). *Identification des composés polyphénoliques dans Gentiana campestris L., Gentiana germanica Willd. et Gentiana ramosa Hegetschw*. Thesis, Université de Neuchâtel, Switzerland.

Kielholtz, P. (1982). Intravenöse Behandlung Depresssiver Zustandsbilder. *Journal Suisse de Pharmacie* **120**, 507-511.

Lins Mesquita, A. A., De Barros Correa, D., Gottlieb, O. R. and Taveira Magalhaes, M. (1968). Methods for the structural investigation of xanthones: Part II. Location of hydroxyl groups by ultraviolet and visible spectroscopy. *Analytica Chimica Acta* **42**, 311-323.

Ma, W.-G., Fuzzati, N., Wolfender, J.-L., Hostettmann, K. and Yang, C. (1994). Rhodenthoside A, a new type of acylated secoiridoid glycoside from *Gentiana rhodentha*. *Analytica Chimica Acta* **77**, 1660-1671.

Markham, K. R. (1965). Gentian pigments (II); Xanthones from *Gentiana bellidifolia*. *Tetrahedron* **21**, 1449-1452.

Marston, A. and Hostettmann, K. (1987). Antifungal, molluscicidal and cytotoxic compounds from plants used in traditional medicine. In *Biologically Active Natural Products, Proceedings of the Phytochemical Society of Europe*, Vol. 27 (eds. K. Hostettmann and P.J. Lea), pp. 65-83. Oxford Science Publication, Oxford.

Marston, A., Hamburger, M., Sordat-Diserens, I., Msonthi, J. D. and Hostettmann, K. (1993). Xanthones from *Polygala nyikensis*. *Phytochemistry* **33**, 809-812.

Mott, K. E. (1987). *Plant Molluscicides*. Wiley, Chichester.

Okpanyi, S. N. and Weischer, M. L. (1987). Tierexperimentelle Untersuchungen zur psychotropen Wirksamkeit eines Hypericum- Extraktes. *Arzneimittelforschung* **37**, 10-13.

Rodriguez, S., Wolfender, J.-L., Hakizamungu, E. and Hostettmann, K. (1995a). An antifungal naphthoquinone, xanthones and secoiridoids from *Swertia calycina*. *Planta Medica* **61**, 362-364.

Rodriguez, S., Wolfender, J.-L., Odontuya, G., Purev, O. and Hostettmann, K. (1995b). Xanthones, secoiridoids and flavonoids from *Halenia corniculata*. *Phytochemistry* **40**, 1265-1272.

Schaufelberger, D. and Hostettmann, K. (1988). Chemistry and pharmacology of *Gentiana lactea*. *Planta Medica* **54**, 219-221.

Smallcombe, S. H., Patt, S. L. and Keiffer, P. A. (1995). WET solvent suppression and its application to LC NMR and high-resolution NMR spectroscopy. *Journal of Magnetic Resonnance* **Series A 117**, 295-303.

Sparenberg, B. (1993). *MAO-inhibierende Eigenschaften von Hypericum-inhaltsstoffen und Untersuchungen zur Analytik und Isolierung von Xanthonen aus Hypericum perforatum* . PhD Thesis, University of Marburg.

Spjut, R. W. and Perdue, R. E. (1976). Plant folklore: A tool for predicting sources of antitumor activity? *Cancer Treatment Reports* **60**, 979-985.

Spraul, M., Hoffmann, M., Lindon, J. C., Nicholson, J. K. and Wilson, I. D. (1993). Liquid chromatography coupled with high field proton nuclear magnetic resonance spectroscopy: Current status and future prospects. *Analytical Proceedings* **30**, 390-392.

Stroelin-Benedetti, M. and Dostert, P. (1992). Monoamine oxidase: from physiology and pathophysiology to the design and clinical application of reversible inhibitors. In *Advances in Drug Research* (ed. B. Testa), pp. 65-125. Academic Press, London.

Suffness, M. and Douros, J. (1982). Current status of the NCI plant and animal product program. *Journal of Natural Products* **45**, 1-14.

Suzuki, O., Katsumata, Y., Oya, M., Chari, V. M., Klapfenberger, R., Wagner, H. and Hostettmann, K. (1978). Inhibition of type A and type B monoamine oxidase by isogentisin and its 3-O-glucoside. *Biochemical Pharmacology* **27**, 2075-2078.

Suzuki, O., Katsumata, Y., Oya, M., Chari, V. M., Klapfenberger, R., Wagner, H. and Hostettmann, K. (1980). Inhibition of type A and Type B monoamine oxidase by isogentisine and its 3-O-glucoside. *Planta Medica* **39**, 19-23.

Suzuki, O., Katsumata, Y., Oya, M., Chari, V. M., Klapfenberger, R., Wagner, H. and Hostettmann, K. (1981). Inhibition of type A and type B monoamine oxidases by naturally occurring xanthones. *Planta Medica* **42**, 17-21.

Suzuki, O., Katsumata, Y., Oya, M., Bladt, S. and Wagner, H. (1984). Inhibition of monoamine oxidase by hypericin. *Planta Medica* **50**, 272-274.

Thomson, R. H. (1987). *Naturally Occurring Quinones*. University Press, Cambridge.

Thull, U. (1995). *Monoamine oxidase inhibitors of natural and synthethic origin: Biological assay and 3D-QSAR*. Thesis, Université de Lausanne.

Whitehouse, R. C., Dreyer, R. N., Yamashita, M. and Fenn, J. B. (1985). Electrospray interface for liquid chromatographs and mass spectrometers. *Analytical Chemistry* **57**, 675-679.

Wolfender, J.-L., Hamburger, M., Msonthi, J. D. and Hostettmann, K. (1991). Xanthones from *Chironia krebsii*. *Phytochemistry* **30**, 3625-3629.

Wolfender, J.-L. and Hostettmann, K. (1993). Liquid chromatographic-UV detection and liquid chromatographic-thermospray mass spectrometric analysis of *Chironia* (Gentianaceae) species. A rapid method for the screening of polyphenols in crude plant extracts. *Journal of Chromatography* **647**, 191-202.

Wolfender, J.-L., Maillard, M. and Hostettmann, K. (1994). Thermospray liquid chromatography-mass spectrometry in phytochemical analysis. *Phytochemical Analysis* **5**, 153-182.

Wolfender, J.-L., Rodriguez, S., Hostettmann, K. and Wagner-Redeker, W. (1995). Comparison of liquid chromatography/electrospray, atmospheric pressure chemical ionisation, themospray and continuous-flow fast atom bombardment mass spectrometry for the determination of secondary metabolites in crude plant extracts. *J. Mass Spectrom.* , S35-S46.

3. International collaboration in drug discovery and development. The United States National Cancer Institute experience

G.M. CRAGG[1], M.R. BOYD[1], M.A. CHRISTINI[2], T.D. MAYS[2], K.D. MAZAN[2] AND E.A. SAUSVILLE[1]

[1]Developmental Therapeutics Program, Division of Cancer Treatment, Diagnosis and Centers and [2]Office of Technology Development, National Cancer Institute, Bethesda, Maryland 20892

Introduction

In 1937, the United States National Cancer Institute (NCI) was established with its mission being "to provide for, foster and aid in coordinating research related to cancer". The NCI is the largest of seventeen Institutes which comprise the National Institutes of Health (NIH), which are components of the Federal Government's Department of Health and Human Services. The NCI and NIH are entirely funded through appropriations from the U.S. Congress and, as such, are entirely non-commercial and non-profit. Thus, while the NCI will attempt to license drugs discovered through its screening programs to pharmaceutical companies for advanced development and marketing, chemotherapeutic agents not licensed, but considered by NCI to be clinically effective, will be distributed by the NCI at no cost to the patients. An example is the antileukemic agent, Erwinia L-asparaginase, which was procured by the NCI and distributed free of charge until its production and marketing were transferred to the private sector.

In 1955, NCI set up the Cancer Chemotherapy National Service Center (CCNSC) to promote a cancer chemotherapy program, involving the procurement, screening, preclinical development, and clinical evaluation of new agents. All aspects of drug discovery and preclinical development are now the responsibility of the Developmental Therapeutics Program (DTP), a major component of the Division of Cancer Treatment, Diagnosis and Centers (DCTDC). During the past 40 years, over 300,000 chemicals submitted by investigators and organizations worldwide, have been screened for antitumor activity, and NCI has played a major role in the discovery and development of almost all of the available commercial and investigational anticancer agents (Boyd 1993).

Naturally-derived anticancer agents

From 1960 to 1982, over 180,000 microbial fermentation products, and over 114,000 plant-derived and 16,000 marine organism derived extracts were tested for *in vivo* antitumor activity, mainly using the L1210 and P388 mouse leukemia models. Extracts showing significant activity were subjected to bioassay-guided fractionation, and the isolated, active agents were submitted for secondary testing against panels of animal tumor models and human tumor xenografts. Those agents showing significant activity were assigned priorities for preclinical and clinical development.

Much of the drug discovery effort was carried out through collaborations with research organizations and the pharmaceutical industry, which either submitted compounds on a voluntary basis or were supported by NCI through contract or grant funding mechanisms. A large number of novel agents belonging to a wide variety of chemical classes were isolated and characterized (Cassady and Douros 1979), but few of these new agents satisfied the stringent preclinical development requirements and advanced to clinical trials (Cragg *et al.* 1993a). Commercial agents of microbial origin include actinomycin D, bleomycin and doxorubicin (adriamycin) (Fig. 3.2), while plant-derived commercial drugs include vinblastine, vincristine, etoposide and teniposide (semisynthetic derivatives of epipodophyllotoxin, an epimer of podophyllotoxin), and taxol (Fig. 3.3). While no marine organism-derived agent has yet been approved for commercial use, bryostatin isolated from the bryozoan, *Bugula neritina*, is showing promise in clinical trials (Fig. 3.4) (Philip *et al.* 1993). Over ten natural product agents are in various stages of clinical development and, of these, the semisynthetic derivatives of camptothecin, irenotecan (CPT-11), topotecan and 9-aminocamptothecin (Fig. 3.5), are showing promising clinical activity against a variety of cancer disease types (Wall and Wani 1993). Over 60% of the compounds currently in preclinical and clinical development by NCI are of natural product origin (Fig. 3.1).

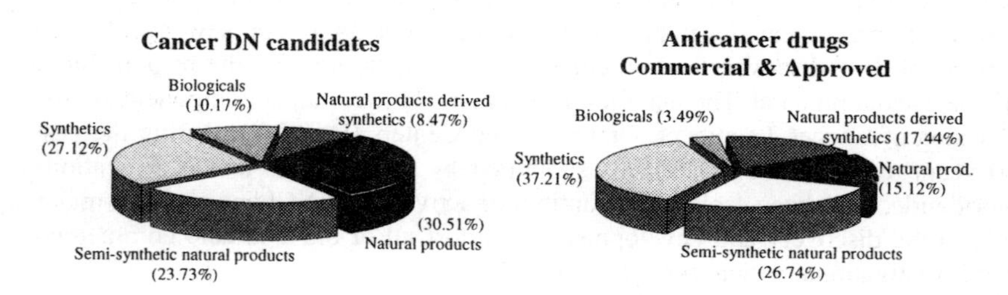

Fig. 3.1. Natural products in drug discovery.

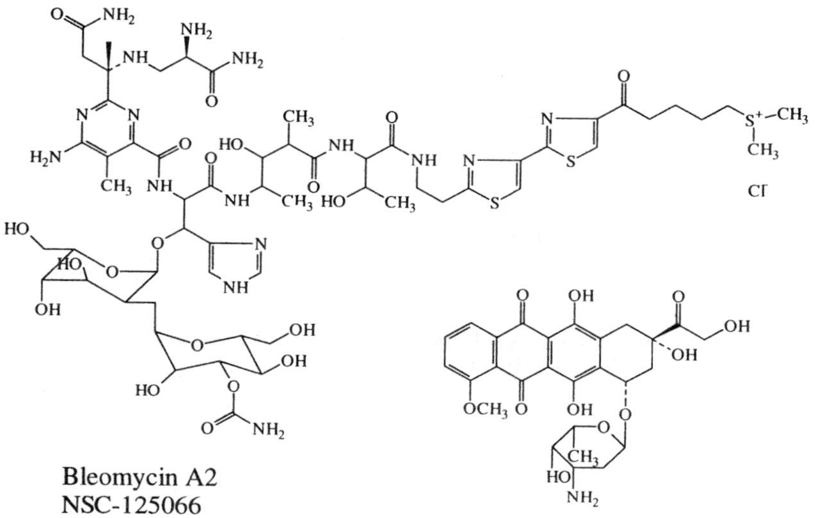

Bleomycin A2
NSC-125066

Adriamycin hydrochloride
NSC-123127

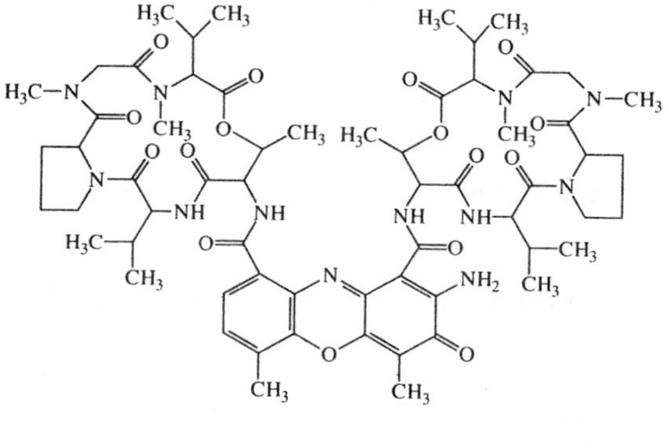

Actinomycin D
NSC-3053

Fig. 3.2. Commercial agents of microbial origin.

Drug discovery at the NCI: Current status

While the primary focus of the NCI drug discovery program remains the search for new anticancer agents, in 1987 the mission of the NCI was expanded to

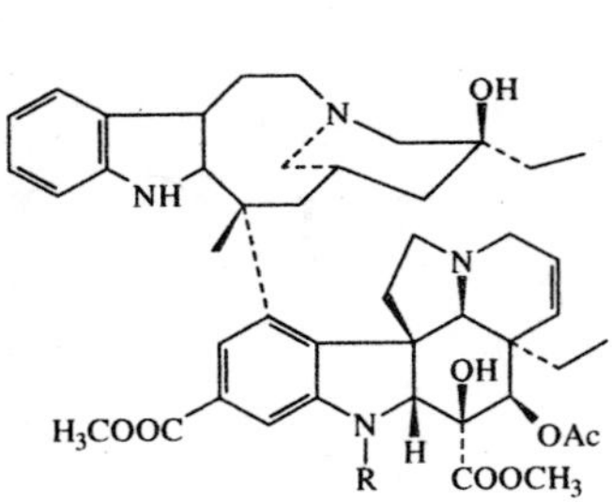

Vincristine: R = CHO
Vinblastine: R = CH_3

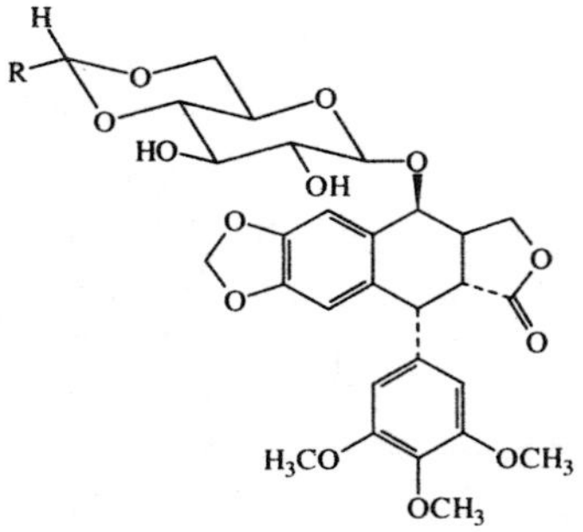

Etoposide: R = CH_3
Teniposide: R =

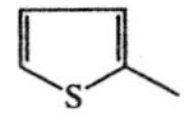

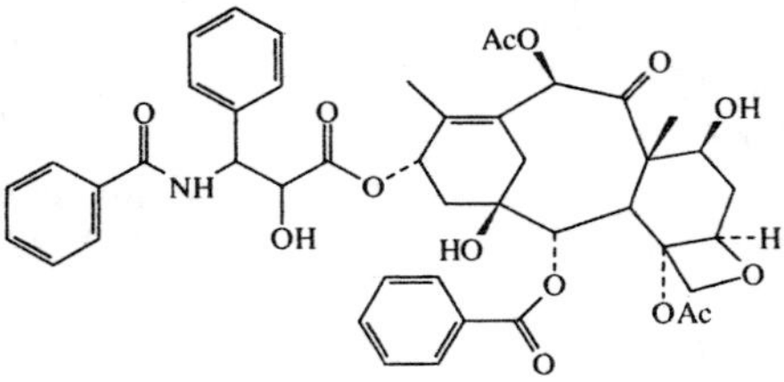

Taxol

Fig. 3.3. Plant-derived commercial drugs.

include the discovery and preclinical development of agents for the treatment of acquired immunodeficiency syndrome (AIDS) (Boyd 1988).

The NCI places a major emphasis on natural products as a source of potential new anticancer and anti-HIV agents, and continues to investigate all natural sources. A contract for the cultivation and extraction of fungi is being performed through a contract with Science Applications International Corporation (SAIC) in NCI facilities at Frederick, Maryland, while the collection of marine invertebrates is being carried out in the Indo-Pacific region through a contract with Coral Reef Research Foundation (CRRF). Collections have been concentrated in the south western Pacific Ocean, but negotiations for expanding the collection program to Madagascar and Tanzania are currently in progress. Terrestrial plant collections have been carried out in over 25 countries in tropical and subtropical regions worldwide through contracts with Missouri Botanical Garden (Africa and Madagascar), New York Botanical Garden (Central and South America), and the University of Illinois at Chicago (Southeast Asia).

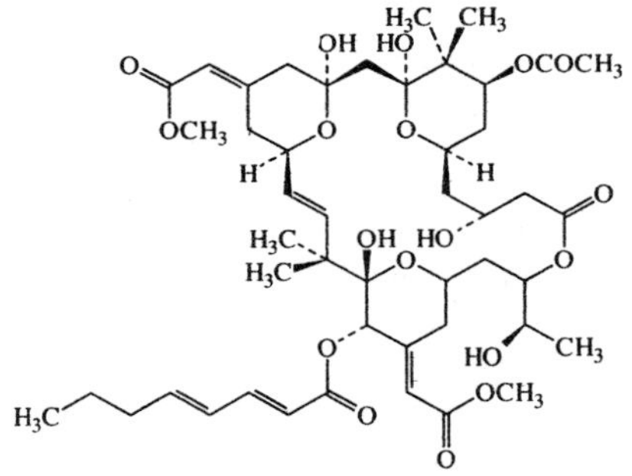

Fig. 3.4. Bryostatin.

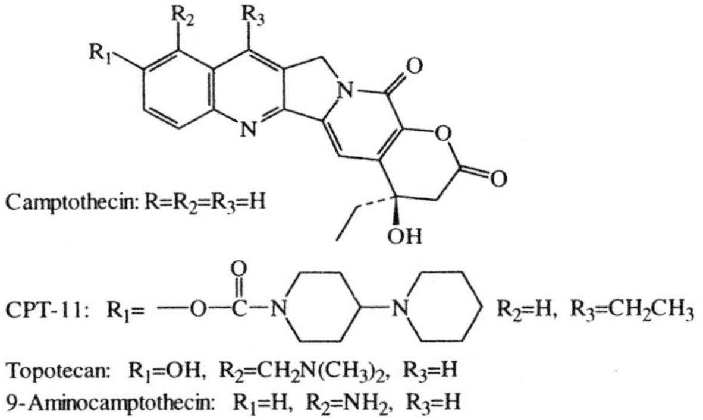

Camptothecin: $R=R_2=R_3=H$

CPT-11: $R_1=$ —O—C(=O)—N ... N ... $R_2=H$, $R_3=CH_2CH_3$

Topotecan: $R_1=OH$, $R_2=CH_2N(CH_3)_2$, $R_3=H$

9-Aminocamptothecin: $R_1=H$, $R_2=NH_2$, $R_3=H$

Fig. 3.5. Semisynthetic derivatives.

In carrying out these collections, the NCI contractors work closely with qualified organizations in each of the source countries. Organizations which have collaborated with the NCI contractor in Africa and Madagascar are listed in Table 3. 1.

Table 3.1. Plant Collection Collaborations in Africa and Madagascar

Cameroon	University of Yaounde; Committee for the Follow-up of the exploration and protection of *Ancistrocladus korupensis*
Gabon	Centre National de la Recherche Scientifique et Technologique
Ghana	University of Ghana, Legon; Ghana Herbarium
Madagascar	Centre National d'Application des Recherches Pharmaceutiques
Tanzania	Institute of Traditional Medicine, Muhimbili University College of Health Sciences.

To date, botanists and marine biologists from source country organizations have collaborated in field collection activities and taxonomic identifications, and their knowledge of local species and conditions has been indispensable to the success of the NCI collection operations. Source country organizations provide facilities for the preparation, packaging, and shipment of the samples to the NCI natural products repository in Frederick, Maryland. The collaboration between the source country organizations and the NCI collection contractors has, in turn, provided support for expanded research activities by source country biologists, and the deposition of a voucher specimen of each species collected in the national herbarium or repository is expanding source country holdings of their biota. When requested, NCI contractors also provide training opportunities for local personnel through conducting of workshops and presentation of lectures. In addition, through its Letter of Collection (LOC) and agreements based upon it, the NCI invites scientists nominated by Source Country Organizations to visit its facilities, or equivalent facilities in other approved U.S. organizations, to participate in collaborative natural products research. To date, scientists (usually chemists) from all the countries listed in Table 3. 1, with the exception of Gabon, have carried out joint research projects for 3-12 months in U.S. laboratories, sponsored by NCI, while representatives of each country have visited NCI and Missouri Botanical Garden for shorter periods.

Dried plant samples (0.3-1kg dry weight) and frozen marine organism samples are shipped to the NCI Natural Products Repository (NPR) in Frederick where they are stored at -20°C prior to extraction with a 1:1 mixture of methanol:dichloromethane and water to give organic solvent and aqueous extracts. All the extracts are assigned discrete NCI extract numbers and returned to the NPR for storage at -20°C until requested for screening or further investigation.

Extracts are tested *in vitro* for selective cytotoxicity against panels of human cancer cell lines representing major disease types, including leukemia, melanoma, lung, breast, colon, central nervous system, ovarian, prostate and renal cancers (Boyd 1989). *In vitro* anti-AIDS activity is determined by measuring the survival of virus-infected human lymphoblastoid cells in the presence or absence of the extracts (Boyd 1988). Extracts showing significant selective cytotoxicity or anti-AIDS activity are subjected to bioassay-guided fractionation by chemists and biologists to isolate the pure chemical constituents responsible for the observed activity (Fig. 3.6). Bioassay-guided fractionation is essential since, in most instances, the active constituents are present in only small amounts in the crude extracts, and are generally isolated in yields of 0.01% or less, based on the mass of raw material. After the active constituent is isolated from an extract, its complete chemical structure is elucidated using modern spectroscopic techniques, and, if necessary and possible, X-ray crystallography.

This isolation and structural elucidation of a potential new agent is but the first phase in a lengthy process of development towards clinical trials, and possible general clinical use.

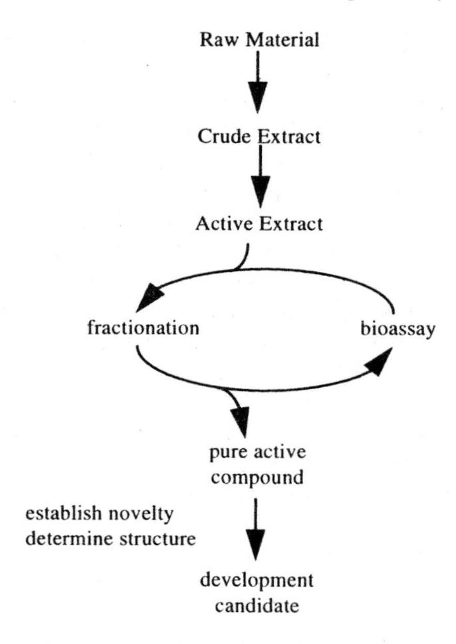

Fig. 3.6. Discovery of new natural products.

Drug development

Agents showing significant activity in the primary *in vitro* human cancer cell line or anti-AIDS screens are entered into various stages of preclinical development to determine their suitability for eventual advancement to clinical trials with human patients.

Preclinical development

Large-scale production of natural products - The initial plant sample (0.3-1.0 kg) collected by the contractor generally yields enough extract (10-40 g) to permit isolation of the pure, active constituent in sufficient milligram quantity for complete structural elucidation. Subsequent secondary testing and preclinical development, however, might require gram or even kilogram quantities, while approval for clinical development could require multi-kilogram quantities.

To obtain sufficient quantities of an active agent for preclinical development, recollections of 5 to 100 kg of dried plant material, preferably from the original collection location, might be necessary; considerably larger quantities (sometimes exceeding 1000 kg) would be required for subsequent clinical development. The performance of large recollections necessitates surveys of the distribution and abundance of the plant, as well as determination of the variation of drug content in different plant parts and the fluctuation of content with the season of harvesting. The potential for mass cultivation of the plant would also need to be assessed. If problems are encountered due to scarcity of the wild plant or inability to adapt it to cultivation, a search for alternative sources would be necessary. Other species of the same genus, or closely related genera, can be analyzed for drug content, and techniques, such as plant tissue culture, can be investigated. While total synthesis must always be considered as a potential route for bulk production of the active agent, it should be noted that the structures of most bioactive natural products are extremely complex, and laboratory bench-scale syntheses often are not readily adapted to large-scale economic production.

The investigation of methods for the large-scale production of an active agent needs to be initiated well before it can be determined whether or not it will become a successful drug. Failure to address this issue could result in a supply crisis should an agent prove to be effective in clinical trials and advance to commercial use. The preclinical development of the *in vitro* active anti-HIV agent, michellamine B (see Section 5), illustrates this strategy.

Formulation - Formulation studies involve the development of a suitable vehicle to solubilize the drug to enable administration to patients, generally by intravenous injection or infusion in the case of cancer. The low solubility of many natural-product agents in water poses considerable formulation problems, but these can often be overcome by use of co-solvents or emulsifying agents (surfactants) (Davignon and Craddock 1987). In the case of the anticancer drug, taxol, a vehicle was developed using ethanol and the emulsifying agent, Cremophore EL (polyoxyethylated castor oil), which is diluted with saline solution prior to administration to patients. Cremophore EL, however, elicits a strong allergic reaction from some patients, and considerable care has to be exercised in the intravenous administration of the solubilized drug.

Pharmacological evaluation - Pharmacological evaluation involves the study of various drug parameters and properties in suitable animal models. Data are developed to determine the route and schedule of administration which give optimal activity. Sensitive, selective analytical techniques are developed to enable the quantitative determination of the drug and its metabolites in biological fluids, such as blood, plasma, and urine; analytical methods used during the chemical isolation and purification procedures might not be suitable, since the compounds of interest often only occur in trace quantities in the biological fluids. Pharmacokinetic studies determine the half-lives, bioavailability, and effective concentrations of the drugs in blood and plasma, as well as their rates of clearance, excretion routes and metabolism. The identity and rates of formation of metabolites can provide insight into the possible mechanisms of action whereby the drugs exert their therapeutic and toxicological effects (Cragg and Suffness 1988). Pharmacokinetic and metabolism data can be used to design analogs and congeners of lead compounds with the aim of enhancing activity and/or decreasing toxicity.

Animal studies with taxol determined that the periodic administration of the drug in small quantities provided greater efficacy in the treatment of various tumors compared to administration of a single large dose (bolus injection); these observations have been confirmed in clinical studies, where a slow infusion over several hours has proved to be more effective in terms of tumor regression and reduction of allergic reactions from the patients.

Toxicological evaluation - Detailed toxicological evaluation is required to determine the type and degree of major toxicities in rodent and dog models, and to develop data for the determination of safe starting doses in humans. Studies are designed to determine the relationship of toxicity to dose and schedule of administration, and to establish the reversibility of observed toxic effects. In the case of taxol, the toxicity was most evident in tissues with high cell turnover, such as bone marrow and the gastrointestinal tract, but all these toxicities were reversible.

Clinical development - On completion of preclinical studies and favorable review by the NCI staff, all the necessary data are collated and submitted to the FDA as an Investigational New Drug Application (INDA). Once the U.S. Food and Drug Administration (FDA) has approved an INDA, the various phases of clinical development may begin. **Phase I** clinical trials are conducted to determine the maximum tolerated dose (MTD) of a drug in humans, and observe the sites and reversibilities of toxic effects. The starting doses administered are generally well below the LD_{10} determined for mice, and doses are gradually escalated until toxic effects are observed. Phase I trial patients are most often terminally ill and, as with all trials, they are entered on a voluntary basis. Due to the advanced state of disease, meaningful responses of the patients to drug treatment in Phase I trials may not occur, though instances of partial and occasionally complete remissions of various cancers have been noted with certain drugs, such as taxol (Arbuck and Blaylock 1995).

Once the MTD has been determined, and NCI staff are satisfied that no insurmountable problems exist with toxicities, the drug advances to **Phase II** clinical trials. The trials are generally conducted to test the efficacy of the drug against a range of different cancer disease-types. In Phase II trials, doses at, or close to, the MTD level are administered, and patients are evaluated for meaningful response to the drug treatment. Additional confirmatory Phase II trials may be conducted against those disease-types showing meaningful responses. **Phase III** clinical trials are conducted against those disease-types responding to the new drug treatment, and the efficacy of the drug is compared with that of the best chemotherapeutic agents currently available for those disease-types. In addition, the new drug may be tried in combination with other effective agents to determine if the efficacy of the combined regimen exceeds that of the individual drugs used alone.

In Phase I trials of taxol, the major toxicity observed was neutropenia, with nausea and alopecia (hair-loss) also evident, but all of these were reversible on cessation of drug treatment. As mentioned earlier, severe allergic reactions were observed with some patients, and these were determined to be due to the large amount of the emulsifying agent, Cremophore EL, in the formulation vehicle. These adverse reactions were overcome by administering the drug as a slow infusion, and pretreatment of patients with anti-allergic regimens. In Phase II trials, significant responses were observed in the treatment of patients with ovarian and breast cancers, while promising results have been obtained against other cancers, such as lung and head-and-neck cancers. Trials are continuing against other cancer disease-types, and Phase III studies are being conducted to define the activity of taxol in combination with cisplatin, and other drugs active in ovarian cancer.

Once sufficient evidence has been accumulated indicating that the new drug is effective for a particular disease type, all the necessary information is assembled and filed as a New Drug Application (NDA) with the FDA. The NDA generally will apply only to the particular responsive disease-type, and approval by the FDA usually only permits marketing of the drug for use in the treatment of that disease-type. Taxol is currently approved for the treatment of breast and ovarian cancers.

Analogue development - In some instances, the active natural product originally isolated from the plant or other organism might not prove to be suitable for development as the final clinical drug. Reasons for not selecting the natural product for clinical development could include insurmountable formulation problems, unacceptable toxicity, or metabolism *in vivo* to inactive metabolites. In such cases, medicinal chemists will attempt to overcome the problem(s) by the synthesis of active derivatives and analogues, and the determination of structure-activity relationships (SAR). A notable example of this strategy is the semisynthesis of the anticancer agent, etoposide (Fig. 3.2) from epipodophyllotoxin, an epimer of podophyllotoxin (Jardine 1979). Podophyllotoxin, isolated from *Podophyllum peltatum* or *P. emodii*, proved to be unacceptably toxic in early clinical trials; extensive research led to the development of etoposide. Etoposide shows clinical activity against small-cell lung and testicular cancers, as well as lymphomas and leukemias (O'Dwyer *et al.* 1985). A more recent example is that of camptothecin (Fig. 3.4), isolated from *Camptotheca acuminata* (Suffness and Cordell 1985b). Clinical trials of a soluble salt of camptothecin in the U.S. in the 1970s were discontinued due to observation of severe toxic effects, but recently several new camptothecin derivatives have entered clinical development in the U.S., Europe and Japan. These derivatives, which are showing activity against leukemias, lymphomas, ovarian cancer, and various forms of lung cancer, are prepared by semi-synthesis from natural camptothecin obtained from Chinese and Indian sources (Wall and Wani 1993).

Costs and timespans of drug discovery and development - The preceding discussion illustrates the complexities of the drug discovery and development process, with particular reference to anticancer drugs. Though the percentage of natural product extracts showing preliminary activity in an *in vitro* screen might vary from less than 1% to 5%, the number of potentially valuable "leads" from plant and animal sources is more likely to be one in 5,000 to 10,000. Such "leads" will undergo extensive research and development, and probably less than 50% of those will advance to commercial drug status. Considering the NCI anticancer screening program from 1960 to 1982, of the 114,000 plant extracts screened, only taxol has advanced to final FDA approval, while camptothecin has yielded several semisynthetic derivatives which show clinical promise and might advance to commercial status. Homoharringtonine is showing efficacy against certain leukemias, and eventually might be developed as a second-line anticancer agent. Meanwhile, considerable resources were devoted to the development to clinical trials of nine other agents, including bruceantin, camptothecin, indicine N-oxide, maytansine, and thalicarpine, only to have trials terminated due to lack of efficacy or unacceptable toxicity; many other agents were entered into preclinical development but dropped for various reasons (for examples see Suffness and Cordell 1985a). The chances of developing an effective commercial anticancer drug, such as taxol, therefore, are of the order of one in 40,000 to 50,000, based on number of plant extracts screened. The timespan for development can vary considerably, and can be from 10-20 years for anticancer drugs. Research on the isolation of taxol started in the mid-1960s and its structure was first published in 1971 (Wani *et al.* 1971). Its development was delayed for various reasons (Cragg *et al.* 1993b), but, once efficacy against refractory ovarian cancer was observed in late 1988 (Arbuck and Blaylock 1995), advancement to final FDA approval was relatively rapid.

In the light of the time and resources which are required for development of a commercial drug, and the resources devoted to eventual failed candidates, it is not surprising that cost estimates for drug discovery and development exceed U.S. $230 million (DiMasi *et al.* 1991).

Discovery and development of Michellamine B

Michellamine B (Fig. 3.7) was isolated as the main active agent from the leaves of the liana, *Ancistrocladus korupensis*, collected in the Korup region of southwest Cameroon. Initially the plant was tentatively identified as *A. abbreviatus*, but collections of this and all other known *Ancistrocladus* species failed to yield any michellamines or show any anti-HIV activity. Subsequent detailed taxonomic investigation of the source plant compared to authentic specimens of *A. abbreviatus* revealed subtle but distinctive morphological differences, and the species was determined to be new to science, and officially named *Ancistrocladus korupensis* (Thomas and Gereau 1993). Michellamine B shows *in vitro* activity against a broad range of strains of both HIV-1 and HIV-2, including several resistant strains of HIV-1 (Boyd *et al.* 1994). The species appears to be mainly distributed within the Korup National Park, and vine densities are of the order of one large vine per hectare. Fallen leaves collected from the forest floor do contain michellamine B, and current collections of these leaves should provide sufficient biomass for the isolation of enough drug for completion of preclinical development and possible preliminary clinical evaluation. It is clear, however, that extensive collections of fresh leaves could pose a possible threat to the wild

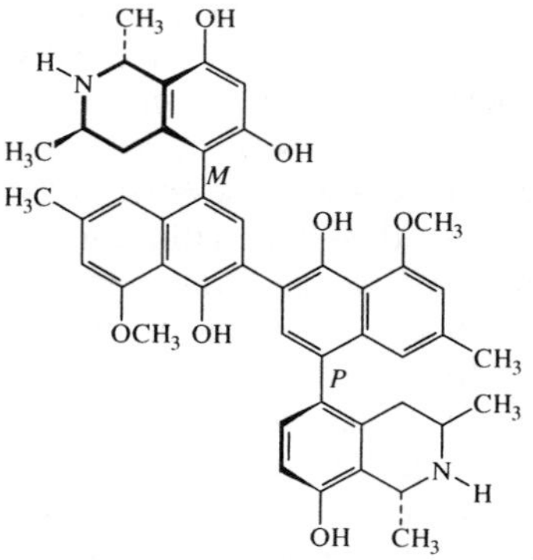

Fig. 3.7. Michellamine B.

source. Thus far, no other *Ancistrocladus* species has been found to contain michellamine B, and investigation of the feasibility of cultivation of the plant as a reliable biomass source was initiated in 1993. An extensive botanical survey has been undertaken, and the range and distribution of the species has been mapped out. Dried leaf samples from representative vines were shipped to NCI for analysis of michellamine B content. Plants indicating high concentrations were re-sampled for confirmatory analysis, and those showing repeated high concentrations were targeted for cloning via vegetative propagation. A medicinal plant nursery has been established to hold and maintain the *A. korupensis* collection at the Korup Park headquarters in Mundemba. The cultivation project has been coordinated by the Center for New Crops and Plant Products of Purdue University, working in close collaboration with the University of Yaounde 1, the World Wide Fund for Nature Korup Project, Missouri Botanical Garden, Oregon State University and the NCI contractor, SAIC. In keeping with the NCI policies of collaboration with source countries, all the cultivation studies are being performed in Cameroon, and involve the local population, particularly those in the regions adjacent to the Korup National Park. In performing this project the cooperation and support of the scientific community and the Government of the Republic of Cameroon has been indispensable, and is greatly appreciated by the NCI and its collaborating organizations.

Based on the observed activity, the NCI has committed michellamine B to INDA-directed preclinical development. Unlike many natural products, formulation presents no problem since the drug is readily water-soluble as its diacetate salt. Continuous infusion studies in dogs indicate that *in vivo* effective anti-HIV concentrations can be achieved at nontoxic dose levels. However, despite these observations and the *in vitro* activity against an impressive range of HIV-1 and HIV-2 strains, there are some serious disadvantages which could preclude advancement of michellamine B to clinical trials. The difference between the toxic dose level and the anticipated level required for effective antiviral activity is small, indicative of a very narrow therapeutic index. In view of some of the toxicities observed in the toxicology studies this narrow therapeutic index is a concern to clinicians considering the drug as a candidate for preliminary clinical trials. In addition, administration by continuous infusion over a period of several weeks is a decided disadvantage compared to oral administration, but, unfortunately, michellamine B is not orally bioavailable.

Even if michellamine B does show some activity in a preliminary clinical trial (assuming it advances that far), it is clear that extensive research will be necessary to determine if the pharmacological and toxicological profiles can be improved through analogue synthesis. Such studies could require substantial quantities of the natural product, or the successful synthetic studies of Professor Bringmann and his group reported in these Proceedings could provide a satisfactory solution (see chapter 1). The isolation of the novel antimalarial compounds, the korupensamines, from *A. korupensis*, provides another class of potential medicinal agents from this plant (Hallock *et al.* 1994). The korupensamines, which are equivalent to the "monomeric" units of the michellamines, are essentially inactive against HIV, whereas the michellamines exhibit only very weak antimalarial activity. The development of the michellamines and/or the korupensamines as effective medicinal agents will require true international collaboration between all parties if Cameroon, as the original source country, is to derive optimal benefits from these significant discoveries.

Potential for international collaboration

The discovery and development of michellamine B illustrates the potential for international collaboration resulting from the contract collection programs supported by the NCI. Further examples of productive collaborations are the development of the calanolides, isolated from the Sarawak (Malaysia) plants, *Calophyllum lanigerum* and *C. teysmanii* (Kashman *et al.* 1992), and conocurvone, isolated from the Western Australian *Conospermum* species (Decosterd *et al.* 1993). In these instances, NCI is collaborating respectively with the Sarawak State Government, and the Western Australian Conservation and Land Management agency (CALM) and the Australian pharmaceutical company, AMRAD.

In addition to the contract acquisition programs, direct collaborations have been established between the NCI and research organizations in countries not covered by the present collection contracts, or organizations studying organisms not included in the NCI program. Medicinal plants from Yunnan Province of the Peoples' Republic of China, Korea, India and Russia are being studied in collaboration with the Kunming Institute of Botany, the Korean Research Institute of Chemical Technology in Seoul, the Central Drug Research Institute in Lucknow, and the Cancer Research Center in Moscow, respectively. Collaborations have also been established with Instituto Nacional de Biodiversidad (INBio) in Costa Rica, the South American Organization for Anticancer Drug Development in Brazil, the Institute of Chemistry at the National University of Mexico, the HEJ Research Institute in Karachi, Pakistan, and the Zimbabwe National Traditional Healers Association (ZINATHA) and the University of Zimbabwe. In establishing these collaborations, NCI undertakes to abide by the same policies of collaboration and compensation as apply to source countries participating in the contract collection programs.

The terms of collaboration are generally stated in a Memorandum of Understanding (MOU) signed by the organization and the NCI. The terms, based on the policies of the NCI Letter of Collection, are summarized in a schematic diagram (Fig. 3.8) and are presented in detail in thé "generic" MOU (Appendix 3.1).

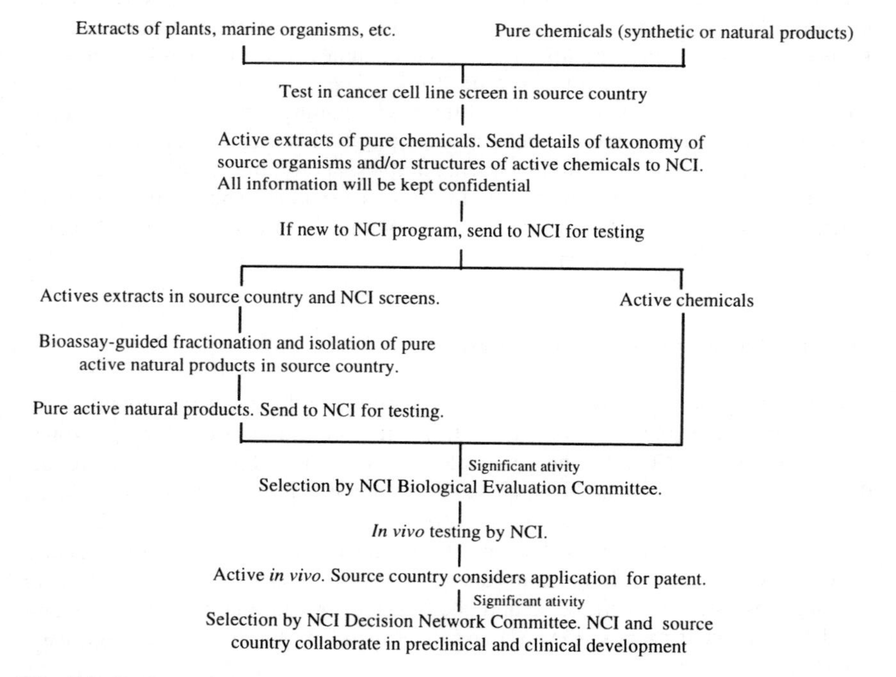

Fig. 3.8. Policy of the NCI in term of possible collaboration between NCI and other organization.

When suitable screening facilities are available in-country, the organization tests pure compounds and/or extracts and provides a list of the identified active materials to the NCI in order that the NCI databases may be checked for earlier submissions. This is a critical requirement, since the NCI has received thousands of compounds and raw materials (plants, marine organisms, etc.) from suppliers worldwide, either on a voluntary basis or through contract mechanisms. It is most important to avoid duplications, since the NCI has legal obligations to suppliers and source countries to protect their rights and intellectual property which could be crucial factors in determining terms of collaboration and compensation should a commercial product be developed. In the case of raw materials, or extracts thereof, if the prior submissions were inactive in the NCI screens, consideration may be given to accepting additional samples since the chemical constituents and bioactivity of plants and other organisms are known to vary depending on the location and time of collection. Those materials new to the NCI program (or extract samples of duplicate organisms originally found to be inactive) are submitted to the NCI *in vitro* screens, and the results provided exclusively to the relevant suppliers. In the case of active pure compounds, additional samples may be requested for *in vivo* testing. Significant *in vivo* activity, and selection by the NCI Decision Network Committee for preclinical development, would merit serious consideration of patent application by the supplier organization to cover the active compound and related derivatives and analogues thereof. In the case of active extracts, bioassay-guided fractionation and isolation of the active constituent(s) may be pursued independently by the organization using in-country screens, or the organization may designate a scientist to visit the NCI facilities in Frederick or equivalent facilities at an approved U.S. institution (*e.g.* a university) to participate in joint isolation studies. The development of any isolated active constituent then follows the same path as described for pure compounds above. If the isolation of an active constituent was performed independently by the supplier organization, patent application would be pursued by the organization at its own cost, and patent rights would belong exclusively to the organization. If, however, the isolation studies were performed jointly by the supplier organization and the NCI or an approved U.S. organization, patent application would be pursued jointly by the collaborating parties with shared costs, and patent rights would be shared by these parties. All information generated and exchanged during the collaboration is considered confidential, with presentation and publication being subject to mutual agreement of the parties involved.

Once an active compound is selected by the NCI Decision Network Committee for preclinical and clinical development, it enters into drug development. The Decision Network Process divides the drug development process into stages designated as DNIIA, DNIIB and DNIII (Fig. 3.9). The NCI Developmental Therapeutics Program (DTP/NCI) collaborates closely with the supplier organization in this process, and the division of tasks depends on the capabilities of the supplier organization. Most pharmaceutical companies can support all aspects of the drug development process, but other suppliers organizations generally do not have the expertise and resources to accomplish most phases of drug development. In such instances, the supplier organization may provide additional quantities of the active agent, while the DTP/NCI will devote its resources to the remaining phases. DTP/NCI can also provide support to the supplier organization to produce additional quantities of the active agent, if necessary. Should the organization be unable to provide the additional quantities in a reasonable period of time, the DTP/NCI will require authorization to prepare them utilizing procedures provided by the organization.

As mentioned earlier, drug development is an extremely costly process, and, once the DTP/NCI has committed itself to the development of a drug, it may expend considerable resources amounting to millions of dollars. Therefore, should a supplier organization decide to cease collaboration at any stage, the DTP/NCI reserves the right to proceed with the development utilizing whatever means it considers appropriate. Even in such an unlikely event, the supplier organization and the corresponding source country still stand to benefit should a commercial product eventually result.

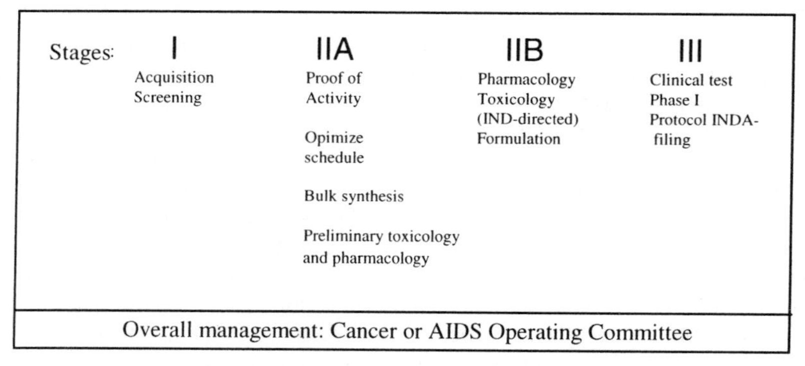

Fig. 3.9. NCI Decision Network Process.

The above discussion primarily applies to research organizations in genetically-rich source countries wishing to collaborate in the investigation of their natural resources. In the case of organizations only wishing to have pure compounds tested, such as pharmaceutical and chemical companies or university research groups, the DTP/NCI has formulated a screening agreement which includes terms stipulating confidentiality, patent rights, routine and non-proprietary screening and testing versus non-routine and proprietary screening and testing, and levels of collaboration in the drug development process. Individual scientists at universities and pure research organizations wishing to submit pure compounds for testing would probably consider entering into this agreement as opposed to the MOU discussed earlier.

Routine/non-proprietary screening and testing versus non-routine/proprietary screening and testing (see Appendix 3.2)

This issue, referred to above, designates those NCI operations which could result in intellectual contributions in the development of a compound by NCI scientists, and which may rise to the level of inventorship as determined under United States patent law. Routine/non-proprietary screening and testing (Step 1) generally refers to the standard NCI *in vitro* and *in vivo* anticancer and anti-HIV screens, and preliminary formulation and toxicology studies necessary to perform the animal *in vivo* screens. Non-routine/proprietary screening and testing (Step 2) encompasses more advanced formulation, pharmacological and toxicological studies aimed at determining optimal parameters for clinical trials, as well as analogue development and mechanism of action studies. These latter operations require significant research involving intellectual input which could result in sole NCI scientist inventorship. In signing either an MOU or a screening agreement, NCI agrees that it will not proceed with Stage II operations without the prior written consent of the supplier organization. In addition, the supplier organization may designate which of the Stage II operations it wishes to delete from the scope of the MOU or screening agreement prior to execution, and DTP/NCI could consider amending the agreement to incorporate them at a later stage. If no limitation in the scope of testing is requested by the supplier organization, it is assumed that DTP/NCI may proceed directly from Stage I to Stage II testing. Should a compound show promising anticancer or anti-AIDS activity through Stage I and designated Stage II testing, the NCI will propose the establishment of a more formal collaboration.

Conclusions

The NCI has the full capability to advance compounds showing promising preliminary anticancer or anti-AIDS activity through all the phases of preclinical and clinical development. The NCI welcomes the opportunity to collaborate with organizations in the discovery and/or development of anticancer and anti-AIDS agents, and has formulated agreements (Memorandum of Understanding or Screening Agreement) outlining terms of collaboration covering issues of confidentiality, inventorship and patent rights, and compensation should a commercial product be developed. The NCI, as a U.S. taxpayer supported, non-profit institution, is firmly committed to protecting the rights of suppliers and Source Countries during all phases of drug discovery and development.

Acknowledgments

NCI gratefully acknowledges the collaboration and support of the many individuals and organizations worldwide which make these programs possible. From the collection of organisms in over 25 countries to the clinical trials of new drugs, and the studies of occurrence and prevention, this is truly an international effort in the fight against the scourges of AIDS and cancer. NCI recognizes the indispensable contributions being made through the provision of valuable natural resources, expertise, knowledge, and skills; through policies of collaboration and compensation, as stated in the Letter of Collection, NCI wishes to assure participating countries of its commitment to working with them in a fair and equitable manner.

References

Arbuck, S.G. and Blaylock B.A. (1995). Taxol: Clinical results and issues in development. In *Taxol®: Science and Applications* (ed. M. Suffness), pp. 379-415. CRC Press, Inc., Boca Raton, Florida, U.S.A.

Boyd, M.R. (1988). Strategies for the identification of new agents for the treatment of AIDS: A national program to facilitate the discovery and preclinical development of new drug candidates for clinical evaluation. In *AIDS: Etiology, Diagnosis, Treatment and Prevention* (eds V.T. Devita, S. Hellman and S.A. Rosenberg), pp. 305-319. J.B. Lippincott, Philadelphia.

Boyd, M.R. (1989). Status of the NCI preclinical antitumor drug discovery screen. In *Principles and Practice of Oncology Updates, Vol 3, No. 10 Prevention* (eds V.T. Devita, S. Hellman and S.A. Rosenberg), pp. 1-12. J.B. Lippincott, Philadelphia.

Boyd, M.R. (1993). The future of new drug development. In *Current Therapy in Oncology. Section I. Introduction to Cancer Therapy* (ed. J. Niederhuber), pp. 11-22. B.C. Decker, Inc., Philadelphia.

Boyd, M.R., Hallock, Y.F., Cardellina II, J.H., Manfredi, K.P., Blunt, J.W., McMahon, J.B., Buckheit Jr, R.W., Bringmann, G., Schaffer, M, Cragg, G.M., Thomas, D.W., and Jato, J.G. (1994). Anti-HIV michellamines from *Ancistrocladus korupensis. Journal of Medicinal Chemistry* **37**, 1740-1745.

Cassady, J.M. and Douros, J.D.(eds.). (1979). *Anticancer agents based on natural product models.* Academic Press, New York.

Cragg, G.M. and Suffness, M. (1988). Metabolism of plant-derived anticancer agents. In *Pharmacological. Therapy. 37* (G. Powis (ed.), pp. 425-461. Pergamon Press, Oxford.

Cragg, G.M., Boyd, M.R., Cardellina II, J.H., Grever, M.R., Schepartz, S.A., Snader, K.M., and Suffness, M. (1993a). Role of plants in the National Cancer Intitute Drug Discovery and Development Program. In *Human Medicinal Agents from Plants. ACS Symposium Series* 534 (eds. A.D. Kinghorn and M.F. Balandrin), pp. 80-95. American Chemical Society, Washington, D.C.

Cragg, G.M., Schepartz, S.A., Suffness M., and Grever, M.R. (1993b). The taxol supply crisis. New NCI policies for handling the large-scale production of novel natural product anticancer and anti-HIV agents. *Journal of Natural Products* **56**, 1657-1668.

Davignon, J.P., and Cradock, J.C. (1987). *The Formulation of Anticancer Drugs*, pp. 212-227. McGraw-Hill, Inc., New York.

Décosterd, L.A., Parsons, I.C., Gustafson, K.R., Cardellina II, J.H., McMahon, J.B., Cragg, G.M., Murata, Y., Pannel, L.K., Steiner, J.R., Clardy, J., and Boyd, M.R. (1993). Structure, absolute sterochemistry, and synthesis of conocurvone, a potent, novel HIV-inhibitory naphthoquinone trimer from a *Conospermum* species. *Journal of the American Chemical Society* **115**, 6673-6679

DiMasi, J.A., Hansen, R.W., Grabowski, H.G., and Lasagna, L. (1991). Cost of innovation in the pharmaceutical industry. *Journal of Health Economics* **10**, 107-142.

Jardine, I. (1979). Podophyllotoxins. In *Anticancer Agents Based on Natural Product Models* (eds. J.M. Cassady and J.D. Douros), pp. 319-351. Academic Press, New York.

Kashman, Y., Gustafson, K.R., Fuller, R.W., Cardellina, II, J.H., McMahon, J.B., Currens, M.J., Buckheit, R.W., Hughes, S.H., Cragg G.M., and Boyd, M.R. (1992). The Calanolides, a novel HIV-inhibitory class of coumarin derivatives from the rainforest tree, *Calophyllum lanigerum. Journal of Medicinal Chemistry* **35**, 2735-2743.

O'Dwyer, P., Leyland-Jones, B., Alonso, M.T., Marsoni, S., and Wittes, R.E. (1985). Etoposide (VP-16-213). Current status of an active anticancer drug. *New England Journal of Medicine* **312**, 692-700.

Philip, P.A., Rea, D., Thavasu, P., Carmichael, J., Stuart, N.S.A., Rockett, H., Talbot, D.C., Ganesan, T., Pettit, G.R., Balkwill, F., and Harris, A.L. (1993). Phase 1 study of Bryostatin 1: assessment of interleukin 6 and tumor necrosis factor a induction *in vivo*. *Journal of the National Cancer Institute* **85**, 1812-1818.

Suffness, M. and Cordell, G.A. (1985a). Academic Press, New York.

Suffness, M. and Cordell, G.A. (1985b). Camptothecin. In *The Alkaloids, Vol. 25* (ed. A. Brossi), pp. 73-89. Academic Press, New York.

Thomas, D.W. and Gereau, R.E. (1993). *Ancistrocladus korupensis* (Ancistrocladaceae): A new species of liana from Cameroon. *Novon* **3**, 494-498.

Wall, M.E. and Wani, M. (1993). Camptothecin and analogues: Synthesis, biological *in vitro* and *in vivo* activities, and clinical possibilities. In *Human Medicinal Agents from Plants. ACS Symposium Series* 534 (ed. A.D. Kinghorn, and M.F. Balandrin), p. 149. American Chemical Society, Washington, D.C.

Wani, M.C., Taylor, H.L., and Wall, M.E. (1971). Plant Antitumor Agents. VI. The isolation and structure of taxol, a novel antileukemic and antitumor agent from *Taxus brevifolia*. *Journal of the American Chemical Society* **93**, 2325-2327.

Appendix 3.1.

Memorandum of understanding (MOU) between Organization and Developmental Therapeutics Program, Division of Cancer Treatment, Diagnosis and Centers and National Cancer Institute

The Developmental Therapeutics Program (DTP), Division of Cancer Treatment, Diagnosis, and Centers (DCTDC), National Cancer Institute (NCI) is currently screening synthetic compounds and natural product materials derived from plants, marine macro-organisms and microbes as potential sources of novel anticancer and AIDS-antiviral drugs. The DTP is the drug discovery program of the NCI which is an Institute of the National Institutes of Health (NIH), an arm of the Department of Health and Human Services of the United States Government. While investigating the potential of natural products in drug discovery and development, NCI wishes to promote the conservation of biological diversity, and recognizes the need to compensate source country **organizations** and peoples in the event of commercialization of a drug developed from an organism collected within their borders.

As part of the drug discovery program, DTP has contracts with various organizations for the collection of synthetic compounds, plants and marine macro-organisms worldwide. DTP has an interest in investigating plants, terrestrial and marine microorganisms and marine macroorganisms from source country, and wishes to collaborate with the **organization** in this investigation. The NCI will make sincere efforts to transfer knowledge, expertise, and technology related to drug discovery and development to **organization**, subject to the provision of mutually acceptable guarantees for the protection of intellectual property associated with any patented technology. **Organization**, in turn, desires to collaborate closely with the DCT/NCI in pursuit of the investigation of its plants, terrestrial and marine microorganisms and marine macroorganisms, and selected synthetic compounds subject to the following conditions and stipulations of this MOU.

1) On the basis of in-house screening results in their anticancer and/or antiviral screens, **organization** may select both synthetic compounds and extracts of plants, marine macroorganisms and microorganisms for confirmatory anticancer testing at DTP/NCI.

2) Until such time as a suitable anti-HIV screen is available to **organization**, **organization** will select both synthetic compounds and extracts of plants for anti-HIV testing at DTP/NCI.

3) Prior to submission of the materials, **organization** will send a data sheet, on each material so that DTP may check its databases for records of prior submission to DTP. **All information provided by organization will be kept strictly confidential**.

4) For pure compounds, the data sheet(s) will give pertinent available data as to chemical constitution, structure, biological data, solubility, toxicity and any precautions which need to be followed in handling, storage and shipping them.

   For crude extracts, data will be provided as to the source organism taxonomy, location and date of collection, any hazards associated with the organism, available biological data and any known medicinal uses of the organism/extracts. **All data provided will be kept strictly confidential** (see section 6, below).

5) DTP will inform **organization** which of the materials are new to the program, and such materials will be shipped to DTP for screening. DTP will provide a record of the accession number for the materials. Quantities of materials required for initial testing are 50 mg for pure compounds and 100 mg for crude extracts.

6) All test results will be provided to **organization** as soon as they are available, but not later than 270 days (nine months) from the date of receipt of the sample. If available, *in vitro* test results will be delivered within 90 days. **Organization** will be informed in writing of any delays beyond this period (270 days) together with an explanation of the reason(s) for delay.

It is clearly understood that no data about the materials will be kept in files open to the public either by DTP, the testing laboratories, or the data processing facilities, all of whom are U.S. government contractors. Only those employees directly engaged in the operation of DTP will have access to the files of information regarding source and nature of confidential materials and results of testing, unless the release of data about the materials or the results of the testing are required under statute or by court order.

7) Any extracts exhibiting significant activity will be further studied by bioassay-guided fractionation in order to isolate the pure compound(s) responsible for the observed activity. Such fractionation will be carried out in **organization** laboratories. If **organization** has no available bioassay, DTP/NCI will assist **organization** in establishment of the necessary bioassay systems or suitably qualified scientists designated by **organization scientists** will be sent to DTP/NCI for the isolation studies subject to the terms stated below in Clause 8. In addition, during the course of this agreement, DTP/NCI will assist the **organization** to develop the capacity to undertake drug discovery and development.

8) Subject to the provision that suitable laboratory space and other necessary resources are available, DTP/NCI agrees to invite senior technician(s) and/or scientist(s) designated by **organization** to work in the laboratories of DTP/NCI or, if the parties agree, in laboratories using technology which would be useful in furthering work under this agreement. The duration of such visits would not exceed one year except by prior agreement between **organization** and DTP/NCI. The designated "Visiting Scientist(s)" will be subject to provisions usually governing Guest Researchers at NIH, except when carrying out research on materials collected in **source country**, Costs and other conditions of visits will be negotiated in good faith prior to the arrival of the scientist(s).

9) In the event that an agent isolated and purified from materials provided by the **organization** from **source country**, and/or in the event that a synthetic compound provided by **organization** meets the criteria established by the Decision Network Committee of NCI's Division of Cancer Treatment (DTP's parent organization), which would include, but not be limited to, *in vivo* activity in rodent models, further development of the agent will be undertaken by DTP/NCI in collaboration with **organization**. Once an active agent is approved by DTP/NCI for preclinical development (*i.e.*, has passed the Decision Network at Stage IIA), DTP/NCI will collaborate with **organization scientists** in the development of the specific agent.

10) The rights of ownership of inventions discovered or made solely in connection with work covered by this MOU, are retained by the **organization** that is the employer of the inventor. Both **organization** and DTP/NCI recognize that inventorship will be determined under patent law.

All inventions developed under this MOU from the materials provided by **organization** shall be the sole property of **organization**, if such inventions are made by any member of **organization** alone. Application for patent protection on inventions made by **organization** employees alone will be the responsibility of **organization**.

For those drugs used for development toward clinical trial by the DTP/NCI, the Government shall have a royalty-free, irrevocable, nonexclusive license under any patent which **organization** may have or obtain on such compound or on a process for use of such compound, which patent relies upon data generated by DTP/NCI or DTP/NCI testing laboratories, to manufacture and/or use by or for the Government the invention(s) claimed in the patent(s). Such license shall be only for medical research purposes related to or connected with the therapy of cancer or AIDS. It is understood that the term "medical research purposes" herein shall not include commercial treatment of patients or commercial distribution of the compound.

All inventions developed under this MOU from the materials provided by **organization** shall be the sole property of DTP/NCI, if such inventions are made by any member of DTP/NCI alone. Application for patent protection on inventions made by DTP/NCI employees alone will be the responsibility of DTP/NCI.

DTP/NCI and **organization** will, as appropriate, jointly seek patent protection on all inventions jointly developed under this MOU by DTP/NCI and **organization** employees, and will seek appropriate protection abroad, including in **source country**, if appropriate. With respect to the inventions jointly owned, any income derived from the sale or licensing of such inventions shall be shared equally between the parties.

11) DTP/NCI will make a sincere effort to transfer any knowledge, expertise, and technology developed during such collaboration in the discovery and development process to **organization**, subject to the provision of mutually acceptable guarantees for the protection of intellectual property associated with any patented technology.

12) All licenses granted on any patents arising from the collaboration conducted under the terms of this MOU shall contain a clause referring to this MOU and shall indicate that the licensee has been apprised of this MOU.

13) Should the agent eventually be licensed to a pharmaceutical company for production and marketing, DTP/NCI will require the licensee to negotiate and enter into agreement(s) with **organization**. The agreement(s) will address the concern on the part of the **source country** that pertinent agencies, institutions and/or persons receive royalties and other forms of compensation, as appropriate.

Such terms shall apply equally to instances where the invention is directed to an actual synthetic compound, an isolated natural product, a product structurally based on the isolated natural product (*i.e.* where the natural product provides the lead for development of the invention) or a derivative of the synthetic compound, though the percentage of royalties negotiated as payment might vary depending upon the relationship of the marketed drug to the originally isolated product. It is understood that the eventual development of a drug to the stage of marketing is a long term process which may require 10-15 years.

14) In obtaining licensees, DTP/NCI will require the applicant for license to seek as its first source of supply the natural products available from **source country**. If no appropriate licensee is found who will use natural products available from **source country**, or if **organization** or its suppliers cannot provide adequate amounts of raw materials at a mutually agreeable fair price, the licensee will be required to pay to the **organization** an amount of money (to be negotiated) to be used for expenses associated with cultivation of medicinal plant species that are endangered by deforestation, or for other appropriate conservation measures. Such terms will also apply to instances where the active agent is prepared by total synthesis.

15) Clause 14 shall not apply to organisms which are freely available from different countries (i.e., common weeds, agricultural crops, ornamental plants, fouling organisms) unless information indicating a particular use of the organism (e.g. medicinal, pesticidal) was provided by local residents to guide the collection of such an organism from **source country**, or unless other justification acceptable to both **organization** and DTP/NCI is provided. In the case where an organism is freely available from different countries, but a genotype producing an active agent is found only in **source country**, Clause 14 shall apply.

16) Publication of data resulting from the collaboration under this MOU will be undertaken at times determined by an agreement between **organization** and DTP/NCI.

17) It is the intention of NCI that **organization** not be liable to DTP/NCI for any claims or damages arising from NCI's use of the Research Material; however, no indemnification for any loss, damage, or liability is intended or provided by any party under this MOU. Each party shall be liable for any loss, claim, damage or liability, that said party incurs, as a result of said party's activities under this MOU, except that the NCI, as an agency of

the United States, assumes liability only to the extent as provided under the Federal Tort Claim Act (28 USC § 171).

18) DTP/NCI will not distribute materials provided by **organization** to other **organization**s. However, should **organization** wish to consider collaboration with **organization**s selected by NCI for distribution of materials acquired through NCI collection contracts, DTP/NCI will establish contact between such **organization**s and **organization**.

19) **Organization** scientists and their collaborators may screen additional samples of the same materials for biological activities other than anti-cancer and anti-AIDS activity and develop them for such purposes independently of this MOU.

This MOU shall be valid as of July 28, 1995 for an initial period of three (3) years, after which, it can be renewed by mutual agreement of **organization** and DTP/NCI are confident that this MOU will lay the basis for a mutually successful cooperation in discovering and developing new therapies in the treatment of cancer and AIDS.

______________________________ ______________________________

Director, **Organization** Director, National Cancer Institute

______________________________ ______________________________

Printed Name Printed Name

_______________ ______________________________

Date Date

Appendix 3.2.

Step 1: Routine or Non-Proprietary Screening and Testing

1. Entry of the structure of a pure synthetic compound or natural product into NCI's confidential data base of structures.
2. Testing of the pure synthetic compound or natural product in NCI's *in vitro* Cancer Drug Screen in a two-day format and a six-day format AND/OR testing of the pure synthetic compound or natural product in NCI's *in vitro* AIDS Drugs Screen against wild type and mutant strains of HIV.
3. Comparison of the pattern of activity in the Cancer or AIDS Screen of the pure synthetic compound or natural product against the pattern of activity of other compounds in NCI's data base, and with the expression of molecular targets in NCI's Cancer Screen in the case of compounds tested in the Cancer Screen.
4. For compounds with evidence of antiproliferative activity in the primary screen, initiation of
   a. Preliminary toxicology with determination of Maximal Tolerated Dose (MTD) according to routine iv and ip protocols.
   b. initiation of formulation studies to determine appropriate vehicle to allow *in vivo* testing.
   c. assessment of *in vivo* activity against athymic mouse xenografts of human tumors dosed according to standard protocols for the tumor types chosen. The tumors will represent those cell types suggested to be sensitive by the *in vitro* screening data. In the case of anti-AIDS drugs, study in *in vivo* mouse models bearing HIV-infected cells to observe evidence of anti-retroviral effect.
5. For compounds with evidence of anti-proliferative activity in the primary screen, study in selected routine subscreens to allow assessment of activity in *in vitro* and *in vivo* models of
   a. AIDS related lymphoma
   b. prostate carcinoma (primary)
   c. breast carcinoma (metastatic)
   d. growth of tumor vasculature (as they may be available to DCTDC)
   e. differentiation

Step 2: Proprietary or Non-Routine Screening and Testing

1. Determine the optimal schedule for routine demonstration of anti-tumor or anti-retro viral activity in selected *in vivo* models.
2. Produce structural analogs of the initial compound that would optimize pharmacologic and or pharmaceutic features.
3. Determine or develop the optimal formulation for the agent to allow clinical use, including verification of
   a. stability of drug as bulk substances
   b. dilution stability in common diluents
   c. stability as aliquoted in clinical dose forms

4. Develop a detailed assay for the bulk drug as well as drug dissolved in vehicles and body fluids.

5. Determine the range of doses that would cause toxic effect in two species according to a dosing schedule that would mimic the proposed clinical use. Also:
   a. determine effect on marrow progenitor growth from at least two species
   b. determine initial clinical pathology parameter changes in response to drug

6. Perform detailed toxicologic evaluation in two species at three dose levels to allow estimation of a "low toxic" dose from which clinical starting doses for a Phase I protocol might be estimated. These studies would include.
   a. detailed gross anatomic and histopathologic studies
   b. correlation of plasma drug concentration with pathology
   c. correlation of clinical pathology with 6a and 6b.

7. If the drug and formulation appears practicable, to file an Investigational New Drug Application (INDA) with the U.S. Food and Drug Administration with NCI as holder of the INDA.

8. Verification that the compound or its analog possess a mechanism of action suggested by comparison of its pattern of activity with known agents, or determine its mechanism of anti-neoplastic or anti-retroviral effect by detailed evaluation in non-routine assays.

4. The search for, and discovery of, two new antitumor drugs, Navelbine and Taxotere, modified natural products

P. POTIER*, F. GUÉRITTE-VOEGELEIN AND D. GUÉNARD

Institut de Chimie des Substances Naturelles du C.N.R.S., F-91198 Gif-sur-Yvette Cedex, France

---

Natural products, a never out-dated subject

Man, since time immemorial, has always sought to extract subsistence and medicines from his environment. It is the accumulation of this experience acquired over several millennia in the selection of remedies, many of which have now been forgotten, which constitutes, even today, the basis of therapeutics. Names such as *Colchicum*, *Digitalis*, *Belladonna*, poppies and opium, *Cinchona*, strychnine, etc. are very current.

Natural products still represent today more than 80% of pharmacological and therapeutic lead compounds. This high percentage is the result not only of what I call archeopharmacology (that based on ancient remedies) but also of the discovery, every year, of several natural substances having major biological and therapeutic interest. These include new antibiotics, of course, antiparasitics, compounds acting on the nervous and cardiovascular systems, etc.

A remarkable advance was the discovery, accidental as is often the case, of cyclosporin, a widely-used drug for the prevention of transplanted organ rejection. This also allowed the development of screening tests for substances of this type and led to the discovery of other immunomodulating compounds such as FK 506, discovered in Japan, as well as rapamycin. Investigation of the mechanism of action of these substances has very recently revealed some of the mysteries of immunology.

Natural products thus constitute, today and for a long time to come, a practically endless source of novel substances able to enrich therapeutics. I would

---

* Member of the Academy of Sciences, Director of the "Institut de Chimie des Substances Naturelles", Centre National de la Recherche Scientifique at Gif-sur-Yvette, former Professor at the Muséum National d'Histoire Naturelle in Paris.

like to give you two examples chosen from the many research programs conducted over twenty years in our Institute at Gif and whose goal is the discovery of new substances of therapeutic interest.

The discovery of Navelbine®, an antitumor drug.

The first example is that of substances isolated from the Madagascan periwinkle, *Catharanthus roseus* G. Don (Apocynaceae). The first sample of this plant brought to France is conserved at the National Museum of Natural History in Paris. It was provided by Etienne de Flacourt (1607-1660), a colonizer sent by the Maréchal de la Meilleraye to pacify the south of Madagascar where the colonial residents of Fort-Dauphin, who worked for the East India Company, had become agitated. Monsieur de Flacourt made use of his mission of pacification to study this country and its nature. As a result, he wrote several books, including a "Dictionary of the Language of Madagascar". In fact, even at that time, numerous dialects existed in Madagascar and the language spoken at Fort-Dauphin (in the Southeast of the island) was not the same as that spoken on the high merina plateaus or in the north (where Arab influence was considerable).

On his before-last return to France, Etienne de Flacourt brought back plants, seeds, and animals from Madagascar. Among these samples was the now famous Madagascan periwinkle which was planted, probably in the form of seeds, in the King's Garden, now known as the Garden of Plants of the National Museum of Natural History in Paris. Like many plants, the periwinkle may have first been subjected to a quarantine in a French Mediterranean or Atlantic port. This general procedure, in fact, explains the present richness and diversity of the botanical gardens in Bordeaux, Nantes, Brest, Le Havre, Marseille, Toulon, etc. Flacourt, however, did not bring back any empirical information concerning the medicinal usage of this plant.

It was already the custom, at that time, that the botanical gardens of the colonizing countries exchanged their collected samples. It is for this reason that *Catharanthus roseus*, the Madagascan periwinkle, reached Chelsea Gardens in London and the Botanical Gardens in Leyden (The Netherlands). From here, the plant was sent to English and Dutch colonies in the West Indies, Indonesia, India, Africa, etc. As a result, the Madagascan periwinkle is now widely disseminated throughout the world where it grows in regions spreading from the equator to the Mediterranean countries.

It was not until 1955 that Canadian scientists, successors of the famous Banting and Best group that discovered insulin at the University of Western Ontario in London (Canada), became by chance interested in the Madagascan periwinkle (Noble *et al.* 1958). This team was looking for substances, natural or synthetic, which could be used to control diabetes. The results of an ethnopharmacological survey in Costa Rica and in the West Indies indicated that the Madagascan periwinkle was used by the locals to treat diabetes. These Canadian scientists thus prepared an extract of this plant and evaluated its antidiabetic activity in rats rendered diabetic by ingestion of alloxan. Not only was the periwinkle extract found to be inactive, but most of the treated animals died of septicemia. An autopsy of these animals revealed that they were all in a state of leucopenia (a large decrease in the number of white blood cells) which, as everyone knows, leads to collapse of an organism's defenses against microbial and viral infections.

It became immediately evident to these investigators that, if in fact the Madagascan periwinkle extracts were responsible for this leucopenia then it should be possible to isolate the "active principles" and perhaps use them for the treatment of leukemias, a disease characterized by a much higher than normal white blood cell number. It was, in fact, in this manner that the beneficial effects of "alkylating agents" such as "mustard gases", used during the First World War and, unfortunately, more recently by Irakian troops against the Kurds and Iranian soldiers, were discovered. Thus, with the end of hostilities in 1918, it became necessary to destroy remaining stocks of mustard gas. During this destruction, accidental leakage led to serious intoxications and here again, pronounced leucopenia was observed in the victims, whence the idea of using these chemicals to lower the number of white blood cells in leukemia patients.

The reasoning was identical for the Madagascan periwinkle. However, before continuing with this story, it must be added that an unexpected event occurred at this time. On the other side of the border, in the United States, scientists at Eli Lilly in Indianapolis had undertaken systematic screening of plant extracts in their search for compounds displaying antitumor activity (Svoboda and Blake 1975). And this is how they observed that the Madagascan periwinkle contained substances active against several experimental murine tumors (L 1210, P388, etc.). The necessary contacts were established with the Canadian scientists and so began the long years of research which led to the isolation of vincaleukoblastine (vinblastine) and leurocristine (vincristine) eventually commercialized under the names of Velbe® and Oncovin®, respectively. These two drugs have now been used for over thirty years in the chemotherapy of cancers and leukemias. Their complex chemical structures led to some remarkable structural chemistry studies around 1957 in which all of the most sophisticated physical methods available at that time (mass spectrometry, nuclear magnetic resonance and, finally, X-ray crystallography) were used to solve this problem. This type of "secondary metabolite" is still, thirty years later, one of the most complex natural molecules known. Certain toxins from marine organisms as well as some antibiotics have since taken first place for structural complexity, but the vinblastine group of molecules still hold an honorable position !

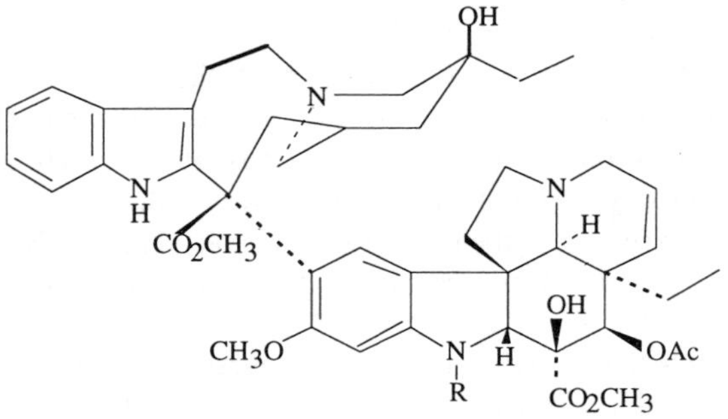

Vinblastine (R= CH_3), Vincristine (R= CHO)

Another characteristic of these drugs is that they are among the most expensive in our therapeutic arsenal [113.20 FF (*ca.* US$ 22.-) for 10 mg of Velbe® and 85.60 FF (*ca.* US$ 17.-) per mg of Oncovin®]. This is due to the very low concentrations of each drug in the plant (several grams per ton of the dried air-exposed parts of the plant). The synthesis of these types of molecules resisted the numerous onslaughts of chemists for seventeen years. Finally, in 1974, my collaborator, Dr. Nicole Langlois and myself managed to synthesize these molecules by using a "biomimetic" approach (that is one imitating Nature). This success not only allowed us to synthesize most of the compounds of this family known at that time (vincristine, vinblastine, leurosine, leurosidine, etc.) but also, and more importantly, to synthesize a completely original type of substance, Navelbine® (nor-anhydro-vinblastine) (Mangeney *et al.* 1979).

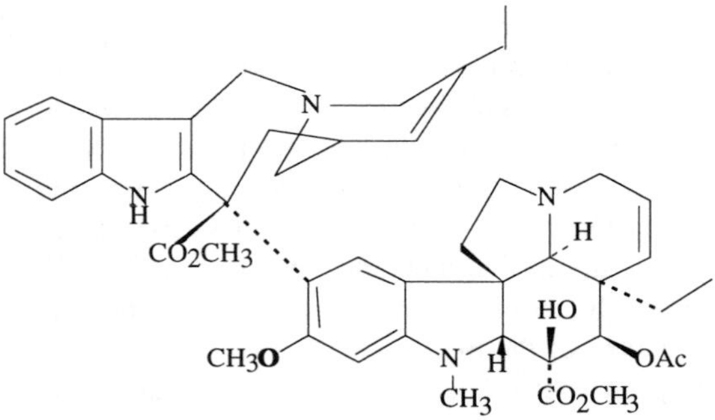

Vinorelbine (Navelbine®)

This compound exhibits therapeutic properties superior to those used up until now, in particular, for the treatment of non-small cell lung cancer (or smoker's cancer) (Vokes *et al.* 1994) and, more recently, of breast cancer. It is altogether probable that Navelbine's spectrum of activity is even wider, particularly if it is associated with other anticancer chemotherapeutic agents.

This substance, a product of French public research, is commercialized by Pierre Fabre Laboratories in cooperation with Burroughs-Wellcome (USA) and Kyowa-Hakko (Japan). Navelbine represents a major advance in the area of anticancer chemotherapy ; it will no doubt have successors.

The discovery of Taxotere® (docetaxel) :

I shall now summarize the history of the discovery of another antitumor drug, Taxotere, the fruit, once again, of research conducted at our research Institute at Gif-sur-Yvette in close collaboration with Rhône-Poulenc Rorer research laboratories in Vitry.

The yew (*Taxus* sp., Taxaceae) is a tree well known for its beauty, its longevity, the quality of its wood and also its toxicity. It grows in temperate and sub-temperate countries, resisting well to both cold and heat. The yew is only rarely attacked by insects, fungi or other parasites.

In 1971, Monroe Wall and his group at the Research Triangle Institute in the United States were involved in systematic screening of extracts prepared from plants collected not only in the USA but also from all over the world. This screening program, on a typically huge, American scale, was financed by the National Cancer Institute. Wall discovered that a substance, which he called taxol, present in the trunk bark of the American West Coast yew tree, presented interesting cytotoxic properties (Wani *et al.* 1971). Although he managed to determine the structure of this very complex molecule, his results remained, for reasons I don't understand, for the most part unexploited. One of the reasons is no doubt because taxol could be extracted only from the bark of these multi-centuried trees and that the yield was very low (0.1 to 0.2 g per kg of bark, which corresponds, as we now know, to one therapeutic dose). Since yews are very slow-growing trees, the problem of obtaining sufficient quantities of taxol for testing was an immediate problem.

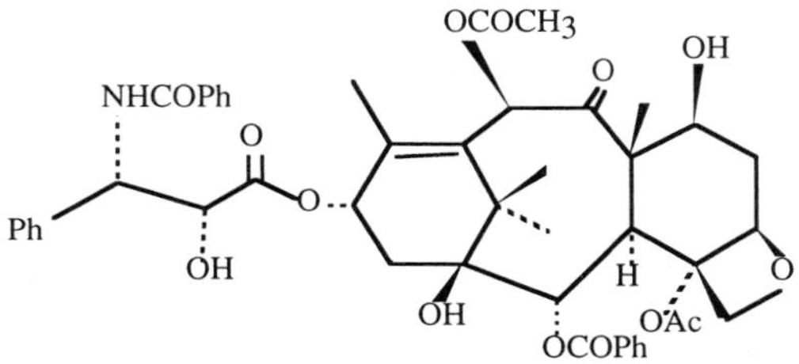

Taxol®

A second reason is the low solubility of taxol which required the implementation of galenical subterfuges such as the use of "Cremophor", an excipient which allows administrable suspensions of taxol to be made but which is not without its own side-effects.

Several years later, the biological mode of action of taxol was discovered by Susan Horwitz in New York (Schiff *et al.* 1979). Taxol belongs to the group of substances called "spindle poisons". These are compounds which inhibit the formation (or the disappearance) of the spindle during the course of cell division.

Derivatives of vinblastine, whose history I have just evoked, but also colchicine and podophyllotoxine, are spindle poisons. They inhibit the polymerization of tubulin, a ubiquitous protein present in all eucaryotic organisms. This polymerization is the basis for the formation of the mitotic spindle, an event that is essential for cell division. This spindle disappears after cell division has terminated. Taxol, on the other hand, inhibits this disappearance. For this reason, then, taxol is a novel compound.

It must be noted that, here again, it is the discovery of a natural product, colchicine, that led to isolation of its biological target, tubulin. The extreme importance of this protein was revealed as a result. The structure of tubulin has not yet been established, though several of these proteins have been cloned.

But let us return to the yew. At our Institute in Gif, our success with Navelbine encouraged us to examine substances which had been described as possessing activity on the tubulin-microtubule (the product of tubulin polymerization) system. This is how our attention was drawn in 1979 to taxol's reported action on tubulin as discovered by Susan Horwitz. The first thing that had to be done was to verify that taxol was present in the trunk bark of the European yew (*Taxus baccata* L.), different from the American yew (*Taxus brevifolia*). This is indeed the case. We were thus able to verify the reported biological properties of taxol, especially with respect to its ability to inhibit depolymerization of the spindle during cell division. However, the major problem which Americans faced, obtaining enough taxol to satisfy therapeutic needs, remained unresolved. The useful therapeutic dose being estimated at 170 mg/patient/month, 2g/patient/year of taxol would be necessary. The number of patients who could benefit from treatment by taxol or better, Taxotere®, as we will see, can be estimated at 250,000. Approximately 500 kg/year of taxol (half as much of Taxotere) would thus have to be made available. Thus, in order to supply sufficient taxol to allow preclinical and clinical studies in the United States, the National Cancer Institute established contracts with West Coast forestry companies to fell 30,000 trees. One can easily imagine the furor of American ecologists on the West Coast, exacerbated by the fact that an endangered species of spotted owl had elected the yew tree as its home !

The decision to cut down these trees became, of course, a major topic of the American press. The conservation of nature versus the treatment of patients suffering, in particular, from ovarian cancer was a dilemma that was widely discussed. All this at a time when it would have sufficed to leave a little room for mathematics : it was quite clear in 1979 when we began studying this problem that if it were necessary to rely on the extraction of the trunk bark of the yew as the sole source of taxol, therapeutic needs would never be satisfied (see above). The complex chemical structure of taxol, moreover, eliminated the possibility of an inexpensive, industrial-scale total synthesis.

It was thus essential to find a reasonable solution to this problem. We thus proceeded to investigate the chemical composition of other parts of the yew. The felling of several of these trees near the C.N.R.S. château at Gif-sur-Yvette, made necessary for the construction of a roadway, furnished us with the raw material needed. It was in this manner that we discovered, in the leaves, a biosynthetic precursor of taxol, that is, 10-deacetylbaccatin III (Chauvière *et al.* 1981), at a concentration of the order of 1 g/kg of leaves (compared to 0.1 - 0.2g/kg of taxol in the trunk bark). It was then simply necessary to do a little chemistry to transform this natural precursor not only into taxol but also into a variety of derivatives. Based on the use of a simple test for biological activity, we were able to select among these derivatives a highly active semi-synthetic compound which we named Taxotere® (Guéritte-Voegelein *et al.* 1990)

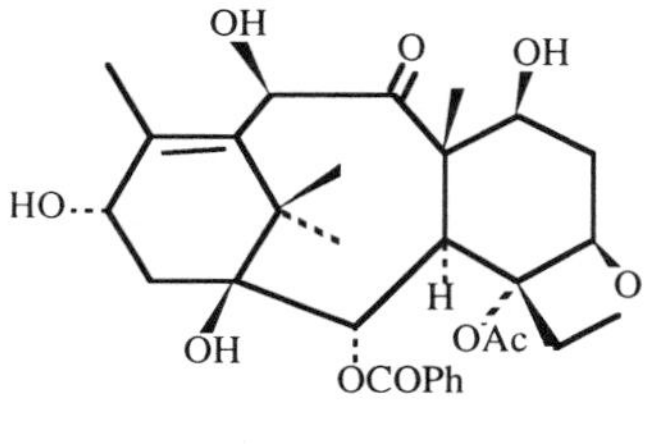

10-deacetylbaccatin III 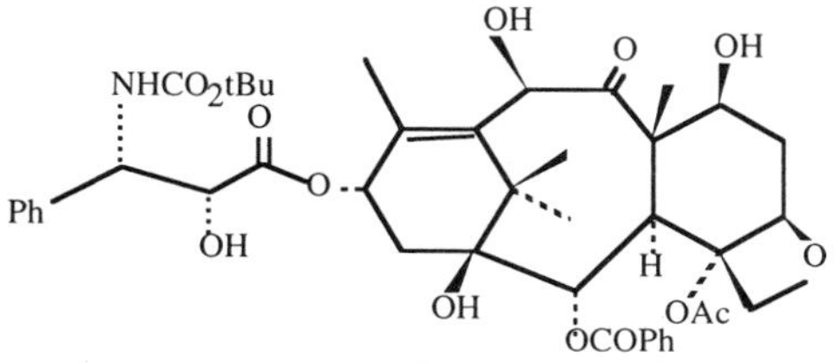 **Taxotere®**

Taxotere has now been launched in most of the world (except Japan which will come later) for the treatment of lung cancers resistant to other chemotherapies, but also of breast and ovarian cancers.

The two drugs (Navelbine and Taxotere) are synergistic.

It is necessary to insist on an important, but unfortunately still quite rare, occurrence in our country - the shoulder-to-shoulder cooperation between our public laboratory and industrial biology and chemistry laboratories, in this case, Rhône-Poulenc Rorer. Taxotere will, as a result of this collaboration, is now available world-wide except Japan.

Present knowledge indicates that Taxotere constitutes, like Navelbine, another French discovery of an important antitumor drug. Clinical specialists are more and more convinced that taxol and related compounds such as Taxotere constitute the most important discovery in the field of antitumor drugs of the last twenty years. Let us hope so, for the sake of patients afflicted with tumors which do not respond to other therapies, for instance, certain ovarian tumors for which taxol and Taxotere have been shown to be effective.

Several lessons may be drawn from these successes. The study of natural substances must continue because, despite the considerable progress that has been made in therapeutic research in the design of new drugs, discoveries made by "guided chance" can bring great rewards. The recent discoveries of certain immunomodulators (cyclosporin, FK 506, rapamycin), of inhibitors of cholesterol synthesis (via inhibition of hydroxymethylglutarate CoA-reductase), of antiparasitic substances of the avermectin family, and of many more, are there to attest.

After all, this is not very surprising. The "Creator" has had more than 4.8 billion years to provide us with all kinds of compounds. Much remains to be explored in the garden of Eden. We have contributed a little to this exploration and we will continue despite, as usual, the vaticinations of a few ill-informed people.

References

Chauvière, G., Guénard, D., Picot, F., Senilh, V., and Potier P. (1981). Structural analysis and biochemical study of isolated products of the yew: Taxus baccata L. *Comptes Rendus de l'Académie des Sciences* **293**, 501-503.

Guéritte-Voegelein, F., Guénard, D., Lavelle, F., Le Goff, M-T., Mangatal, L., and Potier, P. (1990) Relationships between the structure of taxol analogues and their antimitotic activity. *Journal of Médicinal Chemistry* **34**, 992-998.

Mangeney, P., Andriamialisoa, R.Z., Lallemand, J-Y., Langlois, N., Langlois, Y., and Potier, P. (1979). 5'-Nor-Anhydrovinblastine, prototype of a new class of vinblastine derivatives. *Tetrahedron* **35**, 2175-2180.

Noble, R.L., Beer, C.T., and Cutts, J.H. (1958). Role of chance observations in chemotherapy: *Vinca rosea. Annals of the New York Academy of Sciences,* 882-894.

Schiff, P. B., Fant, J., and Horwitz, S. B. (1979). *Promotion of microtubule assembly in vitro by taxol. Nature* **277**, 665-667.

Svoboda, G.H. and Blake, D.A. (1975) The phytochemistry and Pharmacology of *Catharanthus roseus* (L.) G. DON. In *The Catharanthus Alkaloids*, (eds. W.I. Taylor and N.R. Farnsworth) pp. 45-83. Marcel Dekker, Inc. New York.

Vokes, E.E., Rosenberg, R., Jahanzeb, M., Craig, J., Gralla, R., Belani, C., Jones, S., Bigley, J., and Hohneker, J. (1994). Oral Vinorelbine (Navelbine) in the treatment of Advanced Non-Small Cell Lung Cancer: A Preliminary Report. In *Seminar in Oncology*, (eds. J.W. Yarbro, R.S. Bornstein and M.J. Mastrangelo), pp. 35-40. W.B. Saunders Company, Philadelphia.

Wani, M. C., Taylor, H. L., Wall, M. E., Coggon, P., and McPhail. A.T. (1971). Plant antitumor agents VI. The isolation and structure of taxol, a novel antileukemic and antitumor agent from *Taxus brevifolia. Journal of the American Chemical Society* **93**, 2325-2327.

5. Wound healing with plants: the African perspective

O. A. ONAYADE[1], A. A. ONAYADE[2] AND A. SOFOWORA[1]

[1]Department of Pharmacognosy and [2]Department of Community Health, Obafemi Awolowo University, Ile-Ife, Nigeria.

Introduction

A vast range of aspects have been covered already on wounds. These include: definitions (Ellis and Calne 1977), types, extent (Macfarlane and Thomas 1972), the socio-economic implication and the comfort, well being, ambulation as well as restoration of the function of wound sites. There is similarly a good coverage in the scientific press on wound healing including the historical (Fish and Owen Dawson 1967), medical and clinical aspects, the complications of wounds (Macfarlane and Thomas 1972), chemical substances responsible for and factors influencing wound healing (Macfarlane and Thomas 1972; Fish and Owen Dawson 1967; Elliot 1994; Schilling 1968; Douglas 1963).

A brief discussion of the aspects already covered is presented here with a view to familiarizing the reader with the subject matter. Emphasis will be placed on wound healing agents (WHAs) some of which are either natural, synthetic or derived products which can be sourced from plants or plant products. A list of such plants used for wound healing in Africa including the families, active morphological parts, their specific roles (modes/mechanism of their action), isolated active chemical components responsible for wound healing activity as well as their structure-activity relationships (SAR) where known will be discussed.

Wounds

Wound is a collective term for conditions in which there is interruption or damage to the structural integrity of the skin or the underlying tissues (Ellis and Calne 1977). It includes abrasions, abscesses, bites, burns, blisters, boils, bruises (contusions), clean cuts (incisions), fractures, gunshot, injuries, punctures, skin lesions, sores, scalds, sprains, tears (lacerations) and ulcers.

Wounds resulting from injuries, bruises, scalds, cuts, abrasions from road traffic accidents etc., and burns constitute the majority of cases reported at the "Casuality" or "Accidents and Emergency" Units of most hospitals. Patients with intentional or planned wounds *i.e.* surgical wounds *e.g.* excision, orthopedic and plastic surgery, as well as wounds resulting from other already existing medical problems *e.g.* abscesses, boils, ulcers, sores, blisters also constitute a large number of the complaints of in-patients of many hospitals. The occurrence or infliction of wounds is usually of great concern. Chronic wounds, particularly in the elderly (Frantz and Gardner 1994; Lau *et al.* 1994) represent a worldwide problem. The failure of chronic wounds to heal remains a major medical problem (Bhora *et al.* 1995).

The covering of wounds has of necessity always been practiced by man. It was, for a long time, the only branch of the healing arts (Fish and Dawson 1967). As a matter of must, wounds have to be treated, and in good time, to avoid any grave complications such as hemorrhage, infection, development of contracture with subsequent deformity, limitation of joint use, scar formation, scar breakdown and adhesions to underlying tendon or bone (Macfarlane and Thomas 1972). Delayed healing or failure to heal due to poor skin integrity or a patient's multisystemic disorder (Patel and Mach 1994) can also result in keloid formation.

Treatment/management of wounds

The proper way to manage wounds is summarized as follows:

(1) The injured part should be handled gently (First Aid).
(2) The bleeding must be controlled.
(3) The pain must be relieved.
(4) Any shock must be prevented or controlled.
(5) Prophylactic measures against superimposed infections including tetanus and gangrene need to be instituted.
(6) A sterile, protective dressing should be applied immediately.
(7) The wound area or site must be cleaned and shaved. The cleaning can be done under local or general anesthetics, and irrigated with normal saline to remove all foreign bodies (debridement), damaged fat, fascia and muscle; but all essential structures such as blood vessels and nerves should be carefully preserved.
(8) All the edges of the wound must be brought together (Macfarlane and Thomas).

Wound care should generally provide support for the natural healing process and progress (Rakel 1993) and reduce cost. It must considerably decrease the patient's discomfort, morbidity and prevent prolonged hospitalization (Goldenheim 1993).

Cost containment as well as the provision of effective care are the principal goals of wound management (Ratliff and Rodeheaver 1995). Wounds must be attended to very early because the cost of managing chronic wounds in terms of staff time and dressing materials is phenomenal (Benbow 1995). The type of wound, its subsequent treatment and the defense mounted by the injured individual, all affect or mediate microbial infection of wounds. The pathogenicity of the microbes is also of great importance. Wounds can be protected from exogenous microfloral contamination with dressing *e.g.* hydrocolloid dressing with the right moisture content (Mertz and Ovington 1993).

The nutritional status, presence of underlying diseases or co-administration of drugs *e.g.* antineoplastics, anticoagulants and high-doses of steroids, all affect wound repair (Telfer and Moy 1993). Prevention of fibrosis and scar formation are also important clinical considerations in wound healing.

Wound repair

Wound repair is a part of the wound healing process and the stages involved are integrated. In spite of continued research concerning wound repair, it is still not clear precisely what begins the wound healing process and what ultimately stops it (Esterhai (Jr.) and Queenan 1991). It is, however, well known that certain chemicals appear only at the onset of wound, some persist throughout, while others are found only at the end of the wound healing process. Some aspects of the cellular basis and the involvement of intermolecular macromolecules in wound repair have been reviewed (Schilling 1968). After cell injury and following several other occurrences, in parallel or in series, concentric lamellar configurations occur. Some lamellar arrangements and proliferations also occur in the cell and cell contents. Certain chemical substances *e.g.* acid hydrolases are released, though this may be restricted to the wound site (Schilling 1968).

The basic science and clinical aspects of wound repair and healing consist of the following three phases:

(1) The inflammatory phase is marked by platelet accumulation, coagulation and leukocyte migration.
(2) The proliferative phase is characterized by re-epithelialization, mitosis, angiogenesis, synthesis, fibroplasia and wound contraction.
(3) Finally, the remodeling of the extracellular matrix phase takes place over a period of months during which the dermis responds to injury with the production of collagen and matrix proteins and then returns to its preinjury phenotype.

Wound healing

Wound healing is the art of restoring the structural integrity of a disrupted skin or underlying tissues of a deep cut. It comprises:

(1) contraction (gradual approximation of the wound edges),
(2) restoration of the patient to health and
(3) return of body function to normal (Elliot 1994).

Although the wound healing processes are not yet fully understood, they are known to be chemically mediated (Douglas 1963). This is corroborated by the fact that pouring, smearing or impregnating active medicinal compounds (such as gums and other cellulose materials, capsicum, white lead, nettle seed, saline ammoniac and mustard) on the surface of wounds and binding up such in wound protectant materials (like silk or linen) aids the healing of wounds. Antimony litharge, oil, water and wines were also amongst items used as dressings for wounds. Lion grass and oil of roses, honey, starch, aromatic herbs and substances such as borax, salt, spiders webs and boiled snails have also been used. Frankincense, aloes, white of egg and hare's fur were used as styptic for some wounds (Elliot 1994).

The healing of wounds is by the deposition of connective tissue - which is the primary unit of repair. It is a form of a universal cement which appears whenever tissues are wounded (Peacock and van Winkle 1976). It is a fluid, distributed throughout the body, which keeps the various tissues in good health and repairs them when they are injured (Elliot 1994). The connective tissue is concerned with supporting the framework of the body (Peacock and van Winkle 1976). It is found in virtually every part of the organism. Wound healing could also be effected by preventing the alteration of this fluid as well as by maintaining its balance (Elliot 1994).

Wound healing research has produced some startling discoveries during the past decades. For instance, cutaneous wounds created *in utero* are histologically indistinguishable from intact, unwounded tissue (Broker and Reiter 1994). Topical application of hyaluronic acid to wounds in adult diabetic rats leads to enhanced epithelial migration, whilst the addition of cytokine transforming growth factor (TGF-B) to fetal wounds causes an adult like healing reponse with fibrosis and inflammation. When neutralizing antibody to TGF-B was used in adult wounds, it caused an enhanced healing with a more normal dermal architecture with few macrophages, few blood vessels and less collagen (Bleacher *et al.* 1993).

In addition to the local response to injury, there are also sequences of clinical, metabolic and hormonal events of a general nature, called 'recovery or convalescence' which are designed for an ultimate return to full health (Macfarlane and Thomas 1972).

Wound healing does not take a single course of event, but rather a synopsis of events and changes that occur at the cellular and subcellular levels involving purely chemical, biochemical, physical, immunological and other processes, either one after the other or simultaneously. The networks are integrated and sequential, as well as tightly controlled (Clark 1993).

Factors influencing wound healing

Wound Healing is influenced by many factors (Macfarlane and Thomas 1972) including local and general (systemic) factors.

Local factors - These modify the healing process, prevent rapid and effective healing which requires prompt and complete healing or they interfere with the process of repair. They include:

- Inaccurate skin apposition, large tissue defects (creation of gaps).
- Foreign bodies - dead or damaged tissues, blood clots, etc.
- Impaired blood supply- slows healing and inhibits fibroplasia which can cause death of tissues and inadequate defense against infection. The impairment may be due to damage to an artery.
- Venous and lymphatic stasis may cause edema and as such lead to deficient tissue perfusion.
- Local infection - this must be prevented, as subsequent treatment with antibiotics/ antiseptics may interfere with wound healing.
- Complete immobilization of the wounded area may accelerate repair by avoiding damage to the delicate capillaries and regenerating cells. For instance, this value is apparent in dealing with fractures. It may not even be of value in dealing with soft tissues injury, and it may not even be achievable in wounds of heart, lung and arteries.
- Hemorrhage.
- Suture materials.
- Irradiation.

General (systemic) factors - These may be grouped as follows:

(A) Nutritional disturbance which includes:

- Protein deficiency - this is due to lack of collagen in the wound and dehiscence is common in the malnourished, the essential amino acids- cysteine and methionine, being of particular importance.
- Lack of Vitamin C - this leads to defective formation and maintenance of collagen

(B) Others, which include:

- Lathyrism
- Methionine
- Hormones (hypophysis, adrenals, thyroid and gonads). It has not been proved that cortisone or other adrenal steroids will, even in high therapeutic dosage, impair wound healing.
- Anemia - there is little evidence that anemia alone interferes with wound healing, but problems caused can be associated with hypo-proteinemia.
- Age - the young heal better than the elderly but the factors involved are not obvious. This may, however, be due to the fact that the young is still growing,

and certainly more rapidly when compared with the elderly. Therefore, growth in wound healing may be more enhanced in the young.

- Other diseases - systemic disorders, such as diabetes, uremia, jaundice, Cushing's disease and disseminated neoplasia, but no other specific causes are obvious other than hypoproteinemia and lack of vitamin C.

Precautions for appropriate wound healing

Wounds should not be allowed to: get infected, as infected wounds scar more severely (Ward and Saffle 1995), heal more slowly, and re-epithelialization is more prolonged. There is also the risk of systemic infection and eventually death through an infected open wound (Schilling 1968). In addition, wounds should not dry out, as desiccation kills healthy cells. Wound healing is mainly a prerogative of the body, the conditions for adequate healing, However, must be provided by the patient (Rakel 1993).

Chemical wound healing agents (WHAs)

There is a host of agents that are employed in wound healing, namely, chemical, electrical, surgical, thermal agents etc. However, chemical agents, mainly as dressings, out number other agents used in wound healing.

Dressings are materials applied to protect a wound and favor its healing (Fish and Dawson 1967). Much thought has been given to ways of increasing the efficiency of dressings as wound healing agents and means of facilitating their quicker and neater application (Fish and Dawson 1967). Fibrous dressings are known to be fibrous on the macroscale, the microscale and at the molecular level, being composed of flexible, linear macromolecules (polymers) with regular repetition of chain building units (Fish and Dawson 1967). These are of great advantage in the process of sealing up wounds. Some dressings interact with hemoglobin to form a coagulum. This is attributable to the presence of polyanhydro-glucuronic acid *e.g.* in oxidizable cellulose. For this reason, uronic acids and their derivatives are preferred to oxidizable cellulose (Fish and Dawson 1967).

The benefits and advantages of dressings over other well known WHAs, including antibiotics, are:

(1) If wound infection is adequately prevented, there will be no need for wound healing agents (WHA) such as antibiotics, antiseptics, vaccination, surgery etc., since normal wound healing is primarily by deposition of connective tissue (meshwork of fibrils etc.) - the micellar seals up the wounds in plants too (Ikan 1969). However, sutures may be inevitable in cases of deep/wide wounds, the edges of which have to be appropriated.

(2) If such complications as arise from the use of these groups of WHAs is prevented, probably through the use of appropriate dressings, a lot of

improvements would be achieved in wound healing and there will be no need for many of the procedures observed in the management of wounds. The physical aspects of WH, such as aesthetics, lack of deformity and eventual handicap (Elliot 1994) also would be preserved. Extended time of healing, a great disadvantage, would be minimized.

The need to close or cover up wounds to prevent infection cannot be over emphasized, though the presence of some infection may not necessarily prevent the healing of the wound *i.e.* open wounds do not need to be sterile to heal (Clark and Sherman 1993). Wound closure is the ultimate goal of burn care (Smith (Jr.) *et al.* 1994) and the art of closing up wounds has been practiced for long (Fish and Dawson 1967).

Even in clinical practice of wound management, moist occlusive dressings with or without medication, are preferred to many other forms of wound management, since occlusion enhances wound healing primarily by preventing wound desiccation. As a result, epidermal necrosis or eschar formation do not occur thus enabling the wound to re-epithelialize more quickly (Kannon and Garrett 1995). Occlusive dressings have been found to reduce inflammation and subsequent scarring (Hulten 1994).

Moist wound healing agents are associated with hydrocolloid dressings which may provide an alternative treatment modality for certain "partial - unclean" injuries. These dressings produce good functional and cosmetic results, rapid re-epithelialization, and improve patient comfort (Smith (Jr.) *et al.* 1994). Hydrocolloid dressings facilitate the healing of wounds via granulation tissue formation, they keep wound moist and solve many of the practical problems associated with the traditional method of keeping the metal plates, exposed bone and tissues moist (Sochen 1994).

Initial results of a pilot study based on the application of newer dressing materials to animal wounds was reported to support their recommendation for use in human wounds (Cockbill and Turner 1995). Use of collagenous matrices in wounds, made of native bovine collagen I fibres, hyaluronic acid, fibronectin or elastin was added and covered with a protective semi-permeable urethane membrane (De vries *et al.* 1995). Calcium alginate dressing significantly reduced the pain severity and it is easy to use for the nursing personnel (Bettinger *et al.* 1995).

"Chemical component impregnated dressings" *e.g.* saline solution - impregnated dressing followed by hydrocolloid dressing is also advocated (Morris *et al.* 1994). The use of dressing of conventional gauze in wound care and management has such disadvantages as increased frequency of change; harm and associated pain while changing; it also affects the function, freedom and hygiene of the patient (Wijetunge 1994).

Intrasite gel is a hydrogel designed for the debridement of necrotic tissue and effective desloughing, clearing the way for effective wound healing. It is also designed for wounds that are granulating and epithelializing (Williams 1994).

Duoderm - Is an hydrocolloid dressing (Hulten 1994) which lyses fibrin more effectively than others. Pain is significantly reduced when wounds are covered with occlusive dressings (Field and Kerstein 1994).

Hydrocolloid polypeptide biomaterials, *e.g.* Procuren - (William and Da Camara 1993) and synthetic polymers including Polyvinylpyrollidone Iodine (PVP-I) *e.g.* Betadine preparations have also been used (Rakel 1993; Goldenheim 1993).

Some chemicals involved in wound healing

The connective tissue is composed of fibrils, cells, amorphous element or ground substance. The fibrils in turn contain chemical compounds such as collagen (albuminous), reticulin, elastin, proteoglycans and glycosaminoglycans (Lorenz and Adzick 1993; Meyer 1958). These substances intended to heal the wounds are usually applied (Elliot 1994) either to:

- make the wound cicatrize,
- make the flesh grow,
- make the growing flesh firm,
- arrest hemorrhages, or
- remove foreign bodies.

The profiles of fetal proteoglycans, collagens and growth factors are different from those in adult wounds. High level of hyaluronic acid and its stimulation is more rapid and there is highly organized collagen deposition (Bleacher *et al.* 1993; Broker and Reiter 1994). Growth factors are characteristically less prominent in fetal wounds. Topical application of wound healing promoting agents *e.g.* regulated amounts of growth factor to wounds may soon be possible.

All these three compounds viz;- proteoglycans, collagens and growth factors characterize fetal wound which heals without scar formation. These findings are of advantage in the clinical application in the modulation of adult fibrocytic disease and of abnormal scar forming conditions.

Biomaterials *e.g.* biodegradable biopolymers such as some acids and peptides have been reported to enhance dermal and corneal wound healing. This acceleration improves the quality of regenerated tissue, restricts the extent of fibrosis and reduces the risk of microbial infection (Sochen 1994).

Topical therapeutic agents have also been shown to be quite effective in the management of open skin wounds (Ward and Saffle 1995) as or in addition to moist dressings and protectants (Leitch 1995). Growth factors (Falanga 1993), moist bio-occlusive dressings with or without medication *e.g.* alginates etc. (Piacquadio and Nelson 1992) are also used in wound treatment as they provide moist wound healing environment. Pretreatment of skin with all-*trans* retinoic acid (tretinoin) has been shown to enhance wound healing dramatically in photodamaged skin (Popp *et al.* 1995). Acidic firoblast growth factor (aFGF) has been shown to be a potent mitogenic and chemotactic agent for the principal cellular constituents of the skin. It increases wound closure in a dose-dependent manner. It increased granulation tissue formation and re-epithelialization throughout healing. It has potential therapeutic applications for promoting healing of dermal ulcers, especially in healing-impaired individuals (Gerstein *et al.* 1993).

Some wound healing chemical compounds are enzyme inhibitors or antienzymes *e.g.* Echinacin (antihyaluronidase) from *Echinacea* species, lysyl oxidase in *Tridax* spp., and Aloe carboxypeptidase from *Aloe* sp., as well as kauranoic acid which inhibits Bovine serum albumin coagulation. Others are hormones *e.g.* traumatic acid, a straight chain dicarboxylic acid. It is a specific wound hormone which is very active in inducing meristematic activity.

These wound healing chemical compounds have been classified as wound healing accelerators. Their biological activities vary (just as do their structures) and include: anti-inflammatory, antipyretic, analgesic, antimicrobial (antibacterial, antifungal and antiviral) detoxicant, deobstruent, hemostatic, anti-enzyme, antiseptic, anesthetic, nutrient, immunosupressive, peripheral stimulant, astringent and cicatrizant.

Acidic firoblast growth factor (aFGF) has also been found to be one of the most effective wound healing components of the human system. It has potential therapeutic applications for promoting the healing of dermal ulcers, especially in healing-impaired individuals (Popp *et al.* 1995) and has been recommended for that use.

Mucopolysaccharides, which are the major components of the connective tissue and the ground substance are also acidic in nature, examples include: hyaluronic acid, chondroitin, keratan, dermatan and heparin sulfate (Peacock and van Winkle 1976). Some of these compounds have been applied directly as dressings to wounds and hyaluronic acid, chondroitin sulfate (De vries *et al.* 1995; Pruden 1964) have shown good results. They contain at least one uronic acid moiety. Generally, polyuronides are the chief constituents of most mucilages (Trease and Evans 1989) which have also been used in wound healing. In plants, mucilages and gums are well known to bathe cells and keep them healthy as well as repair them when wounded. Alginic acid is a related compound which is also in use (as alginate) for wound healing (Bettinger *et al.* 1995).

Wound healing situation in plants

In plants, gums and related compounds effect wound healing, by acting as protective agents that cover accidental wounds (Ross and Brain 1977). The cellulose fibrils are arranged in a multidimensional net in the primary cell wall whilst the middle lamella contains a plastic cementing layer that holds the adjacent cells together. The cellulose fibrils are arranged with specific orientations such that adjacent layers re-inforce one another (Ross and Brain 1977). This deposition pattern seals wounded cells (Ikan 1969) and could be likened to mucopolysaccharides (glycosaminoglycans) which are biological polymers acting as the flexible connecting matrix between the tough protein filaments in cartilage to form a polymeric system similar to re-inforced rubber (Meyer 1958).

When a plant is wounded, the P-protein, a cell wall material called calose, produces plugs that seem to block the sieve platter. Normal cells do not develop such plugs. The major function of the P-protein is to seal off the sieve cell by blocking the sieve plates thereby eliminating leakage of the assimilates when the plant is wounded (Devlin and Witham 1986). The slime layer is a secretion of complex polymers around the outside cell wall. It may be diffuse or thick. Usually, these polymers are polysaccharides but polypeptides are found in some species (The Pharmaceutical Handbook, 19th edn. 1980).

Plants possess the groups of chemical compounds implicated in the primary repair of tissues which basically include: polysaccharides and specifically, glycosaminoglycans, polyuronides (Ikan 1969), some of which are found as main constituents of mucilages and gums (Ross and Brain 1977; MacGregor and Greenwood 1980). Some of these compounds which also possess antibiotic and/or antimicrobial property would be of double advantage in wound treatment.

A group of chemical compounds which are released by plants after and in response to injury is referred to as phytoalexins. They combat infections and have been tested on zoological pathogens, some of which may be implicated in wounds (Adesanya and Pais 1995).

If latex, waxes, resins, balsams, mucilages and gums are produced, they function for the purpose of wound repair in some plants. They may well function similarly in some animals and humans, especially since some of their current uses have extended to those related to, or primarily intended to achieve wound healing, such as in the cases of cellulose and its derivatives (Fish and Dawson 1967; Ikan 1969; Burkill 1985).

Propolis, a resinous substance found in beehives collected by bees from buds, contains caffeic acid and cinnamyl alcohol. It has been used in the treatment of wounds (Magro-Filho and de Carvalho 1994) just as honey is used (Komolafe 1996). In fact, in some parts of western Nigeria, it is believed that any wound that does not heal with the application of honey may never heal (Komolafe 1996). The future use of plant extracts externally in the management of wounds, is bright because there is a growing interest in the clinical practice of wound management with the use of chemical component - impregnated dressings (Morris *et al.* 1994).

Criteria for selecting prospective wound healing agents

There is a pointer in the scientific literature to the fact that with respect to prospective wound healing agents, emphasis should be on ways of increasing the efficiency of dressings as wound healing agents since they can be applied quickly and neatly (Fish and Dawson 1967). This is a result of the fact that the regeneration of tissues, in the last analysis must be a matter of the synthesis of new chemical substances in the wound, since the processes involved in wound healing are chemical in nature. It is in this area that most advances are to be hoped for in the future (Douglas 1963).

The utilization of medicinal and aromatic plants and plant constituents in the acceleration or promotion of wound healing seems to be the ultimate, and is actively advocated, especially because wound healing has been very successful with the use of natural products (Elliot 1994).

Plants used in wound healing in Africa

Some of the clinical practices of wound management can be likened to the procedures in African traditional medicine and Chinese traditional medicine *e.g.* in bone setting, fracture management, uvulectomy, abdominal surgery, trephination and circumcision (Sofowora 1982, Sofowora 1996).

The African traditional medical practitioners have, over the years, also been treating various wounds with herbal remedies. A compilation of such plants which have been or are in use for the treatment of different types of wounds in Africa is presented in Table 5.1. The plants are used as first aids, in cleansing, washing of wounds, in the extraction of pus, as well as for infected and festering wounds. Other uses include the treatment of boils, abscesses, cuts, skin lesions, wounds, snakes and dog bites, insect stings, bruises, pains, soothing of burns, ulcers, fractures, trauma, sprains, aches, suppurations, inflammations, wounds, scabies, rabies and to dress wounds.

Many of these plants are reported to be effective for diverse activities but all directly or indirectly culminate in the cure or healing of wounds. This implies that their mechanism or mode of action varies. Some act as cicatrizants, antiseptic, antifungal, antiviral, antibacterial, antipyretic, anesthetic, analgesic, hemostatic, antimicrobial, anti-inflammatory, growth promotor, collagen synthesis/fibroblast formation enhancer; whilst others enhance the closing up of wounds with or without scar formation. Yet others are glycosaminoglycan synthesis stimulators.

Most medicinal plants used in the healing of wounds, specifically accelerate the process of wound healing. A majority of these plants have been in use for the purpose of wound healing for a long time now, and a number of them are within reasonable reach of and can be identified with relative ease by the villagers. Some of the plants are reputed to be quite effective and to yield instant results. Most of the plants are used fresh either as juice or sap or latex, while a few are used as extracts against the various types of wounds mentioned above. The plant part(s) are usually rubbed in between the palms and the juice squeezed onto the wound surface with or without tying or binding up with protective bandage.

Mucilages are used as a soothing application to the mucous membranes (Ellis and Calne 1977). The mucilage of Slippery elm *Ulmus rubra* bark is used for its soothing effect on inflamed tissues either in the crude state or in the form of lozenges.

Table 5.1. African plants used in wound healing

Family/Plant Name	Part used	Wound Type/Uses	Ref.
Acanthaceae			
Brillantaisia lamium	juice	sores	Dalziel (1956, p. 450)
Elytraria marginata	leaves	fresh wounds	Dalziel (1956, p. 11)
Graptophyllum pictum	leaves	ulcers, abscesses, cuts, broken bones	Holdsworth and Rali (1989), Ozaki *et al.* (1989)
Hypoestes verticillaris	plant sap	sores	Dalziel (1956, p. 15)
Justicia shimperi	leaves	fresh wounds	Dalziel (1956, p. 18)
Phaulopsis falcisepala		wound dressing, sores	Adesomoju and Okogun (1985)
Phaulopsis falcisepala	whole plant, juice, root	fresh wounds, sores	Dalziel (1956, pp. 23, 24, 452)
Thomandersia hensii	plant sap, leaves	external ulcers, sores	Dalziel (1956, p. 28)
Amaranthaceae			
Amaranthus caudatus	seed	sores, antifungal	Watt and Breyer-Brandwijk (1962, p. 14), Kubas (1972).
Amaranthus paniculatus	whole plant	sores	Watt and Breyer-Brandwijk (1962, p. 16)
Boophone disticha	fresh/dry leaves	fresh wounds	Watt and Breyer-Brandwijk (1962, p. 23)
Cyathula postrata	leaves, twigs	fresh wounds, burns, sores	Dalziel (1956, p. 58)
Cyathula spathulifolia	stem fruit, seed	sores	Watt and Breyer-Brandwijk (1962, p. 18)
Grinum kirkii	whole plant	sores	Kokwaro (1993, p. 248)
Haemanthus coccineus	fresh leaves	ulcers	Watt and Breyer-Brandwijk (1962, p. 33)
Hypoxis rooperi	root juice	anti-burn	Watt and Breyer-Brandwijk (1962, p. 41)
Anacardiaceae			
Lannea barteri	bark	sores, ulcers	Dalziel (1956, p. 76)
Lannea velutina	bark	fresh wounds, ulcers	Dalziel (1956, p. 80)
Schinus molle	bark, leaves	sores	Dalziel (1956, p. 87)
	oleoresin	used as a cicatrizant	Martinez (1984)
Sorindeia longifolia	bark, leaves	sores, ulcers	Dalziel (1956, p. 341)
Trichoscypha longifolia	bark, leaves	fresh wounds, sores, ulcers	Dalziel (1956, p. 96)
Annonaceae			
*Annona senegalensis***	bark, root, leaves	used as dressing, sores	Burkill (1985); Adesakin , personal communication
Enantia chlorantha	bark, leaves	fresh wounds, ulcers, sores	Dalziel (1956, pp. 4, 111, 112)
Hexalobus crispiflorus	bark	fresh wounds	Dalziel (1956, p. 114)
Uvaria chamae	leaves juice	fresh wounds, sores, swelling	Hedberg *et al.* (1983a), Arnold and Gulumian (1984)

Xylopia aethiopica	bark	sores	Dalziel (1956, p. 8)
		used to dress umbilicus after cord stump has dropped off	Ayensu (1978).
Apocynaceae			
Alafia lucida	leaves	sores	Dalziel (1956, p. 137)
Alafia multiflora	latex	fresh wounds	Dalziel (1956, p. 137)
Alstonia bonnei	juice	fresh wounds	Adesakin, personal communication
Conopharyngia usambarensis	latex	fresh wounds	Watt and Breyer-Brandwijk (1962, p. 81)
Diplorhynchus condylocarpon	bark	fresh wounds, snake bites, sore eyes	Watt and Breyer-Brandwijk (1962, p.83), Chhabra and Uiso (1991)
Funtumia africana	leaves	anti-burn	Dalziel (1956, p. 150)
Isonema smeathmanni	latex	sores	Dalziel (1956, p. 157)
Strophanthus preussi	sap	fresh wounds, sores	Dalziel (1956, p. 183)
Araceae			
Colocasia esculenta[**]	rasping	applied to maturate boils	Burkill (1985)
Araliaceae			
Polyscias fulva	leaves	external ulcers	Adjanohoun *et al.* (1993, p. 45, 321)
Asclepiadaceae			
Calotropis procera[**]	juice	used as dressing	Burkill (1985)
Kanalua laniflora	latex	sores	Dalziel (1956, p. 229)
Tylophora conspicua	leaves	fresh wounds, ulcers	Dalziel (1956, p. 390-391)
Asteraceae			
Ageratum conyzoides	latex	fresh wounds, antimicrobial	Watt and Breyer-Brandwijk (1962, p. 197), Kokwaro (1993, p. 69)
Anisopappus africanus	leaves	sores	Kokwaro (1993, p. 69)
Artemisia tridentata	leaf oil	wounds	Northway (1975)
Bidens bipinnata	juice	fresh wounds	Dalziel (1956, p. 416)
Bidens pilosa	leaves	external ulcers, antiseptic, skin lesions, anti-inflammatory, wounds, cuts	Adjanohoun *et al.* (1993, p. 61, 321), Gonzalez *et al.* (1993)
Blumea aurita	leaves	ulcers	Dalziel (1956, p. 452-3)
Calendula officinalis[a]	flowers, tincture	gunshot wounds	Reynolds (1886), Livezey (1868), Rao *et al.* (1991).
Crassocephalum picridifolium	leaves	fresh wound	Dalziel (1956, p. 462); Kokwaro (1993, p. 74)
Echinacea angustifolia[a,c]	aq. extract (echinacin)	leukocytes stimulation, activation of the reticulo-endothelial system	Nikol'skaya (1954), Tunnerhoff and Schwabe (1955a,b, 1956, 1965). Zoutewelle and van Wijk (1990),
Echinacea pallida	extract	antibody formation	Kabelik (1965)
Echinacea purpurea[*]	extract	antibody and fibroblast formation	Kabelik (1965)

Table 5.1. Cont'd

Family/Plant Name	Part used	Wound Type/Uses	Ref.
Asteraceae cont'd			
Gutenbegia fischeri	leaves	sores	Kokwaro (1993, p. 76)
Gynura cernua	fresh juice	sores	Dalziel (1956, p. 418)
Helichrysum appendiculatum	leaves	fresh wounds	Watt and Breyer-Brandwijk (1962, p. 237)
Helichrysum foetidum	leaves	sores	Dalziel (1956, p. 477)
Helichrysum pedunculare	leaves	sores	Watt and Breyer-Brandwijk (1962, p. 239)
Notonia spp.	leaves	fresh wounds	Kokwaro (1993, p. 79)
Senecio fuchsia	extract	wounds	Nikol'skaya (1954)
Tridax procumbens	fresh leaves	fills wounds dead space	Diwan *et al.* (1982), Sarma *et al.* (1990), Udupa *et al.* (1991),
Vernonia aemulans	whole plant	fresh wounds	Kokwaro (1993, p. 83)
Vernonia cinerea	whole plant	fresh wounds, boils, blisters	Kloos (1977), Abebe, (1986), Desta (1993)
Vernonia homilantha	leaves	sores	Kokwaro (1993, p. 85)
Wedelia chinensis	leaves	sutured wounds	Hedge *et al.* (1994)
Basellaceae			
*Basella alba***	stem, leaves	poultices	Burkill (1985)
Bignonaceae			
Kigellia africana	inner bark	wounds, abscesses, anti-inflammatory	Dalziel (1956, p. 256), Khan *et al.* (1978)
Newbouldia laevis	bark	inflamed sores, ulcers, antibacterial	Dalziel (1956, p. 444), Le Grand (1989)
Stereospermum kunthianum	bark	ulcers, skin lesions	Dalziel (1956, p. 265-266), Desta (1993)
Bombacaceae			
*Adansonia digitata***	bark, dry leaves	cleaning sores, 'laali' in West Africa	Burkill (1985)
*Bombax buonopozense***	bark ointment	skin-diseases, ringworm	Burkill (1985)
*Ceiba pentandra***	bark decoction	as topic on swelling and to wash sores	Burkill (1985)
*Rhodognaphalon brevicuspe***	bark and leaves liquor	sores and as poultices on 'blue boil'	Burkill (1985)
Boraginaceae			
Alkanna tinctoria[a]	root	leg ulcers	Papageorgiou (1978a,b)
*Cordia myxa***	extract	emollient plaster to maturate abscesses	Burkill (1985)
Heliotropium indicum[a]	whole plant	ulcers, wound healing acceleration	Dalziel (1956, p. 426), Schoental (1968)
Pulmonaria officianalis	leaves	wounds	Nikol'skaya (1954)
Symphytum officinale	leaves	wounds	Goldman *et al.* (1985)

Capparidaceae			
Gyandropsis gynandra		fresh wounds	Adjanohoun *et al.* (1993, p. 113)
Chrysobalanaceae			
Parinari excelsa	bark	fresh wounds	Dalziel (1956, p. 383-385)
Cochlospermaceae			
*Cochlospermum tinctorium***	leaves pulp	wet dressing to maturate abscesses	Burkill (1985)
Combretaceae			
Combretum cinereopetalum	whole plant	external ulcers	Adjanohoun *et al.* (1993, p. 123, 321)
Combretum gueinzii	leaves	fresh wounds	Watt and Breyer-Brandwijk (1962, p. 193)
Commelinaceae			
Aneilema lanceolatum	juice	sores	Dalziel (1956, p. 465)
Aneilema pomeridianum	whole plant	sores	Dalziel (1956, p. 428)
*Commelina benghalensis***	extract, sap	as poultice for sore feet, burns	Burkill (1985)
Commelina diffusa	whole plant	fresh wounds, ulcers, bruises, sores	Caceres *et al.* (1987), Adjanohoun *et al.* (1993, pp. 127, 345)
Connaraceae			
Roureopsis obliquifoliolata	leaves	fresh wounds	Dalziel (1956, p. 524)
Convolvulaceae			
Bonamia mossambicensis	leaves, root	fresh wounds, antifungal	Sawhney *et al.* (1978), Khan *et al.* (1978), Kokwaro (1993, p. 88)
Ipomea involucrata	leaves	fresh wounds	Watt and Breyer-Brandwijk (1962, p. 337)
Ipomea pes-caprae[a]	leaves	decreases tissue destruction in jelly fish sting	Pongprayoon *et al.* (1987)
Crassulaceae			
Kalanchoe spathulata[a]	fresh leaves	wounds, prevention of scar formation	Yadav and Yadav (1985)
Cucurbitaceae			
Luffa acutangula	leaves juice	sores	Dalziel (1956, p. 61)
Cyperaceae			
Cyperus dilatatus	stem	fresh wounds	Dalziel (1956, p. 613)
Ebenaceae			
Diosyros mespiliformis	leaves, twigs	fresh wounds, burns	Watt and Breyer-Brandwijk (1962, p. 389)

Table 5.1. Cont'd

Family/Plant Name	Part used	Wound Type/Uses	Ref.
Elaeginaceae			
Hipphophae rhamoides	fruits.	wound healing acceleration	Neamtu and Cociu (1982), Mironov *et al.* (1983)
Euphorbiaceae			
	(contains corillagin)		
Aleurites fordii	oil from the plant	ulcers, burns	Watt and Breyer-Brandwijk (1962, p. 395-6), Nonaka *et al.* (1990)
Croton lechleri[c]	juice	wound healing acceleration	Vaisberg *et al.* (1989), Pieters *et al.* (1992)
Croton macrostachyus	leaves juice	fresh wounds	Kokwaro (1993, p. 101)
Euphorbia balsamifera[**]	juice, root	dressing for yaws in horse, sores	Burkill (1985)
Euphorbia grantii	sap	fresh wounds	Kokwaro (1993, p. 103)
Euphorbia hirta	leaves	wound healing acceleration	Santhanam and Nagarajan (1990)
Euphorbia maequilatera	whole plant	fresh wounds	Kokwaro (1993, p. 104)
Jatropha curcas	leaves juice	fresh wounds, sprains, sores, abscesses, mouth/throat sores infected wounds, hemostatic and anti-inflammatory	Watt and Breyer-Brandwijk (1962, p. 420), Dhawan *et al.* 1977), Arnold and Gulumian (1984), John (1984), Martinez (1984), Weniger *et al.* (1986), Kone-Bamba *et al.* (1987), Le Grand and Wondergem (1987), Le Grand (1989), Madulid *et al.* (1989), Muanza *et al.* (1994)
Jatropha multifida		fresh wounds, first aid antiseptic	Adjanohoun *et al.* (1993, p. 153 & 345), Kosasi *et al.* (1987)
Jatropha zeyheri	sap	sores and burns	Watt and Breyer-Brandwijk (1962, p. 422)
Mallotus oppositifolius		fresh wounds	Adesakin, personal communication
Phyllanthus aspreicaulis	leaves	fresh wounds	Kokwaro (1993, p. 107)
Ricinus communis	bark	fresh wounds, sores, sprains, trauma, aches, inflammation, ulcers, strong bactericide, antifungal	Watt and Breyer-Brandwijk (1962, p. 428), Khan *et al.* (1978), Holdsworth (1983), Martinez (1984), Ramirez (1988), Chhabra and Uiso (1991), Anesini and Perez (1993), Desta (1993), Muanza (1994)
	seed oil	burns	Tanaka *et al.* (1980), Adesina (1982), Boukef *et al.* (1982), Sebastian and Bhandari (1984), Samuelsson *et al.* (1992)
Synadenium sp.	sap	fresh wounds	Kokwaro (1993, p. 112)
Synadenium cupulare		infected wounds, boils	Nwude and Ebong (1980)
Graminae			
Cynodon dactylon	extract	wound healing	Subramanian and Nagarajan (1988)
Saccharum officinatum		stomatitis, aches, strings.	Hedberg *et al.* (1983b)
Sasa albomarginata		wound healing, needs Vit C to work	Shibata *et al.* (1980).

Guttiferae			
Hypericum perforatum		wound healing	Fedorchuk (1964), Rao *et al.* (1991)
Icacinaceae			
Rhaphiostylis beniniensis	leaves	sores	Dalziel (1956, p. 292)
Lamiaceae			
Leonotis mollisima	root	wounds, snake bites, festering sores	Hedberg *et al.* (1983a), Kokwaro (1993, p. 121)
Leonotis nepetaefolia	leaves	ulcers	Watt and Breyer-Brandwijk (1962, p. 520)
Plectranthus sp.	leaves juice	fresh wounds, scabies, antibacterial	Chhabra *et al.* (1984), Kokwaro (1993, p. 126)
Salvia haematodes	fruit, root	wound healing	Akbar (1989)
Lauraceae			
Persea americana	fruit pulp	fresh wounds, bruises, sores, analgesic, scar remover growth promotor	Watt and Breyer-Brandwijk (1962, p. 532), Gazit and Blumenfeld (1972), Ortiz de Montellano (1975), Browner (1985), Ramirez *et al.* 1988), Werman *et al.* (1991)
Leguminosae			
Abrus precatorius	seed	fresh wounds	Adjanohoun *et al.* (199 , p. 219, 233)
Acacia farnesiana	bark, leaves	sores	Dalziel (1956, p. 207)
Acacia fischeri	root	fresh wounds	Kokwaro (1993, p. 136)
Acacia polyacantha	leaves	sores	Kokwaro (1993, p. 138)
Cassia acutifolia	leaves	fresh wounds, burns	Dalziel (1956, p. 179)
Cassia alata	leaves	infected wounds	Palanichamy *et al.* (1991)
Cassia nigricans	leaves	external ulcers	Adjanohoun *et al.* (1993, p. 105, 347)
Crotalaria deserticola	leaves	fresh wounds	Kokwaro (1993, p. 144)
Crotalaria pallida	leaves	fresh wounds	Kokwaro (1993, p. 146)
Crotalaria cleomifolia	leaves	anti-burn	Kokwaro (1993, p. 144)
Mimosa tenuiflora	bark	burns, abrasion wounds	Tellez and Dupoy (1990),
Pongamia pinnata	leaves	wound healing	Subramanian and Nagarajan (1988)
Smithia ochreata	whole plant	fresh wounds	Dalziel (1956, p. 261)
Tephrosia paucynga	leaves, root	fresh wounds	Kokwaro (1993, p. 156)
Liliaceae			
Allium sativum[a]	Bulb	wound healing in perforated ear drum	Singh *et al.* (1984).
Aloe aculeata	juice	anti-burn	Suga and Hirata (1983)
Aloe africana	juice	anti-burn	Suga and Hirata (1983)
Aloe ammophila	juice	anti-burn	Suga and Hirata (1983)
Aloe arborescens[a,c,*]	fresh leaves, juice	anti-burn	Kameyama and Shinho (1979)

Table 5.1. Cont'd

Family/Plant Name	Part used	Wound Type/Uses	Ref.
Liliaceae cont'd			
Aloe arenicola	juice	anti-burn	Kameyama and Shinho (1979)
Aloe candelabrum	juice	anti-burn	Kameyama and Shinho (1979)
Aloe castanea	juice	anti-burn	Kameyama and Shinho (1979)
Aloe comosa	juice	anti-burn	Kameyama and Shinho (1979)
Aloe ferox	juice	anti-burn	Kameyama and Shinho (1979)
Aloe gariepiensis	juice	anti-burn	Kameyama and Shinho (1979)
Aloe globuligemma	juice	anti-burn	Kameyama and Shinho (1979)
Aloe lettyal	juice	anti-burn	Kameyama and Shinho (1979)
Aloe nyeriensis	juice	anti-burn	Kameyama and Shinho (1979)
Aloe perryi	juice	anti-burn	Kameyama and Shinho (1979)
Aloe saponaria	fresh leaves	anti-burn	Kameyama and Shinho (1979)
Aloe speciosa[*]		wound healing acceleration	Kameyama and Shinho (1979)
Aloe spectabilis	juice	anti-burn	Kameyama and Shinho (1979)
Aloe species		anti-burn	Anon. (1980), Winters *et al.* (1981), Suga and Hirata (1983)
Aloe transvalensis	juice	anti-burn	Kameyama and Shinho (1979)
Aloe vanbalenii	juice	anti-burn	Kameyama and Shinho (1979)
Aloe vera[a]	leaves	burns, roentgen dermatitis, wound healing after dental surgery complete regeneration of skin, new hair growth, complete restoration of sensation, lack of scar tissue	Collins and Collins (1935), Kesten and Laughlin (1936), Rattner(1936), Loveman (1937), Crewe (1939), Mandeville (1939), Rowe (1940), Rowe *et al.* (1941), Barnes (1947), Lushbaugh and Hale (1953), Goff and Levenstein (1964), El Zawahry *et al.* (1973), Cobble (1975), Northway (1975), Ship (1977), Hegazy *et al.* (1978), Sayed (1980), Davis *et al.* (1987, 1988, 1989, 1991, 1994), Lerner (1987), Rodriguez-Bigas *et al.* (1988), Crowell *et al.* (1989), Kivett (1989), Verma *et al.* (1989), Egawa *et al.* (1990), Fulton (Jr.)(1990), Kaufman *et al.*(1990), Thompson (1991), Heggers *et al.* (1993), Hormann and Korting (1994), Patel and Mach (1994), Udupa *et al.*(1994), Bouthet *et al.* (1995)
Aloe volkensii	extract	anti-burn	Suga and Hirata (1983)
Aloe wickensii	extract	anti-burn	Suga and Hirata (1983)
Loganiaceae			
Strychnos cocculoides	root	sores	Kokwaro (1993, p. 158)

Lythraceae			
Lawsonia inermis	leaves	wound healing acceleration	Malekzadeth and Shabestrari (1989)
Malvaceae			
Abutilon fruticosum	root	fresh wounds	Kokwaro (1993, p. 161)
Hibiscus aponeurus	leaves, stem	anti-burn	Kokwaro (1993, p. 165)
Hibiscus flavifolius	leaves, root	anti-burn	Kokwaro (1993, p. 165)
Sida ovata	bark	fresh wounds, ear inflammation, bactericide	John (1984), Kokwaro (1993, p. 168)
Meliaceae			
*Azadirachta indica**	leaves	wound healing acceleration	Davis *et al.* (1991)
*Trichilia heudelotti***	bark	sores	Dalziel (1956, p. 329)
Menispermaceae			
Cissampelos pareira	root	fresh wounds, boils, burns, insect and snake bites	Hedberg *et al.* (1983a), Jain and Puri (1984), Martinez (1984), Shah and Jain (1988), Anesini and Perez (1993), Kokwaro (1993, p. 172)
Tinospora caffra	leaves	fresh wounds	Kokwaro (1993, p. 172)
Moringaceae			
Moringa pterygosperma	stem	pyoderma	Caceres and Lopez (1991)
Nyctaginaceae			
Boerhavia diffusa	root	abscesses, hemostatic, antibacterial, anti-inflammatory, ulcers	Dalziel (1956, p. 43), Mudgal (1975), Anon. (1978b), Dabral and Sharma (1983), Olukoya *et al.* (1993)
Passifloraceae			
Adenia digitata	root	ulcers	Watt and Breyer-Brandwijk (1962, p. 826)
Pedaliaceae			
Sesamum indicum	leaves	sores, ulcers, dog bites	Watt and Breyer-Brandwijk (1962, p. 832), Ortiz de Montellano (1975), Kumar and Prabhakar (1987)
Phytolaccaceae			
Phytolacca dodecandra	leaves juice	fresh wounds, dog bites, rabies, skin lesions, ulcers, anti-inflammatory	Watt and Breyer-Brandwijk (1962, p. 837), Abebe (1986), Desta (1993)
Piperaceae			
Peperomia polhicida	whole plant	sores	Dalziel (1956, p. 16)
Piper betle	leaves	wounds	Santhanam and Nagarajan (1990)

Table 5.1. Cont'd

Family/Plant Name	Part used	Wound Type/Uses	Ref.
Plantaginaceae			
Plantago lanceolata	leaves	epithelialization	Monastyrkaya and Petropavlovskaa (1953)
Plantago major	leaves	wounds	Nikol'skaya (1954), Mironov *et al.* (1983)
Plantago psyllium	extract	wounds	Nikol'skaya (1954)
Poaceae			
Chloris virgata	leaves	fresh wounds	Kokwaro (1993, p. 253)
Cymbopogon dieterlenni	whole plant	fresh wounds	Watt and Breyer-Brandwijk (1962, p. 470)
Saccharum officinarum	whole plant	fresh wounds	Watt and Breyer-Brandwijk (1962, p. 484)
Polygalaceae			
Polygala arenaria	whole plant	fresh wounds	Dalziel (1956, p. 27)
Heinsia crinata	root, leaves	fresh wounds	Kokwaro (1993, p. 193)
Polygonum cuspidatum		anti-burn	Anon (1978b)
Polyporaceae			
Fomes rimosus	plant ash	fresh wounds	Watt and Breyer-Brandwijk (1962, p. 1113)
Portulacaceae			
Portulaca foliosa	whole plant	fresh wounds	Kokwaro (1993, p. 194)
Rubiaceae			
Chasssalia albiflora	fruit	fresh wounds	Kokwaro (1993, p. 201)
Crossopteryx febrifuga	bark	fresh wounds	Watt and Breyer-Brandwijk (1962, p. 898), Hedberg *et al.* (1983b)
Sarcocephalus esculentus	bark, root	fresh wounds	Dalziel (1956, p. 412)
Scrophulariaceae			
Limnophilia conferta	extract	wound healing acceleration	Reddy (1991)
Rhamphicarpa herzfeldiana	leaves	fresh wounds	Kokwaro (1993, p. 219)
Solanaceae			
Nicotiana rustica	leaves	sores	Adesakin, personal communication
Withania somnifera	leaves	sores, snakebite, scabies, inflammation, festering boils, deobstruent, ulcers, anti-inflammation, analgesic	Anon. (1946), Dalziel (1956, p. 435), Arseculeratne *et al* (1985), Shah and Gopal(1985), Begum and Sadique (1987), Nagaraju and Rao (1990)

Tiliaceae			
Grewia occidentalis	stem	sores	Kokwaro (1993, p. 230)
Grewia similis	bark	sores	Kokwaro (1993, p. 231)
Umbelliferae			
Centella asiatica[a,c,*]	whole plant	external ulcers, postphlebitic syndrome, suppurating wounds, wound healing promotion, analgesic, antifungal, anti-inflammatory, sores, aches, abscesses cuts, fractures	Yantadilaka and Raktavat (1950), Rastogi *et al.* (1960), Burkill (1966), Pasich *et al.* (1968), Poizot and Dumaz (1978), MacGregor and Greenwood (1980), Holdsworth *et al.* (1983), Jain and Puri (1984), John (1984), Singh (1986), Morisset *et al.* (1987), Tenni *et al.* (1988), Arpaia *et al.* (1990), Sakina and Dandiya (1990), Montecchio *et al.* (1991), Adjanohoun *et al.* (1993, p. 39, 321)
Centella coriacea	fresh herb	fresh wounds	Watt and Breyer-Brandwijk (1962, p. 1035)
Ferula pseudooreoselinum	root	anti-burn	Dzhumazhanov (1959)
Verbenaceae			
Clerodendron glabrum	leaves	fresh wounds, fractures	Arnold and Gulumian (1984), Weniger *et al.* (1986),
Priva cordifolia	leaves	anti-burn	Kokwaro (1993, p. 242)
Verbena officinalis	leaves	fresh wounds, ulcers, bleeding, analgesic, insect and snake bites, anti-inflammatory	Watt and Breyer-Brandwijk (1962, p. 1054), Le Grand (1989)
Vitex leucoxylon	stem bark	fills wounds dead space	Sarma *et al.* (1990)
Vitex rufa	bark, leaves	ulcers	Dalziel (1956, p. 458)
Vitaceae			
Cissus quadriangularis	leaves	fresh wounds	Watt and Breyer-Brandwijk (1962, p. 1058), Adjanohoun *et al.* (1991)
	leaves	external ulcers, fractures, aches	Watt and Breyer-Brandwijk (1962, p. 1058), Udupa and Prasad (1964), El-Hamid (1970), Chopra *et al.* (1976), Nagaraju and Rao (1990)
Rhoicissus tridentata	sap	sores, wound healing promotion, cuts, anesthetizer	Al-Yahya (1985), Holdsworth and Rali (1989),Kokwaro (1993, p. 246)
Zingiberaceae			
Aframomum melegueta	leaves	fresh wounds	Adjanohoun *et al.* (1991, p. 301)
Curcuma aromatica[c]	rhizome	granulation tissue	Santhanam and Nagarajan (1990)
Curcuma longa[c]	rhizome	granulation tissue	Kumar *et al.* (1993)
Anchomanes difformis	stem, juice	fresh wounds	Adesakin, personal communication
Sphaerocentrum jollyanum	stem bark	sore	Adesakin, personal communication

Table 5.1. Cont'd

Family/Plant Name	Part used	Wound Type/Uses	Ref.
Miscellaneous			
Achatina fulica[**,***]	slimy juice from giant African snail	heals circumcision, wounds	Sofowora (1996)
Apis mellifera[**,***]	honey from bees	it is believed in some parts of Nigeria that a wound that does not heal with honey may never heal	

* Wound healing plant already in clinical use
** Mucilage-containing plant and carbohydrate derived natural products with wound-healing properties
*** Not of plant origin

[a] Plant already tested on humans
[b] Plant with outstanding wound healing effects on animals but not tested on humans
[c] Plant from which the active wound healing chemical compound has been isolated/characterized/patented

Table 5.1. includes 11 plants that have been tested in humans. Many of the active products and their biological activity have been patented. The most widely used genus of all the plants is *Aloe*. The genera tested in humans are distributed in the families as follows: Liliaceae (3), Asteraceae/Compositae (2), Boraginaceae (2), Convolvulaceae (1), Umbelliferae (1), Leguminosae (1), Crassulaceae (1).

Wound healing after traditional surgery

A few examples of surgical operations carried out in African traditional medicine and the treatment for which involve plants are:

Bone Setting - A traditional bone setter is a specialist skilled in the treatment of fractures. The bone setter ties splints and medicaments to the fracture. The fractured part is laid flat and immobilized and herbal dressings are placed on the fracture, examples of herbs used include a decoction of *Cissus quadrangularis* (Vitaceae) leaf which is drunk three times daily and used to bathe the affected parts. The plant has actually been demonstrated to have wound and fracture healing activity (Udupa and Prasad 1964; Chopra *et al.* 1976).

Treatment of burns - In African traditional medicine burns are treated with herbal preparations which produce a soothing effect. For example papaya juice ointment is applied by some practitioners to produce a gradual removal of dead tissue, after this process is completed and the healthy granulation tissue appears, the burn is treated with a herbal medication especially to promote wound healing (Sofowora 1996).

After some surgical operations certain types of diet are forbidden to ensure that there is rapid healing up of wounds, *e.g.* no "okro" soup prepared from *Hibiscus esculentus* or other slimy foods including *Corchorus olitorius* leaves is permissible after traditional circumcision until the wound is healed (Sofowora 1996).

After the traditional abstraction of bullets from wounds sustained by warriors, herbal medication (usually oily preparations) as dressings and heat therapy are applied to such wounds to aid healing.

Circumcision - Other surgical operations carried out by the traditional medical practitioners include circumcision of the male. This is carried out with care using a cold knife and keeping the penis cold to effect some vasoconstriction and reduce blood loss during the operation. Local or general anesthetics are rarely applied. A few drops of spent traditional (from plant) dye (*i.e.* dye which has already been used for dyeing native cloth) is splashed onto the surface probably for its antimicrobial effect. A little later (this is common in all cases) a giant African snail (*Achatina fulica*) is broken open at the base of its shell to release the slimy juice which is allowed to drop directly onto the cut surface. No more dressing is applied on that day. On the second day a preparation of palm oil containing some herbs is applied to the wound with the aid of a feather. The sore heals within 2-3 weeks, if the wound is not damaged.

Tribal marks - In the treatment of wounds resulting form tribal marks, spent dye is also utilized immediately after the incisions are made. Two days later, palm oil is applied with the aid of a feather and the wound area kept dry. On the third day the patient bathes and rubs the sore with a face towel and native soap (sometimes mixed with maize pap to absorb the fluid oozing from the sores). Maize pap, gum or mucilage may be acting as a wound sealing plug as well. After cleansing the wounds, various herbs which have astringent, hemostatic or antimicrobial properties *e.g. Hoslundia opposita, Dissotis rotundifolia, Ehreta cymosa, Solanum nodiflorum* and *Ocimum gratissimum* are wrapped in banana leaf, heated in hot ash and the leaf juice squeezed onto the sores once or twice daily. Healing is effected within 3 weeks. The tribal marks are often pressed into shape with a thumb during the healing period to avoid keloid (hardened tissue swelling) formation. In traditional medicine herbal preparations are used to dissolve away keloids instead of removal by surgery as in orthodox medicine.

The traditional surgical patients are advised against eating slimy foods *e.g.* vegetable soup made from *Corchorus olitorus* leaves during the healing process because it is believed that such foods when taken internally prevent rapid healing of sores. It is noteworthy to state here that slimy items from natural sources including mucilages, gums and slimes from snails have been used externally in wound healing (Elliot 1994), whereas they are forbidden for internal consumption in Western Nigeria and other parts of Africa (Sofowora 1996).

Possible mode of action of some wound healing plants from Africa

As earlier mentioned, the mechanism or mode of action of wound healing agents in the plants used for wound healing vary. A correlation of the possible mode of action of some of these plants is provided below. The active wound healing agent(s) reported in several wound healing plants and which provide justification for their use is shown in Table 5.2.

1. The juice of *Aristolochia bracteata* (Aristolochiaceae) is used in the treatment of foul and neglected ulcers. The leaves of a related species *A. elegans* was found to contain among other compounds, aristolochic acid and allantoin (Sofowora 1996). Allantoin is one of the wound healing agents listed in Table 5.2. The antimicrobial activity of *A. bracteata* may explain its traditional use for treating sores. The presence of allantoin is contributory.

2. After circumcision, a shoot of *Pergularia daemia* (Asclepiadaceae) is wilted over fire and the warm juice is squeezed onto the circumcision wound, probably as an antiseptic, anesthetic or analgesic. A poultice of the leaves of the plant is applied to boils and abscesses in India and Ghana and it is applied to sore eyes. The paralytic effect of the extract of *P. daemia* on cockroaches within 2-4 days could possibly indicate that the plant has anesthetic properties which will be beneficial as a soothing agent in circumcision (Sofowora 1996). The plant *P. daemia* also possesses antimicrobial activity which could explain its use in the treatment of wounds and abscesses. The plant also has mild analgesic effect when administered intraperitonially or orally.

   Kalanchoe crenata (Crassulaceae). The crushed leaves or the juice expressed after heating the leaves of *K. crenata* are mixed with shea butter or oil and rubbed on abscesses or other swellings or applied to ulcers and burns. Juice from dried leaves are squeezed out and applied to septic wounds (Sofowora 1996). Its main constituents include malic acid and α-tocopherol found in the green callus.

α-Tocopherol has also been reported to possess wound healing activity see Table 5.2.

3. In Ghana, *Euphorbia hirta* leaves are used in sore and wound healing. It is used in East Africa to treat boils (Sofowora 1996). In East Africa, The Malay peninsula and Liberia the latex is used in treating conjunctivitis and ulcerated cornea. The Toukouleurs and Wolofs also use the latex externally as antiseptic and for sore healing. At one time in southern Malawi it was used in eye treatment, but it is no longer so commonly used. The plant contains flavonoids, triterpenes, mucilage and some acids *e.g.* ellagic acid (Sofowora 1996). Ellagic acid may be responsible for the wound healing properties (see Table 5.2). Ellagitannin derivatives have been isolated from the leaf of a Chinese specimen of *E. hirta*. Antiulcer activity was demonstrated by a chromatographic fraction of *E. hirta* from Taiwan. The fraction contains amongst others, protocatechic acid and gallic acid which may be responsible for the antiulcer and wound healing activities.

4. *Moringa pterygosperma* (Moringaceae) root and root bark are used by the Indians to treat mouth sores. The root contains a gum which is made up of bassorin and enzymes. The plant also contains cytokinins, zeatin and zeatin riboside (Sofowora 1996) which have some effects on wound healing activities as normal cell growth promotors. The plant has antimicrobial activity (against a wide range of micro-organisms) which corroborates the traditional use of the plant in gargles for sore mouth. An intramuscular injection or local administration of spirochin (Sofowora 1996) is antiseptic and prophylactic against wound infections, even in patients with already marked infection. It has analgesic and antipyretic effects.

5. In Angola, the crushed bark of *Ximenia americana* (Oleaeceae) is applied to sores of domestic animals and in west tropical Africa the pulverized bark and root are used as a dressing for ulcers, etc. The main constituents of the whole plant include hydrocyanic acid. The bark yields 16-17% of tannins (Sofowora 1996). These tannins may be responsible for its wound healing activity. See Table 5.2. for wound healing tannins.

6. *Borreria verticillata* (Rubiaceae) is used in Casamance for the treatment of whitlow and boils, by applying a paste obtained by pounding the leaves in a mortar with the extract of *Carapa procera*. The volatile oil it contains is rich in terpenes, phenolics and aromatic polycarboxylic acids. Azulene is present in the oil and this compound has also been reputed to possess wound healing activity. See Table 5.2. The high boiling components of the volatile oil showed strong antimicrobial activity against Gram positive and Gram negative bacteria (Sofowora 1996).

Table 5.2. Wound healing active chemical compounds and their sources

Chemical Compound	Wound haling Activity	Plant/Source	References
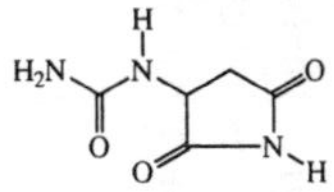 Allantoin	wound healing accelerator	synthesized from uric acid	Thompson (1991)
[a]Aloe Carboxypeptidase	anti-burn	*Aloe arborescens* var. *natalensis*	Obata *et al.* (1993).
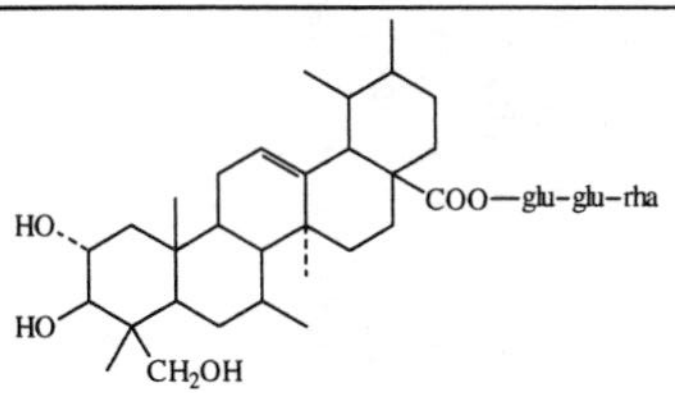 Asiaticoside	wound healing accelerator	*Centella asiatica*	Velasco and Romero (1976)
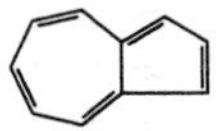 Azulene	anti-inflammatory	*Anthemis nobilis*	Takeda *et al.* (1983)
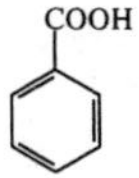 Benzoic acid	antifungal, anti-inflammatory	wide spread in nature	Yamasaki and Saeki (1967)
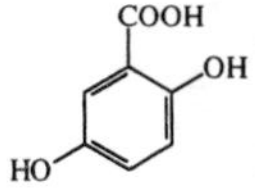 2,5-Dihydroxy-benzoic acid	anti-inflammatory	wide spread in nature	Yamasaki and Saeki (1967)

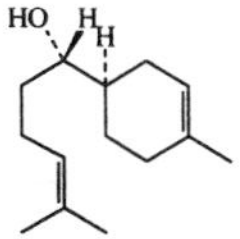 α–Bisabolol	anti-burn	*Anthemis nobilis*	Zita and Steklova (1955)
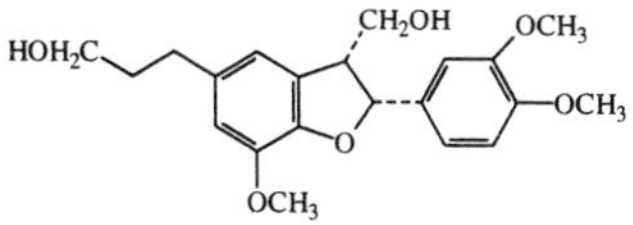 Dimethyl-cedrusin	simulates fibroblast collagen synthesis	*Croton* sp.	Santhanam and Nagarajan (1990). Pieters (1992)
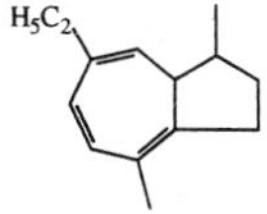 Chamazulene	anti-burn	*Anthemis nobilis*	Zita and Steklova (1955)
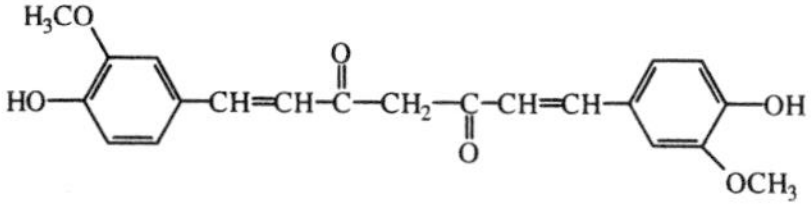 Curcumin = Diferuloyl methane	wound healing accelerator	*Curcuma aromatica* *Curcuma longa*	Deodhar *et al.* (1980), Santhanam and Nagarajan (1990), Kumar *et al.* (1993)
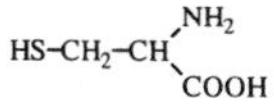 Cysteine	wound healing accelerator	wide spread in proteins	Harvey and Gibson (1984).
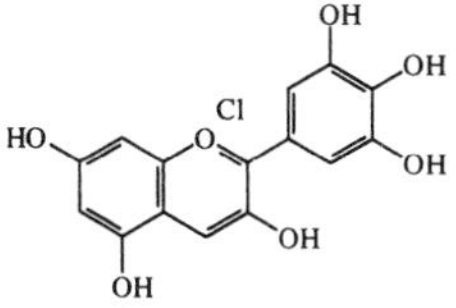 Delphinidin chloride	wound healing accelerator	wide spread in plants	Conti *et al.* (1992).
[a]Echinacin B an extract	anti-hyaluronidase	*Echinacea angustifolia*	Bonadeo *et al.* (1971)

Table 5.2. Cont'd

Chemical Compound	Wound haling Activity	Plant/Source	References
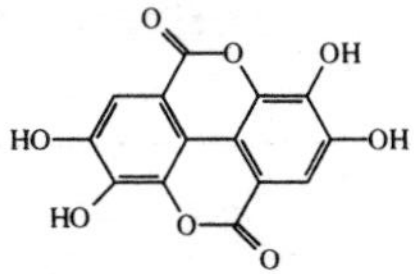 Ellagic acid	hemostatic, anti-inflammatory	*Castanea* sp., *Eucalyptus* sp.	Egawa *et al.* (1990).
Essential oil	wound healing accelerator	*Chromolaena odorata*	George (1974).
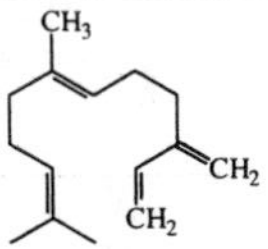 ß-Farnesene	anti-burn	Essential oils (*Chamomile*)	Zita and Steklova (1955)
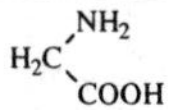 Glycine	wound healing accelerator	Gelatin, silk fibroin	Harvey and Gibson (1984).
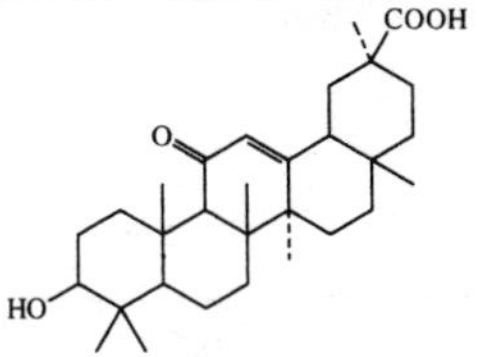 Glycyrrhetinic acid	cicatrizant	*Glycyrrhiza glabra* and its varieties	Vevron and Giustiniani (1988).

Compound	Activity	Source	Reference
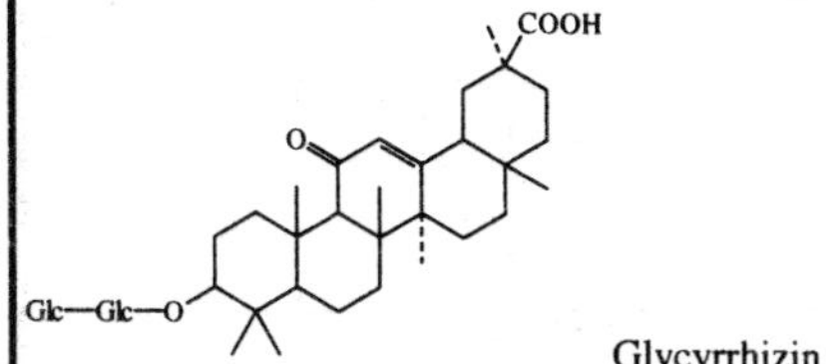 Glycyrrhizin	wound healing accelerator	*Glycyrrhiza glabra* and its varieties	Davydova *et al.* (1992).
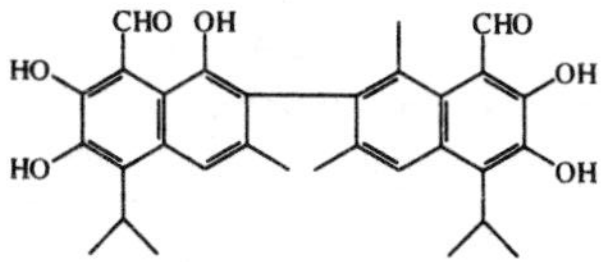 Gossypol	antimicrobial	*Gossypium barbadense*, *Thespesia populnea*	Aizikov *et al.* (1977)
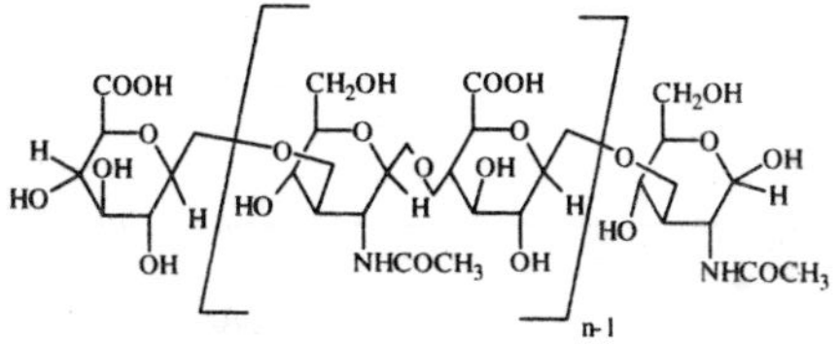 Hyaluronic acid	wound healing accelerator	connective tissue	Peacock, and van Winkle (1976), Bleacher *et al.* (1993)
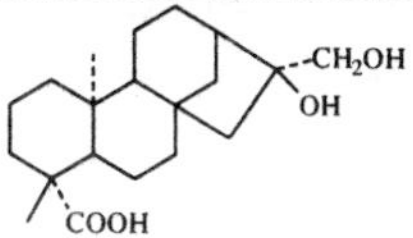 16ß, 17-Dihydroxy-kauran-19-oic acid	wound healing accelerator	*Siegesbeckia pubescens*	Han *et al.* (1975).

Table 5.2. Cont'd

Chemical Compound	Wound haling Activity	Plant/Source	References
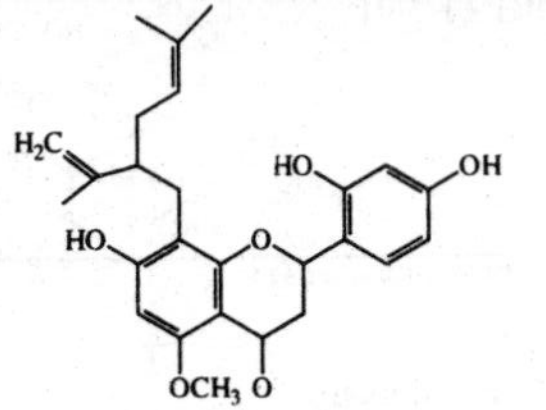 Kurarinone	antiulcer	*Sophora flavescens*	Yamahara *et al.* (1990)
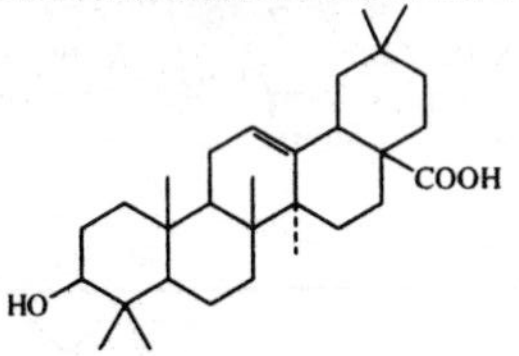 Oleanolic acid	wound healing promotor	Wide spread	Tamai and Yamahora (1992).
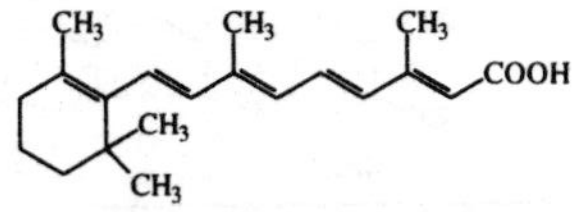 Retinoic acid = tretinoin	keratolytic	synthesized from vitamin A	Trease and Evasns (1989)
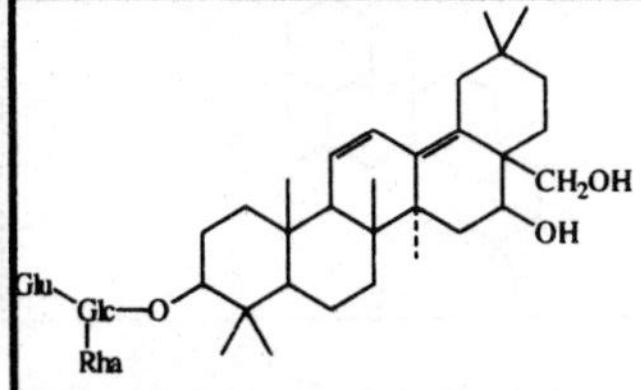 Saikosaponin B1	wound healing accelerator	*Bupleurum falcatum*	Nishiyama and Akutsu (1992), Hostettmann and Marston (1995).

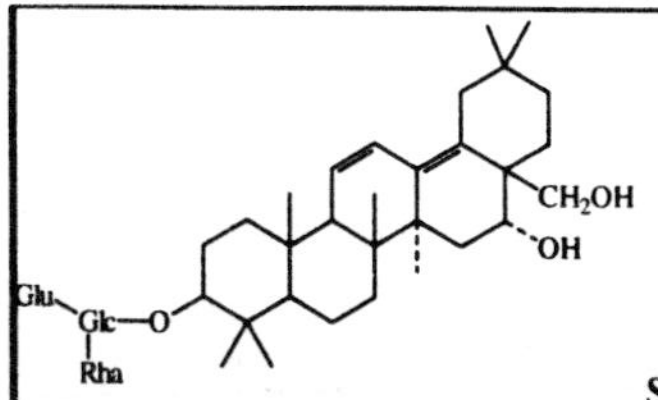 Saikosaponin B2	wound healing accelerator	*Bupleurum falcatum*	Nishiyama and Akutsu (1992) Nishiyama and Akutsu (1993) Hostettmann and Marston (1995)
Tannic acid ($C_{76}H_{52}O_{46}$) = penta-*O*-(m-digalloyl)-β-D-glucose	anti-burn	*Pinus caribaea*	Bope *et al.* (1948)
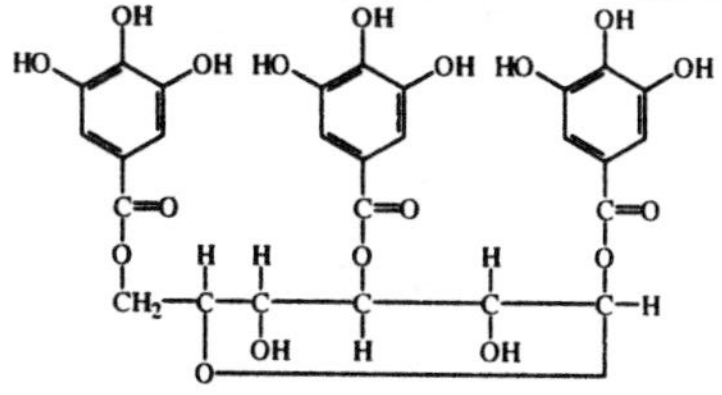 Corilagin (tannin)	anti-burn	*Pinus caribaea* *Aleurites fordii* *Acalypha wilkesiana*	Bope *et al.* (1948) Nonaka *et al.* (1990), Olugbade *et al.* (1996).
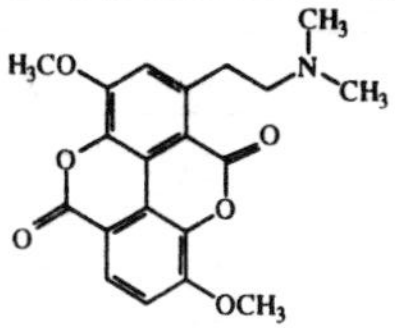 Taspine	anti-burn, anti-inflammatory, cicatrizant	*Croton lechleri*	Vaisberg *et al.* (1989) Pieters (1992) Porras-Reyes *et al.* (1993)
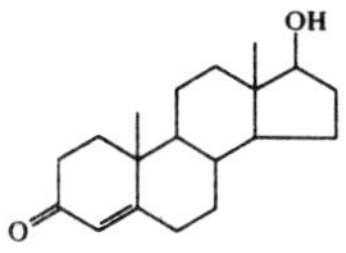 Testosterone	wound closure	testes of bull/synthesis	Morton and Malone (1972)

Table 5.2. Cont'd

Chemical Compound	Wound haling Activity	Plant/Source	References
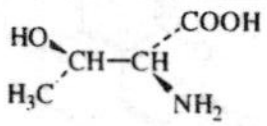 DL-Threonine	wound healing accelerator	wide spread in proteins	Harvey and Gibson (1984)
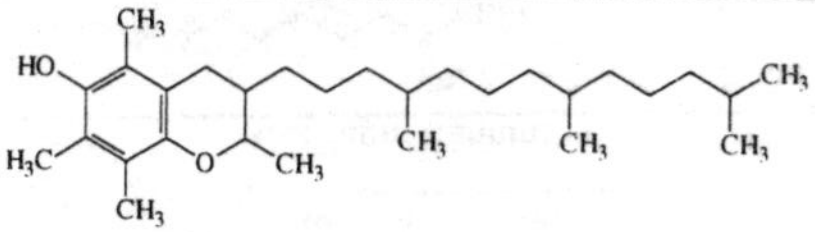 α-Tocopherol	improves skin integrity	embryos of cereals, many seed oils, alfalfa, lettuce	Bernhard (1988).
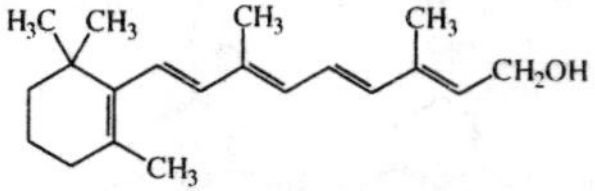 Vitamin A = Retinol	wound healing accelerator, improves skin integrity	Butter/Egg yolk, fish liver (not in plants)	Gao *et al.* (1992)
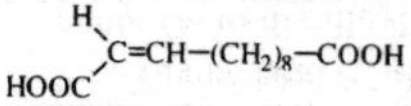 $HOOC-CH=CH-(CH_2)_8-COOH$ Traumatic acid	wound hormone	Bean pod tissue or *Aloe* sp	Freytag 1954)

[a] = Already in clinical use.

The structure-activity relationships (SAR) of wound healing chemical compounds isolated from plants and animals

The chemical compounds with wound healing activity are not restricted to a particular chemical group. They are, however, mostly proteins, amino acids, terpenes, flavonoids, alkaloids, quinonoids, tannins, steroids, carbohydrates or coumarins. Branched chains, uronic acid moiety and pyran rings are fairly common in the diverse structures. So also are polyhydroxyl and acidic functional groups.

Most of the Wound Healing agents are acids, and/or triterpenes. This would suggest a pharmacodynamics based essentially on the presence of at least one carboxylic acidic group, and a structure-activity relationship based on the presence of some terpene and carboxylic acid moieties.

The fact that many growth regulators in plants *e.g.* gibberelins are acidic and terpenoid in nature seem to support this suggestion. The gibberelic acids are biosynthesized via kaurenoic acid which, by a multi-step ring contraction, furnishes the gibbane ring system (Trease and Evans 1989). A kaurenoic acid derivative has been reported to have wound healing properties (Han *et al.* 1975). (see Table 5.2). Abcissic acid (Trease and Evans 1989) and traumatic acid (Devlin and Witham 1986; Freytag 1954) are also involved in cell upkeep, proliferation, elongation phases etc. (Trease and Evans 1989) which are really responsible for re-epithelialization and wound closure. Traumatic acid is abundant in *Aloe* species which are highly reputed for the treatment of wounds and burns (Freytag 1954).

Cell division hormones-cytokinins - have a more specific effect on cell division. These include kinetin (not found in plant, but in the sperm of herrings) and zeatin which has been found in the embryo of maize, at the milky stage. Other derivatives of zeatin have been found in many woody plants (Trease and Evans 1989). Cytokine and TGF (transforming growth factor) play an important role in wound healing (Bleacher *et al.* 1993; Meyer 1958; Sullivan *et al.* 1995). Induced cell cytokinins are believed to be the major promotors in cellular division and hormonal control of morphogenesis (Devlin and Witham 1986). Excessive amounts of these compounds cause scarring in adult wounds, whilst fetal wounds heal without scar due to deficiency or lack of these compounds in the foetus.

Many pentacyclic saponins and their genins have been shown to affect the cell as antiexudative (Hostettmann and Marston 1995) preventing spread. They may be acting as antihyaluronidase or as hyaluronic acid or related compounds since these compounds have been applied to wounds (Bleacher *et al.* 1993). Some of the wound healing compounds listed in Table 5.2. are saponins or aglycones of saponins.

Many triterpene saponins and their aglycones have been reported to possess antiulcerogenic, anti-inflammatory, fibrinolytic, antipyretic, analgesic and antiedematous activities (Hostettmann and Marston 1995). Antiulcerogenics - skin and gastric *e.g.* glycyrrhetinic and glycyrhizic acids, both of which constitute Biogastrone acid (Elks and Ganellin 1990) which is used in the treatment of ulcers and as an anti-inflammatory also in skin diseases are well known. The saponins are reported to act by promoting mucous formation (Hostettmann and Marston 1995).

This property favours wound healing by preventing wound desiccation as well as furnishing important growth factors.

Conclusion

The profiles of fetal proteoglycans, collagens and growth factors are different from those in adult wounds. High level of hyaluronic acid and its stimulation is more rapid and there is highly organized collagen deposition (Bleacher *et al.* 1993; Broker and Reiter 1994). Growth factors are characteristically less prominent in fetal wounds. Topical application of wound healing promoting agents *e.g.* regulated amounts of growth factor to wounds may soon be possible.

The traditional surgical patients are advised against eating slimy foods *e.g.* vegetable soup made from *Corchorus olitorus* leaves during the healing process because it is believed that such foods when taken internally prevent rapid healing of sores. It is noteworthy to state here that slimy items from natural sources including mucilages, gums and slimes from snails have been used externally in wound healing (Elliot 1994), whereas they are forbidden for internal consumption in Western Nigeria and other parts of Africa (Sofowora 1993).

Acidic firoblast growth factor (aFGF) has also been found to be one of the most effective wound healing components of the human system. It has potential therapeutic applications for promoting the healing of dermal ulcers, especially in healing-impaired individuals (Popp *et al.* 1995) and has been recommended for that use.

References

Abebe, W. (1986). A survey of prescriptions used in traditional medicine in Gondar region (Ethiopia),: General pharmaceutical practice. *Journal of Ethnopharmacology* **18**, 147-165.

Adesanya, S.A. and Roberts, M.F. (1995). Inducible compounds in *Phaseolus,Vigna* and *Dioscorea* species. In *Handbook of Phytoalexin Metabolism and Action.* (eds. M. Daniel and R.P. Purkayastha), pp. 333-373. Marcel Dekker, New York.

Adesina, S.K. (1982). Studies on some plants used as anticonvulsants in Amerindian African traditional medicine. *Fitoterapia* **53**, 147-162.

Adesomoju, A.A. and Okogun, J.I. (1985). Phytochemical investigation of *Phaulopsis falcisepala. Fitoterapia* **56**, 279-280.

Adjanohoun, J.E., Ahyi, M.R.P., Ake Assi, L. Dramane, K. Elewude, J.A., Fadoju, S.O., Gbile, Z.O., Goudote, E., Johnson, C.L.A., Keita, A., Morakinyo, O., Ojewole, J.A.O., Olatunji, A.O., and Sofowora, E.A. (1991). *Traditional Medicine and Pharmacopoea. Contribution to Ethnobotanical and floristic studies in Western Nigeria.* Organization of African Unity/Scientific Technical and Research Commission. Uganda

Adjanohoun, J.E., Ahyi, M.R.P., Ake Assi, L., Alia, A.M., Amai, C.A., Gbile, Z.O., Johnson, C.L.A.,Kakooko, Z.O., Lutakome, H.K., Morakinyo, O., Mubiru, N.K., Ogwal-Okeng, J.W., and Sofowora, E.A. (1993). *Traditional Medicine and Pharmacopoea. Contribution to Ethnobotanical and floristic studies in Uganda.* Organization of African Unity/Scientific Technical and Research Commission. Uganda

Aizikov, M.I., Kurmukov, A.G., and Isamukhamedov, I. (1977). Antimicrobial and wound healing effect of gossypol. *Chemical Abstract* **88**, 164324B.

Akbar, S. (1989). Pharmacological investigations on the ethanolic extract of *Salvia haemtodes*. *Fitoterapia* **60**, 270-272.

Al-Yahya, M.A., Hifnawy, M.S., Mossa, J.S., Al-Meshal, I.A., and Mekkawi, A.G. (1985). Aromatic plants of Saudi Arabia, Part 7. Essential oil of *Plectranthus tenuiflorus* (Vatke) Agnew. *Proceedings of the Saudi Biological Society*, 1985, 147-153.

Anesini, C. and Perez, C. (1993). Screening of plants used in Argentine folk medicine for antimicrobial activity. *Journal of Ethnopharmacology* **39**, 119-128.

Anon. (1946). Western Arabia and the Red sea. *Geographical handbook series B.R. 527*, pp. 590-602. Great Britain Naval Intelligence division, London.

Anon. (1978a). Studies on the toxic effects of certain burn escharotic herbs. *Chung-hua I Hsueh Tsa Chih (New Series)* **4**, 388.

Anon. (1978b). Antifertility studies on plants. *Indian Council of Medical Research - Annual Report, Director general* 1978, 63-64.

Anon. (1980). Cosmetics for skin. *Patent-Japan Kokai Tokkyo Koho*-80 104, 205.

Arnold, H.J. and Gulumian, M. (1984). Pharmacopoeia of traditional medicine in Venda. *Journal of Ethnopharmacology* **12**, 35-74.

Arpaia, M. R. Ferrone, R. Amitrano, M., Nappo, C., Leonardo, G., and Del Guercio, R. (1990). Effect of *Centella asiatica* extract on mucopolysaccharide metabolism in subjects with varicose veins. *International Journal of Clinical Pharmacology Research*, **10**, 229-233.

Arseculeratne, S.N., Gunatilaka, A.A.L., and Panabokke, R.G. (1985). Studies on medicinal plants of Sri Lanka. Part 14: Toxicity of some traditional medicinal herbs. *Journal of Ethnopharmacology* **13**, 323-335.

Ayensu, E.S. (1978). *Medicinal Plants of West Africa*. Reference Publications, Inc., Algonac.

Barnes, T.C. (1947). The healing action of extracts of *Aloe vera* on abrasions of human skin. *American Journal of Botany* **34**, 597.

Begum, V.H. and Sadique, J. (1987). Effect of *Withania somnifera* on glycosaminoglycan synthesis in carrageenin-induced air pouch granuloma. *Biochemical Medicine and Metabolic Biology* **38**, 272-277.

Benbow, M. (1995). Extrinsic factors affecting the management of chronic wounds. *British Journal of Nursing* **4**, 534-538.

Bennet, N.T. and Schultz, G.S. (1993). Growth factors and wound healing: part II- Role in normal and chronic wound healing. *American Journal of Surgery* **166**, 74-81.

Bernhard, J. D. (1988). *Aloe vera* and vitamin E as dermatologic remedies. *Journal of the American Medical Association* **259**, 101.

Bettinger, D., Gore, D., and Humphries, Y. (1995). Evaluation of calcium alginate for skin graft donor sites. *Journal of Burn Care and Rehabilitation* **16**, 59-61.

Bhora, F.Y., Dunkin, B.J., Aly, H.M., Bass, B. L. , Sidawy, A.N., and Harmon, J.W. (1995). Effect of growth factors on cell proliferation and epithelialization in human skin. *Journal of Surgical Research* **59**, 236-244.

Bleacher, J.C., Adolph, V.R., Dillon, P.W., and Krummel, T.M. (1993). Fetal tissue repair and wound healing. *Dermatologic Clinics* **11**, 677-683.

Boily, Y. and Van Puyvelde, L. (1986): Screening of medicinal plants of Rwanda (Central Africa) for antimicrobial activity. *Journal of Ethnopharmacology* **16**, 1-13.

Bonadeo, I., Bottazzi, G., and Lavazza, M. (1971). Echinacin B: Active polysaccharides from *Echinacea*. *Rivista Italiana: Essenze, Profumi, Piante Officinale, Aromi, Saponi, Cosmetici, Aersosol* **53**, 281-295.

Bope, F.W., Cranston, E.M., and Gisvold, O. (1948). A preliminary pharmacological investigation of the tannin obtained from *Pinus caribaea*. *Journal of Pharmacology and Experimental Therapeutics* **94**, 209.

Boukef, K. Souissi, H. R., and Balansard, G. (1982). Contibution to the study on plants used in traditional medicine in Tunisia. *Plantes Médicinales et Phytothérapie*, **16**, 260-279.

Bouthet, C.F., Schrif, V.R., and Winters, W.D. (1995). Stimulation of Neuron-like Cell growth by *Aloe* substances. *Phytotherapy Research* **9**, 185-188.

Broker, B.J. and Reiter, D. (1994). Fetal wound healing. *Otolaryngology-Head and Neck Surgery* **110**, 547-549.

Browner, C.H. (1985). Plants used for reproductive health in Oaxaca, Mexico. *Economic Botany*, **39**, 482-504.

Burkill, I.H. (1966). *Dictionary of economic products of the Malay Peninsula*, vol. I. Ministry of Agriculture and Cooperatives, Kuala Lumpur.

Burkill, I.H. (1985). *The Useful Plants of West Tropical Africa. 2nd edn. Vol. 1, Families A-D.* Royal Botanic Gardens Kew, Richmond.

Caceres, A. and Lopez, S. (1991). Pharmacological properities of *Moringa oleifera* - 3. Effect of seed extracts in the treatment of experimental pyodermia. *Fitoterapia* **62**, 449-450.

Caceres, A., Giron, L.M., Alvarado, S.R., and Torres, M.F. (1987) Screening of antimicrobial activity of plants popularly used in Guatemala for the treatment of dermatomucosal diseases. *Journal of Ethnopharmacology* **20**, 223-237.

Chagnon, M. (1984). General pharmacologic inventory of medicinal plants of Rwanda. *Journal of Ethnopharmacology* **12**, 239-251.

Chhabra, S.C. and Uiso, F.C. (1991). Antibacterial activity of some Tanzanian plants used in traditional medicine. *Fitoterapia* **62**, 499-503.

Chhabra, S.C., Uiso, F.C., and Mshiu, E.N. (1984). Phytochemical screening of Tanzanian medicinal plants. I. *Journal of Ethnopharmacology* **11**, 157-179.

Chhabra, S.C., Mahunnah, R.L.A., and Mshiu, E.U. (1987). Plants used in traditional medicine in eastern Tanzania. I. Pteridophytes and angiosperms (Acanthaceae to Canellaceae). *Journal of Ethnopharmacology* **21**, 253-277.

Chopra, S.S., Patel, M.R., and Awadhiya, R.P. (1976). Studies on *Cissus quadrangularis* in experimental fracture repair: A histopathological study. *Indian Journal of Medical Research* **64**, 1365.

Clark N. and Sherman R. (1993). Soft-tissue reconstruction of the foot and ankle. *Orthopaedic Clinics of North America* **24**, 489-503.

Clark, R.A. (1993). Basics of cutaneous wound repair. *Journal of Dermatologic Surgery and Oncology* **19**, 693-706.

Cobble, H.H. (1975). Stabilized *Aloe vera* gel. *Patent-US*-3, 892,853.

Cockbill, S.M. and Turner, T.D. (1995). Management of veterinary wounds. *Veterinary Records*, **136**, 362-365.

Collins, C.E. and Collins, C. (1935). Roentgen dermatitis treated with fresh whole leaf of *Aloe vera. American Journal of Roentgenology and Radum Therapy* **33**, 396.

Conti, M., Cristoni, A., and Magistretti, M.J. (1992). Activity of delphinidin on microvascular damage models in rodents. *Phytotherapy Research* **6**, 99-103.

Crewe, J.E. (1939). Aloes in the treatment of burns and scalds. *Minnesota Medicine* **22**, 538-539.

Crowell, J., Hilsenbeck, S., and Penneys, N. (1989). *Aloe vera* does not affect cutaneous erythema and blood flow following ultraviolet b exposure. *Photodermatology* **6**, 237-239.

Dabral, P.K. and Sharma, R.K. (1983) Evaluation of the role of Rumalaya and Geriforte in chronic arthritis, a preliminary study. *Probe* **22**, 120-127.

Dalziel, J.M. (1956). *Useful Plants of West Tropical Africa*, Crown Agents for Overseas Government, London.

Davis, R.H., Kabbani, J.M., and Maro, N.P. (1987). *Aloe vera* and wound healing. *Journal of American Pediatric Medicine Association* **77**, 165-169.

Davis, R.H., Leitner, M.G., and Russo, .J.M. (1988). *Aloe vera* - A natural approach for treating wounds, edema, and pain in diabetes. *Journal of American Pediatric Medicine Association* **78**, 60-68.

Davis, R.H., Leitner, M.G., Russo, J.M., and Bryne, M.E. (1989). Wound healing, oral and topical activity of *Aloe vera. Journal of American Pediatric Medicine Association* **79**, 559-62.

Davis, R.H., Parker, W.L., Samson, R.T., and Murdoch, D.P. (1991). Isolation of a stimulatory system in an aloe extract. *Journal of American Pediatric Medicine Association* **81**, 473-478.

Davis, R.H., Di Donato, J.J., Hartman, G.M., and Haas, R.C. (1994). Anti-inflammatory and wound healing activity of a growth substance in *Aloe vera*. *Journal of American Pediatric Medicine Association* **84**, 77-81.

Davydova, V.A., Tolstikova, T.G., Baltina, L.A., Zarudii, F.S., Murinov, Y.J., Kondratenko, R.M., and Tolstikov, G.A. (1992). *Pharmaceutical Chemistry Journal* **25**, 309-311.

Deodhar, S.D., Sethi, R., and Srimal, R.C. (1980). Preliminary study on antirheumatic activity of curcumin (diferuloyl methane). *Indian Journal of Medical Practice* **71**, 632-634.

Desta, B. (1993). Ethiopian traditional herbal drugs. Part II. Antimicrobial activity of 63 medicinal plants. *Journal of Ethnopharmacology* **39**, 129-139.

Devlin, R.M. and Witham, F.H. (1986). *Plant Physiology*, 4th edn. CBS publishers, Delhi.

De vries, H.J., Zeegelaar, J.E., Middelkoop, E., Gijsbers, G., Van Marle, J., Wildevuur, C.H., and Westerhof, W. (1995). Reduced wound contraction and scar formation in punch biopsy wounds. Native collagen dermal substitutes. *British Journal of Dermatology* **132**, 690-697.

Dhawan, B.N., Patnaik, G.K., Rastogi, R.P., Singh, K.K., and Tandon, JS. (1977). Screening of Indian plants for biological activity. VI. *Indian Journal of Exerimental Biology* **15**, 208.

Diwan, P.V., Tilloo, L.D., and Kulkarni, D.R. (1982). Influence of *Tridax procumbens* on wound healing. *Indian Journal of Medical Research*, **75**, 460-464.

Douglas, D. M. (1963) *Wound Healing and Management - A monography for Surgeons*. E & S. Livingstone, London.

Dzhumazhanov, O.D. (1959). The pharmacology of *Ferula pseudoreoselinum* roots. *Trudy Instituta Fiziologii Akademija Nauk Kazachskhoj SSR* **1**, 42-65.

Egawa, M., Ishida, K., Maekawa, M., and Sato, Y. (1990). Anti-inflammatory and wound-healing topical skin preparations containing aloe extract and ellagic acids. *Patent-Japan Kokai Tokkyo Koho*,-02 231,408; *Chemical Abstract* **114**, 129142Z.

El-Hamid, A. (1970). Drug plants of the Sudan Republic in native medicine. *Planta Medica* **18** 278.

Elliot, I.M.Z. (1994). *A Short History of Surgical Dressings*. The Pharmaceutical Press, London.

Ellis, H. and Calne, R.Y. (1977). *Lecture notes on General Surgery*, 5th. edn. Blackwell Sientific Publications, Oxford.

Elks, J. and Ganellin, C.R. (1990). *Dictionary of Drugs, Chemical Data, Structures and Bibliographies*. Chapman and Hall, London.

El-Zawahry, M., Hegazy, M.R., and Helal, M. (1973). Use of aloe in treating leg ulcers and dermatoses. *International Journal of Dermatology* **12**, 68-73.

Esterhai, J.L (Jr.) and Queenan, J. (1991). Management of soft tissue wounds associated with type III Open fractures. *Orthopaedic Clinics of North America* **22**, 427-432.

Eszter, T.S. (1992). *Euphorbia hirta* extracts as immunostimulants. *Patent-Ger Offen* **4**, 102,054.

Falanga, V. (1993). Growth Factors and wound healing. *Journal of Dermatologic Surgery and Oncology* **19**, 711-714.

Fedorchuk, A.M. (1964). Effect of *Hypericum performatum* on experimentally infected wounds. *Mikrobiolohicnyj Zurnal(Kiev)* **26**, 32.

Field, F.K. and Kerstein, M.D. (1994). Overview of wound healing in a moist environment. *American Journal of Surgery* **167**, 2S-6S.

Fish, F. and Dawson, J.O. (1967) *Surgical Dressings, Ligatures and Sutures*. Heinemann Medical, London.

Frantz, R.A. and Gardner, S. (1994). Elderly skin care: principles of chronic wound care. *Journal of Gerontological Nursing* **29**, 35-44.

*Freytag, A. (1954). *Pharmazie*, **9**, 705.

Fulton (Jr.), J.E. (1990). The stimulation of postdermabrasion wound healing with stabilized *Aloe vera* gel-polyethylene oxide dressing. *Journal of Dermatolology, Surgery and Oncolology* **16**, 460-467.

Gao, L. X., Xu, Z.Q., Jin, H., Wango, Z.Y., Xu, D., and Gu, J.F. (1992). Study on suitable dosages of vitamin A in wound healing. *Yingyang Xuebao* **14**, 33-37.

Gazit, S. and Blumenfeld, A. (1972). Inhibitor and auxin activity in the avocado fruit. *Physiology of Plant* **27**, 77-82.

George, A. (1974). Essential oil from *Eupatorium odoratum. Patent-Indian-* 121,818; *Chemical Abstract* **82**, 7558Q

Gerstein, A.D., Phillips, T.J., Rogers, G.S., and Gilchrest, B.A. (1993). Wound healing and aging. *Dermatologic Clinics* **11**, 749-757.

Gilliam, A.J. and Da Camara, C.C. (1993). Treatment of wounds with procuren. *Annals of Pharmacotherapy* **26**, 1201-1203.

Goff, S. and Levenstein, I. (1964). Measuring the effects of topical preparations upon the healing of skin wounds. *Journal of the Society of Cosmetics and Chemistry* **15**, 509-518.

Goldenheim, P.D. (1993). An Appraisal of povidone-iodine and wound healing. *Postgraduate Medical Journal* **69** Suppl 3, S97-S105.

Goldman, R.S., Freitas, P.C.D., and Oga, S. (1985). Wound healing and analgesic effect of crude extracts of *Symphytum officinale* in rats. *Fitoterapia* **56**, 323-329.

Gonzalez, A., Ferreira, F., Vazquez, A., Moyna, P., and Paz, E.A. (1993). Biological screening of Uruguayan medicinal plants. *Journal of Ethnopharmacology* **39**, 2187-2220.

Harvey, S.G. and Gibson, J.R. (1984). The effects on wound healing of three amino acids. A comparison of two models. *Chemical Abstract* **101**, 143285 M.

Han, K.D., Kim, J.H., and Oh, S.J. (1975). *Proceedings of the Symposium on Terpenoids 1974*, (ed. W:S. Woo). Natural Products Research Institute, Seoul National Univrsity, Seoul, South Korea, 17-31.

Hedberg, I., Hedberg, O., Madati, P.J., Mshigeni, K.E., Mshiu, E.N., and Samuelsson, G. (1983a). Inventory of plants used in traditional medicine in Tanzania. Part II. Plants of the families Dilleniaceae-Opiliaceae. *Journal of Ethnopharmacology* **9**, 105-127.

Hedberg, I., Hedberg, O., Madati, P.J., Mshigeni, K.E., Mshiu, E.N., and Samuelsson, G. (1983b). Inventory of plants used in traditional medicine in Tanzania. Part III. Plants of the families Papilionaceae-Vitaceae. *Journal of Ethnopharmacology* **9**, 237-260.

Hedge, D.A., Khosa, R.L., and Chansouria, J.P.N. (1994). A study of the effect of *Wedelia calendulacea* Less. on wound healing in rats. *Phytotherapy Research* **8**, 439-440.

Hegazy, M.A., Mortada, A., Hegazy, M.R., and Helal, M. (1978). The use of *Aloe vera* extract in the treatment of experimental corneal ulcers in rabbit. *Journal of Drug Research* **10**, 199-209.

Heggers, J.P., Pelley, P.R., and Robson, M.C. (1993). Beneficial effects of aloe in wound healing. *Phytotherapy Research* **7**, S48-S52.

Holdsworth, D.K. (1990). Traditional medicinal plants of Rarotonga, Cook Islands. Part I. *Inter national Journal of Crude Drug Research* **28**, 209-218.

Holdsworth, D.K. and Rali, T. (1989). A survey of medicinal plants of the southern highlands, Papua New Guinea. *International Journal of Crude Drug Research* **27**, 1-8.

Holdsworth, D.K,. Pilokos, B., and Lambes, P. (1983). Traditional medicinal plants of New Ireland, Papua New Guinea. *International Journal of Crude Drug Research* **217**, 161-168.

Hormann, H.P. and Korting, H.C. (1994). Evidence for the efficacy and safety of topical herbal drugs in dermatology: Part I: Anti-inflammatory agents. *Phytomedicine* **1**, 161-171.

Hostettmann, K. and Marston, A. (1995). *Saponins*. Cambridge University Press, London.

Hulten, L. (1994). Dressings for surgical wounds. *American Journal of Surgery* **167**, 42S-44S.

Ikan, R. (1969). *Natural Products: A laboratory Guide*. Academic Press, London.

Jain, S.P. and Puri, H.S. (1984) Ethmomedicinal plants of Jaunsar-Bawar Hills, Uttar Pradesh, India. *Journal of Ethnopharmacology* **12**, 213-222.

John, D. (1984). One hundred useful raw drugs of the Kani tribes of Trivandrum forest division, Kerala, India. *International Journal of Crude Drug Research* **22**, 17-39.

Kameyama, S. and Shinho, M. (1979). Wound-healing compositions from *Aloe arborescens* extracts. *Patent- Japan Kokai Tokkyo Koho*-79, 151, 113.

Kannon, G.A. and Garrett, A.B. (1995). Moist wound healing with occlusive dressings. A clinical review. *Dermatologic Surgery* **21**, 583-90.

Kaufman, T., Newman, A.R., and Wexler, M.R. (1990). *Aloe vera* and burn wound healing. *Plastic and Reconstructive Surgery* **83**, 1075-1076.

Kesten, B.M.C. and Laughlin, R. (1936) Roentgen ray dermatitis treated with ointment containing viosterol. *Archives of Dermatology and Syphilology* **34**, 901-903.

Khan, M.R., Ndaalio, G., Nkunya, M.H.H., and Wevers, H. (1978). Studies on the rationale of African traditional medicine. Part II. Preliminary screening of medicinal plants for antigonococci activity. *Pakistan Journal of Scientific and Industrial Research* **27**, 189-192.

Kirsner, R.S. and Eaglstein, W.H. (1993). The wound healing process. *Dermatologic Clinics* **11**, 629-640.

Kivett, W.F. (1989). *Aloe vera* for burns. *Plastic and Reconstructive Surgery* **83**, 195.

Kloos, H. (1977). Preliminary studies of medicinal plants and plant products in markets of central Ethiopia. *Ethnomedicine* **4**, 63-104.

Kokwaro, J.O. (1993). *Medicinal Plants of East Africa,* 2nd edn. Kenya Literature Bureau, Nairobi.

Kone-Bamba, D., Pelissier, Y., Ozoukou, Z.F., and Kouao, D. (1987). Hemostatic activity of 216 plants used in traditional medicine in the Ivory coast. *Plantes Médicinales et Phytothérapie* **21**, 122-130.

Kosasi, S.T., Hart, L.A., Fischer, F.C., and Labadie, R.P. (1987). Isolation of two components from latex of *Jatropha multifida* L. which inhibit classical pathway complement activity *in vitro. Pharmazeutische Weekblad* **9**, 224.

Kubas, J. (1972). Investigations on known or potential antitumoral plants by means of microbiological activity of some cultivated plant species in "*Neurospora Crassa* test". *Acta Biologica Cracoviense (Series Botanica)* **15**, 87-100.

Kumar, D.S. and Prabhakar, Y.S. (1987). On the ethnomedical significance of the arjun tree, *Terminalia arjuna* (Roxb.) Wight & Arnot. *Journal of Ethnopharmacology* **20**, 173-190.

Kumar, KA., Sharma, V.K., Singh, H.P., Prakash, P. Singh, S.P. (1993). Efficacy of some indigenous drugs in tissue repair in buffaloes. *Indian Veterinary Journal* **70**, 42-44.

Lau, H.C., Granick, M.S., Aisner, A.M., and Solomon, M.P. (1994). Wound care in the elderly patient. *Surgical Clinics of North America* **74**, 441-463.

Le Grand, A. (1989). Anti-infectious phytotherapy of the tree-savannah, Senegal (western Africa) III: a review of the phytochemical substances and anti-microbial activity of 43 species. *Journal of Ethnopharmacology* **25**, 315-338.

Le Grand, A. and Wondergem, P.A. (1987). Anti-infective phytotherapy of the savannah forests of Senegal (East Africa). I - An inventory. *Journal of Ethnopharmacology* **21**, 109-125.

Leitch, I.O.(1995). New developments in burn management. *Australian Family Physician* **24**, 136-144.

Lerner, F. N. (1987). Investigation of effects of proteolytic enzymes, aloe gel and ionophoresis on chronic and acute athletic injuries. *Chiropatic Sports Medicine* **1**, 106-110.

Livezey, A. (1868). Some observations on our indigenous medical flora. *Medicine and Surgery Report* **19**, 85.

Lorenz, H.P. and Adzick, N.S. (1993). Scarless skin wound repair in the fetus. *Western Journal of Medicine* **159**, 350-355.

Loveman, A.B. (1937). Leaf of *Aloe vera* in treatment of Roentgen ray ulcers: Report on two additional cases. *Archives of Dermatology and Syphilology* **36**, 838.

Lushbaugh, C.C. and Hale, D.B. (1953). Experimental acute radiodermatitis following beta irradiation. B. Histopathological study of the mode of action of therapy with *Aloe vera. Cancer* **6**, 690-8.

Macfarlane, D.A. and Thomas, L.P. (1972). *Textbook of Surgery,* 3rd edn. The English Language Book Society and Churchill Livingstone, London.

MacGregor, E. and Greenwood, C. (1980). *Polymers in Nature.* John Wiley & Sons, Chichester.

Madulid, D.A., Gaerlan, F.J.M., Romero, E.M., and Algoo, E.M.G. (1989). Ethnopharmacological study of the Ati tribe in Nagpana, Barotac viejo, Iloilo. *Acta Manilana* **38**, 25-40.

Magro-Filho, O. and de Carvalho, A.C. (1994). Topical effect of propolis in the repair of sulcoplasties by the modified Kayanjian technique. Cytological and clinical evaluation. *Journal of Nihon University School of Dentistry* **36**, 102-111.

Malekzadeth, F. and Shabestrari, P.P. (1989). Therapeutic effects and *in vitro* activity of an extract from *Lawsonia inermis. Journal of Science (Islamic Republic of Iran)* **1**, 7-12.

Mandeville, F.B. (1939). *Aloe vera* in the treatment of radiation ulcers of mucous membranes. *Radiology* **32**, 598-599.

Martinez, M.A. (1984) Medicinal plants used in a totonac community of the Sierra norte de Puebla: Tuzamapan de Galeana, Puebla, Mexico. *Journal of Ethnopharmacology* **11**, 203-221.

Mertz, P.M. and Ovington, L.G. (1993). Wound healing microbiology. *Dermatologic Clinics* **11**, 739-747.

Meyer, K. (1958). *Polysaccharides in Biology*. Springer, London.

Mironov, V.A., Matrosov, V.S., Zamureenko, V.A., Mairanovskii, V.G., Vasil'ev, G.S., Filipova T.M. Mishchenko, V.V., and Fel'dshtein, M.A. (1983). Physiologically active alcohols from great plantain (*Plantago major*). *Pharmaceutical Chemistry Journal* **17**, 1321-1324.

Monastyrkaya, B.I. and Petropavlovskaa, A.A. (1953). Hemostatic and wound-healing effects of plantain. *Farmakologija i Toksikologija* **16**, 30-32.

Montecchio, G.P., Samaden, A., Carbone, S., Vigotti, M., Siragusa, S., and Piovella, F. (1991). *Centella asiatica* triterpenic fraction (CATTF) reduces the number of circulating endothelial cells in subjects with post phlebitic syndrome. *Haematologica* **76**, 256-259.

Morisset, R., Cote, N.G., Panisset, J.C., Jemni, L., Cmirand, P., and Brodeur, A. (1987). Evaluation of the healing activity of Hydrocotyle tincture in the treatment of wounds. *Phytotherapy Research* **1**, 117-121.

Morris, E.J., Dowlen, S., and Cullen, B. (1994). Early clinical experience with topical collagen in vascular wound care. *Journal of Wound, Ostomy and Continence Nursing* **21**, 247-250.

Morton, J.J.P. and Malone, M. H. 91972). *Archives of International Pharmacodynamics and Therapeutics* **196**, 117-126.

Muanza, D.N., Kim, B.W., Euler, K.L., and Williams, L. (1994). Antibacterial and antifungal activities of nine medicinal plants from Zaire. *International Journal of Pharmacognosy* **32**, 337-345.

Mudgal, V. (1975). Studies on medicinal properties of *Convolvulus pluricaulis* and *Boerhaavia diffusa*. *Planta Medica* **28**, 62.

Nagaraju, N. and Rao, K.N. (1990). A survey of plant crude drugs of Rayalaseema, Andhra Pradesh, India. *Journal of Ethnopharmacology* **29**, 137-158.

Neamtu, G. and Cociu, A. (1982). Treatment of animal wounds with oil extracted from ripe fruits of *Hippophae rhamnoides* L. *Studii si Cercetari de Biochimie* **25**, 30-34.

Nikol'skaya, B.S. (1954). The blood-clotting and wound-healing properties of preparations of plant origin. *Trudy Vsesojuznogo Obscestvo Fiziologov, Biokhimikov i Farmakologov* **2**, 194-197.

Nishiyama, T. and Akutsu, N. (1992). Skin cosmetics containing saikosaponin B1 and/or B2 and cell growth factors. *Patent-Japan Kokai Tokkyo Koho*-04 29,916; *Chemical Abstract* **116**, 241736P.

Nishiyama, T. and Akutsu N. (1993). Topical preparations containing saikosaponins and sugars. *Patent- Japan Kokai Tokkyo Koho*-05 17,332; *Chemical Abstract* **118**, 198249J.

Nonaka, G.I., Hayashi, M., Tanaka, T., Saijo, R., and Nishioka, I. (1990). Tannins and related compounds. XCII. Isolation and characterization of cyanogenic ellagitannins, aleurinins A and B, and a related O-glycosidic ellagitannnin, aleurinin C, from *Aleurites fordii* Hemsley. *Chemical and Pharmaceutical Bulletin* **38**, 861-865.

Northway, R.B. (1975). Experimental use of *Aloe vera* extract in clinical practice. *Veterinary Medicine and Small Animal Clinician* **70**, 89.

Nwude, N. and Ebong, O.O. (1980). Some plants used in the treatment of leprosy in Africa. *Leprosy Review* **51**, 11-18.

Obata, M., Ho, S., Beppu, H., and Fujita, K., and Nagatsu, T. (1993). Mechanism of antiinflammatory and antithermal burn action of carboxypeptidase from *Aloe arborescens* var. *natalensis* in rats and mice. *Phytotherapy Research* **7**, 530-533.

Olugbade, T.O., Adesina, S.K., Ogundaini, A.O., Oladimeji, H., and Onawumi, G.O. (1996). Antimicrobial Ellagitannins of *Acalypha wilkesiana* Muell and Arg. In *Book of Abstracts, IOCD International Symposium on Chemistry, Biological and Pharmacological Properties of African Medicinal Plants*, **P37**. Feb 25-28, 1996, Victoria Falls, Zimbabwe.

Olukoya, D.K., Idika, N., and Odugbemi, T. (1993). Antibacterial activity of some medicinal plants from Nigeria. *Journal of Ethnopharmacology* **39**, 69-72.

Ortiz de Montellano, B. (1975). Empirical Aztec medicine. *Science* **188**, 215-220.

Ozaki, Y., Sekita, S., Soedigdo, S., and Harada, M. (1989). Anti-inflammatory effect of *Graptophyllum pictum* (L.) Griff. *Chemical and Pharmaceutical Bulletin* **37**, 2790-2802.

Palanichamy, S., Amala Bhaskar, E., Bakthavathsalam, R., and Nagarajan, S. (1991). Wound healing activity of *Cassia alata*. *Fitoterapia* **62**, 153-156.

Papageorgiou, V.P. (1978a). Pharmaceutical composition for treating *Ulcus cruris*. *Patent-Ger Offen*- 2, 700, 448.

Papageorgiou, V.P. (1978b). Wound healing properities of naphthaquinone pigments from *Alkanna tincto ria*. *Experientia* **34**, 1499-1501.

Pasich, B., Kowalewski, Z., and Socha, A. (1968). The triterpenoid and Sterol compounds in plant material. XIII. The isolation of asiaticoside from the herb *Centella asiatica* Urb using cationite. *Dissertationes Pharmaceuticae et Pharmacolicae* **20**, 69.

Patel, C.T. and Mach, M.S. (1994). When wounds do not heal: a case study. *Critical Care Nursing Clinics of North America*,

Peacock, E.E. and van Winkle, (1976). *Wound Repair*, 2nd edn. W. B. Saunders Co., London.

Perez, R.M., Ocegueda, G.A., Munoz, J.L., Avila, J.G., and Morrow, W.W. (1984). A study of the hypoglycemic effect of some Mexican plants. *Journal of Ethnopharmacology* **12**, 253-262.

Piacquadio, D. and Nelson, D.B. (1992). Alginates . A "new" dressing alternative. *Journal of Dermatologic Surgery and Oncology* **18**, 992-995.

Pieters, L., de Bruyne, T., Mei, G., Lemiere, G., van den Berghe, D., and Vlietinck, A.J. (1992). *In vitro* and *in vivo* biological activity of south american dragon's blood and its constituents. *Planta Medica* **58**, A582-A583.

Poizot, A. and Dumaz, D. (1978). Modification of the duration of the cicatrization on the healing effect in the rat. Action of triterpenoids on the duration of cicatrization. *Comptes Rendus de l'Académie des Sciences, Série D* **286**, 789.

Pongprayoon, U., Wasuwat, S., Sunthornpalin, P., and Bohlin, L. (1987). Chemical and pharmacological studies of the Thai medicinal plant *Ipomea pes-caprae* (ahakbung tha-le). *Abstracts of the 1st Princess Congress*, **abstr-bp-31**. Bangkok Thailand, 10-13 December 1987.

Popp, C., Kligman, A.M., and Stoudemayer, T.J. (1995). Pretreatment of photodamaged forearm skin with topical tretinoin accelerates healing of full-thickness wounds. *British Journal of Dermatology* **132**, 46-53.

Porras-Reyes, B.H., Lewis, W.H., Roman, J., Simchowitz, L., and Mustoe, T.A. (1993). *Proceedings of the Society for Experimental Biology and Medicine* **203**, 18-25.

Pruden, J.F. (1964). Wound healing produced by cartilage preparations. The enhancement of acceleration, with the report of the use of a cartilage preparation in clinically chronic ulcers and in primarily closed human surgical excisions. *Archives of Surgery* **89**, 1046-59.

Rakel, R.E. ed. (1993). *Conn's Current Therapy (1993)*. W.B. Saunders Co., Pennsylvania.

Ramirez, V.R., Mostacero, L.J., Garcia, A.E., Mejia, C.F., Pelaez, P.F., Medina, C.D., and Miranda, C.H. (1988). *Vegetales Empleados en medicina tradicional Norperuana*. Banco Agrario del Peru & Nacional Universidade de Trujillo, Trujillo, Peru.

Rao, S.G., Udupa, A.L., Udupa, S.L., Rao, P.G.M., Rao, G., and Kulkarni, D.R. (1991). *Calendula* and *Hypericum*: two homeopathic drugs promoting wound healing in rats. *Fitoterapia* **62**, 508.

Rastogi, R.P., Sarkar, B. Dhar, M.L. (1960). Chemical examination of *Centella asiatica* Linn. Part I. Isolation of the chemical constituents. *Journal of Scientific and Industrial Research, Section B* **19**,252-7.

Ratliff, C. and Rodeheaver, G. (1995). The chronic wound care clinic: "one-stop shopping". *Journal of Wound, Ostomy and Continence Nursing* **22**, 77-80.

Rattner, H. (1936). Roentgen ray dermatitis with ulcers. *Archives of Dermatology and Syphilology* **33**, 593-594.

Reddy, G.B.S., Melkhani, A.B., Kalyani, G.A., Venkata, R.A.O. J., Shirwaikar, A., Kotian, M. Ramani, R., Aithal, K.S. Udupa, A. L., Bhat, G., and Srinivasan, K.K. (1991). Chemical and pharma cological investigations of *Limnophila conferta* and *Limnophila heterophylla. International Journal of Pharmacognosy* **29**, 145-153.

Reynolds, R.D. (1886). Calendula. *Pacific Medicine and Surgery Journal* **29**, 720.

Rodriguez-Bigas, M., Cruz, N.I., and Suarez, A. (1988). Comparative evaluation of *Aloe vera* in the management of burn wounds in guinea pigs. *Plastic and Reconstructive Surgery* **81**, 386-389.

Ross, M.S.F. and Brain, K.R. (1977). *An Introduction to Phytopharmacy*. Pitman Medical, Kent.

Rowe, T.D. (1940). Effect of fresh *Aloe vera* jell in the treatment of third degree Roentgen reactions on white rats, a preliminary report. *Journal of the American Pharmaceutical Association, Scientific Edition* **29**, 348-350.

Rowe, T.D., Lovell, B.K., and Parks, L.M. (1941). Further observations on the use of *Aloe vera* leaf in the treatment of third degree X-ray reactions. *Journal of the American Pharmaceutical Association, Scientific Edition* **30**, 266-269.

Sakina, M.R. and Dandiya, P.C. (1990). A psycho-neuropharmacological profile of *Centella asiatica* extract. *Fitoterapia* **61**, 291-296.

Samuelsson, G. , Farah, M.H., Claeson, P. Hagos, N., Thulin, M., Hedberg, O., Warfa, A.M., Hassan, A.O., Elmi, A.H., Abdurahman, A.D., Elmi, A.S. Abdi, Y. A and Alin, M.H. (1992). Inventory of plants used in traditional medicine in Somalia. Part II. Plants of the families Combretaceae-Labiatae. *Journal of Ethnopharmacology* **37**, 47-70.

Santhanam, G. and Nagarajan, S. (1990). Wound healing activity of *Curcuma aromatica* and *Piper betle*. *Fitoterapia* **61**, 458-459.

Sarma, S.P., Aithal, K.S., Srinvassa, K.K., Udupa, A.L., Kumar, V. Kulkarni, D.R., and Rajagopal, P.K. (1990). Anti-inflammatory and wound healing activites of the crude alcoholic extract and flavonoids of *Vitex leucoxylon*. *Fitoterapia* **61**, 263-265.

Sawhney, A.N., Khan, M.R., Ndaalio, G., Nkunya, M.H.H., and Wevers, H. (1978). Studies on the rationale of African traditional medicine. Part III. Preliminary screening of medicinal plants for antifungal activity. *Pakistan Journal of Scientific and Industrial Research* **21**, 193-196.

Sayed, M.D. (1980). Traditional medicine in healing care. *Journal of Ethnopharmacology* **2**, 19-22.

Schilling, J.A. (1968). Wound Healing. *Physiological Reviews*, **48**, 374-423.

Schoental, R. (1968). Toxicology and carcinogenic action of pyrolizidine alkaloids. *Cancer Research* **28**, 2237.

Sebastian, M.K. and Bhandari, M.N. (1984). Medico-Ethnobotany of Mount Abu, Rajasthan, *Indian Journal of Ethnopharmacology* **12**, 223-230.

Shah, G.L. and Gopal, G.V. (1985). Ethnomedical notes from the tribal inhabitants of the north Gujarat (India). *Journal of Economic and Taxonomic Botany* **6**, 193-201.

Shah, N.C. and Jain, S.K. (1988) Ethno-medico-botany of the Kumaon Himalaya, India. *Social Pharmacology* **2**, 359-380.

Sharma, P.K. and Kaul, M.K. (1993). Specific ethnomedicinal significance of *Kigella africana* in India. *Fitoterapia* **64**, 467-468.

Shibata, M., Sato, F., Takeshita, K., and Otani, K. (1980). Pharmacological studies on bamboo grass. V. Combined effects of the extract (F-D) with Vitamin C. *Shoyakugaku Zasshi* **34**, 274-279.

Ship, A.G. (1977). Is topical aloe vera plant mucus helpful in burn treatment? *Journal of the American Medical Association* **238**, 1770.

Singh, V., Kumar, A., and Singh, S.P. (1984). Effect of normal saline, potassium permanganate and garlic extract on healing of contaminated wound in buffalo-calves. *Indian Journal of Animal Science* **54**, 41-45.

Singh, Y.N. (1986). Traditional medicine in Fiji: some herbal folk cures used by Fiji Indians. *Journal of Ethnopharmacology* **15**, 57-88.

Smith, D.J. (Jr.), Thomson, P.D., Garner, W.L., and Rodriguez, J.L. (1994). Burn wounds: infection and healing. *American Journal of Surgery* **167**, 46S-48S.

Sochen, J.E. (1994). Orthopaedic wounds. *American Journal of Surgery*, **167**, 52S-55S.

Sofowora, A. (1982). *Medicinal Plants and Traditional Medicine in Africa.* John Wiley and Sons, Chichester.

Sofowora, A. (1996). *Plantes Médicinales et Médecine Traditionelle d'Afrique.* Karthala, Paris.

Subramanian, S. and Nagarajan, S. (1988). Wound healing activity of *Pongamia pinnata* and *Cynodon dactylon. Fitoterapia* **59**, 43-44.

Suga, T. and Hirata, T. (1983). The efficacy of the aloe plants' chemical constituents and biological activities. *Cosmetics and Toiletries* **98**, 105-108.

Sullivan, K.M., Lorentz, H.P., Muelli, M., Lin, R.Y., and Adzic, N. S. (1995) A model of scarless human foetal wound repair is deficient in transforming growth factor (beta). *Journal of Pediatric Surgery* **30**, 198-203.

Tamai, H. and Yamahora, J. (1992). *Patent-Japan Kokai Tokkyo Koho*-04 26,623; *Chemical Abstract* **116**, 241964M

Tanaka, S., Saito, M., and Tabata, M. (1980). Biossay of crude drugs for hair growth promoting activity in mice by a new simple method. *Planta Medica* **40**, 84-90.

Takeda, S., Tanaka, Y., and Otsuka, M (1983). *Oyo Yakuri*, **25**, 1-6.

Telfer, N.R. and Moy, R.I. (1993). Drug and nutrient aspects of wound healing. *Dermatologic Clinics* **11** , 729-37.

Tellez, P.J. and Dupoy, G.J. (1990). Pharmaceutical preparation containing *Mimosa tenuiflora* extract with skin-regenerating properties. *Patent-Europe Pat Appl*, 349, 469.

Tenni, R., Zanaboni, G., De Agostini, M.P., Rossi, A., Bendotti, C., and Cetta, G. (1988). Effect of the triterpenoid fraction of *Centella asiatica* on macromolecules of the connective matrix in human skin fibroblast cultures. *Italian Journal of Biochemistry* **37**, 69-77.

The Pharmaceutical Handbook, 19th edn. (1980). The Pharmaceutical Press, London.

Thompson, J.E. (1991). Topical use of *Aloe vera* derived allantoin gel in otolaryngology. *Ear Nose Throat Journal* **70**, 119.

Trease, G.E. and Evans, W.C. (1989). *Pharmacognosy*, 13th edn. Bailliere, Tindall.

Tunnerhoff, F.K. and Schwabe, H.K. (1955a). Animal and human studies on tissue changes after gelatin and thrombin implants. Part 1. *Arzneimittel Forschung* **5**, 201-204.

Tunnerhoff, F.K. and Schwabe, H.K. (1955b) Experimental study on the question of issue changes following gelatin and thrombin implants. Part 2. *Arzneimittel Forschung* **5**, 372-376.

Tunnerhoff, F.K. and Schwabe, H.K. (1956). Studies in human beings and animals on the influence of *Echinacea* extracts on the formation of connective tissue following the implantation of fibrin. Part 4. *Arzneimittel Forschung* **6**, 330-334.

Tunnerhoff, F.K. and Schwabe, H.K. (1965). Experimental study on the question of tissue changes following gelatin and thrombin implants: Part 3. *Arzneimittel Forschung* **15**, 520-522.

Udupa, K.N. and Prasad, G. (1964). Biomechanical and 45-CA studies on the effect of *Cissus quadrangularis* in fracture repair. *Indian Journal of Medical Research* **52**, 480-487.

Udupa, S.L., Udupa, A.L., and Kulkarni, D.R. (1991a). Influence of *Tridax procumbens* lysyl oxidase activity on wound healing. *Planta Medica* **57**, 325-627.

Udupa, S.L., Udupa, A.L., and Kulkarni, D.R. (1991b). Influence of *Tridax procumbens* on dead space wound healing. *Fitoterapia* **62**, 146-150.

Udupa, S.L., Udupa, A.L., and Kulkurni, D.R. (1994). Anti-inflammatory and wound healing properites of *Aloe vera. Fitoterapia* **65**, 141-145.

Vaisberg, A.J., Milla, M., Planas, M.D.C., Cordova, J.L., de Agusti, E.R., Ferreyra, R., Mustiga, M.D.C., Carlin, L., and Hammond, G.B. (1989). Taspine is the cicatrizant principle in sangre de grado extracted from *Croton lechleri. Planta Medica* **55**, 140-143.

Velasco, M. and Romero, E. (1976). *Current Therapeutic Research and Clinical Experiments* **19**, 121.

Verma, S.B.S., Schulze, H.J., and Steigleder, G.K. (1989). The effect of externally applied remedies containing *Aloe vera* gel on the proliferation of the epidermis. *Parfumerie und Kosmetik* **70**, 452-459.

Vevron, H. and Giustiniani, V. (1988). Use of glycyrhetinic acid as a cicatrizant. *Patent-Europe Appl*- 275, 222. *Chemical Abstract* **109**, 216041F.

Ward, R.S. and Saffle, J.R. (1995). Topical agents in burn and wound care. *Physical Therapy* **75**. 526-538.

Watt, M.J. and Breyer-Brandwijk, M.G. (1962). The Medicinal and Poisonous Plants of Southern and Eastern Africa, 2nd edn., E. & S. Livingstone Ltd, London.

Weniger, B., Rouzier, M., Daguilh, R., Henrys, D., Henrys, J.H., and Anton, R. (1986). *Journal of Ethnopharmacology* **17**, 13-30.

Werman, M.J., Mokady, S., Nimni, M.E., and Neeman, I. (1991). The effect of various avocado oils on skin collagen metabolism. *Connective Tissue Research* **26**, 1-10.

Whistler, W.A. (1985). Traditional and herbal medicine in the Cook Islands. *Journal of Ethnopharmacology* **13**, 239-280.

Wijetunge, D.B. (1994). Management of acute and traumatic wounds: Main aspects of care in adults and children. *American Journal of Surgery*, **167**, 56S-6OS.

Williams, C. (1994). Intrastite Gel: a hydrogel dressing. *British Journal of Nursing* **3**, 843-846.

Winters, W.D., Benavides, R., and Clouse, W.J. (1981). Effects of aloe extracts on human normal and tumor cells *in vitro*. *Economic Botany* **35**, 89-95.

Yadav, C.L. and Yadav, C.S. (1985). Preliminary clinical study of *Kalanchoe spathulata* DC. On inflammatory wound. *Ancient Sciences and Life* **5**, 30-31.

Yamahara, J., Mochizuki, M., Fujimura, H., Takaishi, Y., Yoshida, M., Tomimatsu, T., and Tamai, Y. (1990). Antiulcer action of *Sophora flavescens* root and an active constituent. I. *Journal of Ethnopharmacology* **29**, 173-177.

Yamasaki, H. and Saeki, K. (1967). Inhibition of mast-cell degranulation by anti-inflammatory agents. *Archives of International Pharmacodynamics and Therapeutics* **168**, 166.

Yantadilaka, P. and Raktavat, S.A. (1950). A preliminary Phytochemical study of *Hydrocotyl asiatica* L. *Journal of the Pharmaceutical Association (Siam)*, **3**, 257-62.

Zamora-Martinez, M.C. and Pola, C.N.P. (1992). Medicinal plants used in some rural populations of Oaxaca, Puebla and Veracruz, Mexico. *Journal of Ethnopharmacology* **35**, 229-257.

Zita, C. and Steklova, B. (1955). The Influence of pure constituents of Camomile oil on thermal burns. *Chemical Abstract* **49**, 11875G.

Zoutewelle, G. and van Wijk, R. (1990). Effects of *Echinacea purpurea* extracts on fibroblast populated collagen lattice contraction. *Phytotherapy Research* **4**, 77-81.

6. Chemistry and biological properties of the African Combretaceae

C.B. ROGERS[1] AND L. VEROTTA[2]

[1]Department of Chemistry, University of Durban-Westville, Private Bag X54001, Durban, Republic of South Africa and [2]Dipartimento di Chimica Organica e Industriale, Unversita degli Studi di Milano, via Venezian 21, 20133 Milano, Italy

---

Introduction

Of the many genera that comprise the African Combretaceae, the two largest, *Combretum* and *Terminalia*, occur in most parts of Africa where they are often the dominant groups as regards numbers. They consist of climbers, shrubs and trees and are readily characterised by fruits with wing-shaped appendages.

Although traditional healers throughout Africa have used species of the Combretaceae for the treatment of a wide range of disorders, only about twenty five out of the approximately ninety nine African species of *Combretum* have been subjected to any form of scientific study; with the exception of a few species of the *Terminalia, Anogeissus* and *Guiera*, virtually nothing has been reported on the phytochemistry of any of the remaining genera. This family thus represents a practically unexplored reservoir of potentially useful substances.

Metabolites isolated so far include alkaloids (*G. senagalensis*), tannins (*A. schimperi*), flavonoids (*C. micranthum*), and amino acids (*C. zeyheri*); substituted phenanthrenes from various heartwoods; a rich variety of triterpenoid acids and their saponins mainly of the cycloartane (*C. molle*) and oleane (*C. imberbe*) types; and a series of unique stilbenes, their glucosides. and macrocyclic lactones called combretastatins (*C. caffrum, C. kraussii*). Many of the *Combretum* species exude gums that are similar in composition and properties to gum arabic.

Electron microscope and chemical investigations have shown that the acidic triterpenoid mixtures isolated from the *Combretum* are secreted onto the surface of the leaves and the fruit through epidermal, scale-like trichomes. The anatomy of the trichomes and the mixtures of triterpenoid acids and saponins that they secrete are both species specific, which raises interesting questions and provides an important combination of taxonomically useful characters for this genus.

Certain of the metabolites show cytotoxic, molluscicidal, anti-HIV, anti-microbial and anti-inflammatory activity, and several of the triterpenoid acid mixtures have strong inhibitory effects on plant germination and seedling growth. Isolation of the toxins that render the fruit of the *Combretum* poisonous to animals and cause violent and prolonged hiccuping in humans (hence the name "hiccup nut" given to many species) has not yet been accomplished.

Taxonomic classification of the Family

The Combretaceae family belongs to the order Myrtales (Fig. 6.1) which is divided into two sub-families of which only the Combretoideae is of interest. Of the two tribes comprising this sub-family, only the Combreteae, which is further divided into three sub-tribes, is of relevance in Africa. The genera from these sub-tribes that are African and have been investigated are *Combretum, Terminalia, Anogeissus,* and *Guiera*; the number of species for each genera is given in

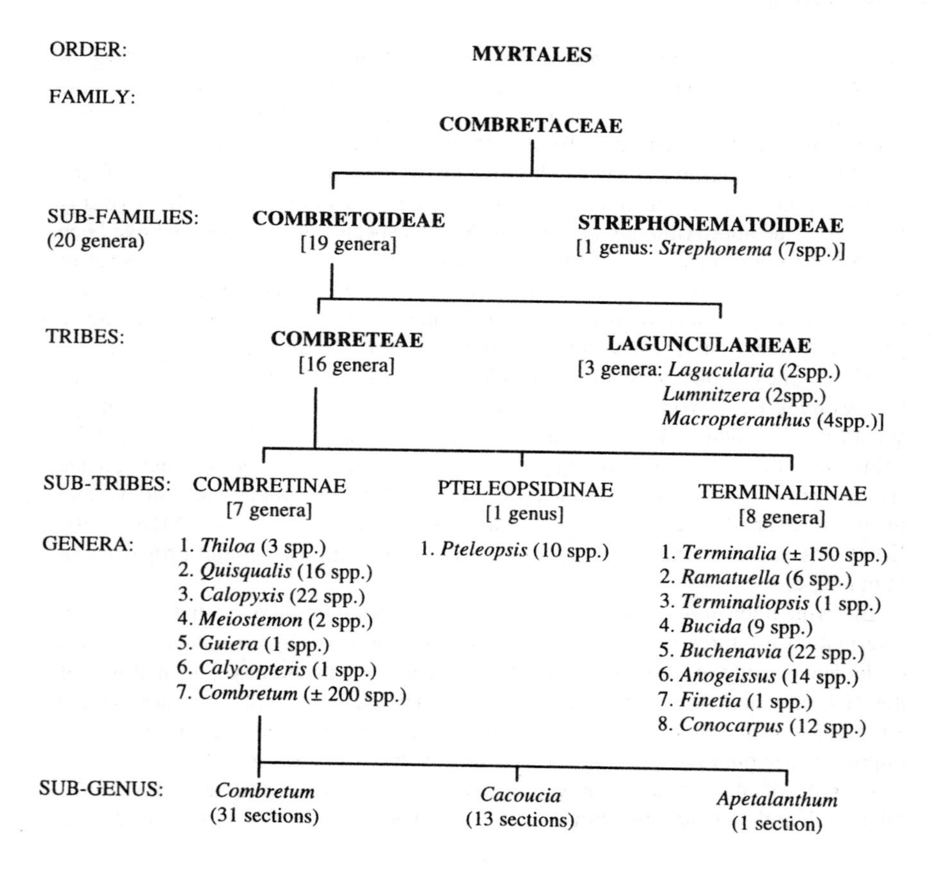

Fig. 6.1. Taxonomic classification of the family Combretaceae.

parenthesis in Fig. 6.1. Of these the genus *Combretum*, which contains about one hundred African species in two sub-genera *Combretum* and *Cacoucia* (*Apatalanthum* is an American genus), has been investigated the most extensively and is the most interesting. Consequently any investigation of the properties of the African Combretaceae is essentially an investigation of the genus *Combretum*.

Use of the Combretaceae in Traditional Medicine

Reports in the literature (Watt *et al.* 1962; Gelfand *et al.* 1985; Kokwaro 1976) indicate that traditional healers throughout Africa have confined themselves almost exclusively to the use of species from the genus *Combretum* and to a lesser extent the *Terminalia* in the treatment of a wide range of maladies (Table 6.1). Although the use of the leaves and bark from *Combretum* species is widespread, the winged fruits, which are produced in great abundance, are never used in medicine (nor are they eaten by wild animals) because of their reported toxicity to humans.

The interest in the use of members of the Combretaceae in traditional medicine has led to a burgeoning interest in this family with an increasing number of species now under investigation for biological activity.

Pharmacological activity in the Combretaceae

Despite the wide use of species of this family by traditional healers, very little of pharmacological importance had been reported until recently. The first scientific study carried out was that on the west-African drug "Kinkeliba" made from the leaves of *C. micranthum* (Paris 1942). This drug used in French Sudan, Senegal and Nigeria for the treatment of biliary fever, colic and vomiting, has a cholagog and diüretic action and is antimicrobial (gram positive and gram negative). In the last two decades a series of stilbenes and dihydrostilbenes (the combretastatins) with potent cytotoxic activity and acidic triterpenoids and their glycosides with molluscicidal, antifungal, antimicrobial and antiinflammatory activity have been isolated from species of *Combretum* (Rogers 1989). These will be dealt with in more detail later.

More recently lignans with HIV-1 reverse transcriptase inhibitory activity have been isolated from *A. acuminata* (Rimondo *et al.* 1994) and 3,4,5-tri-O-galloylquinnic acid, also shown to have anti-HIV activity, has been isolated from *G. senegalensis* (Mahmood 1993). Antimicrobial activity found in *Anogeissus* species in the Sudan has been attributed to 3,3',4'-tri-O-methylflavellagic acid extracted from the bark (Almagboul 1988), and chewing sticks from *A. leiocarpa* are reported to prevent dental caries (Sanui 1983). Leaf decoction extracts from *C. glutinosum* from Senegal strongly inhibit hepatitis B virus antigen (HBsAg) *in vitro* and angiotensin-converting enzyme (Poussel *et al.* 1993).

Table 6.1 : Some Medicinal uses of the Combretaceae

COMBRETUM		
C. apiculatum	E. Africa:	Snake bite, scorpion bite, bloody diarrhoea, leprosy.
	S. Africa:	Abdominal disorders, conjunctivitis.
	Zimbabwe:	Weak body
C. erythrophyllum	S. Africa:	Fattening tonic for dogs.
	Zimbabwe:	To reduce size of vaginal orifice.
C. fragrans	E. Africa:	Chest coughs, syphilis.
	Zimbabwe:	Aphrodisiac.
C. glutinosum	Senegal:	Hepatic disease, antihypertensive, diuretic, bronchial disease.
C. hereroense	E. Africa:	Bilharzias.
	Zimbabwe:	Headache, infertility in women.
C. imberbe	S. Africa:	Coughs, colds
	Zimbabwe:	Diarrhoea, to drive away bad spirits bilharziasis.
C. microphyllum	Zambia:	Lunacy.
	Zimbabwe:	Lucky charm.
C. molle	E. Africa:	Hookworm, stomach ache, snakebite, leprosy, fever, dysentery, chest complaints, anthelmintic.
	Zambia:	Headaches.
	Malawi:	Anthelmintic, snake bite.
	Zimbabwe:	Abdominal pains, diarrhoea, headaches, convulsions, infertility in women, to stop bleeding after childbirth, to fatten babies, as a dressing for wounds.
C. platypetalum	Zambia:	Swelling caused by mumps.
	Zimbabwe:	Pneumonia, abdominal pains, diarrhoea, antiemetic, dysmenorrhoea, infertility in women, earache, epistaxis, haemoptysis.
C. zeyheri	E. Africa:	Toothache, cough
	Tanzania:	Scorpion bite, diarrhoea with blood.
	Zambia:	To arrest menstrual flow, eye lotion, embrocation, diarrhoea.
	Zimbabwe:	Diarrhoea with blood, abdominal disorders.
TERMINALIA		
T. brachystemma	Zimbabwe:	Haematuria, bilious vomiting, constipation, diarrhoea.
T. sericea	E. Africa:	Bilharziasis, stomach troubles.
	Botswana:	To arrest purging.
	S. Africa:	Stomach disorders, bilharziasis, diabetes.
	Tanzania:	Stomach disorders, bilharziasis.
	Zimbabwe:	Diarrhoea, epistaxis, prolapsed rectum of infants, backache, to widen vagina, wounds, abdominal pains, worms on arms, antiemetic, infertility in women, tonic, depressed fontanelle, sore throat, gonorrhoea, bilharziasis, abortion, dilated birth canal.
T. stenostachya	Zimbabwe:	Epilepsy, antidote for poison.
T. brevipes	Somalia:	Hepatitis, malaria.

Toxicity of the Combretum

The "nut" inside the four winged fruit or samara from species such as *C. collinum* and *C. zeyheri* is comparable in size and appearance to a pecan nut yet, despite being produced in vast amounts, it is not consumed by wild animals or the indigenous people. On a continent where starvation was endemic, this is a clear indication that the fruits are toxic, although there is only one case on record where this has been clinically proven (BSA Police 1970).

In an attempt to identify the toxins in the *Combretum* fruit, cytotoxic studies on the fruit of fifteen species where carried out using brine shrimp and other assays (Panzini *et al.* 1993). Compounds isolated included a series of acidic triterpenoids and their glycosides, non-proteinogen aminoacids and several combretastatins and their glucosides-these compounds will be discussed fully later. Although showing other forms of activity, none of these compounds appeared responsible for the toxicity.

More recently it has been reported that five women have died in Zimbabwe after inserting material from *C. erythrophyllum* into their vaginas to reduce the size (Mavi, S., National Herbarium, Harare, Zimbabwe, personal communication, 1996). More unreported deaths have possibly occurred, since this is a widespread practice in rural areas of Zimbabwe. Compounds thus far isolated from *C. erythrophyllum* include a series of unusual cycloartane dienone lactones (**1-3**) from the leaves (Rogers, C.B. University of Durban-Westville, unpublished data) and combretastatin glucosides from the roots (Brookes, B., Mangusuthu Technicon, Durban, personal communication, 1996)

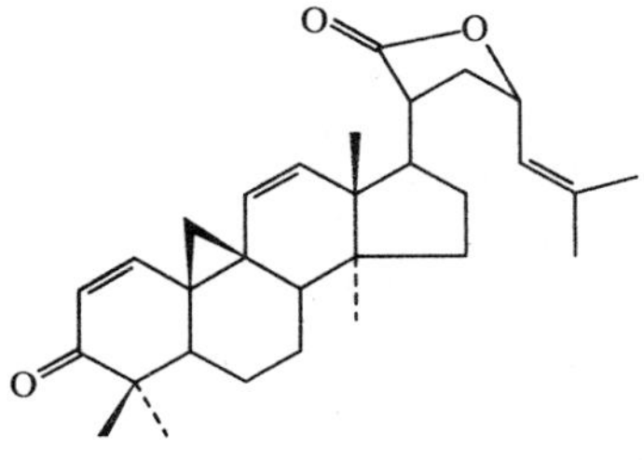

(**1**)

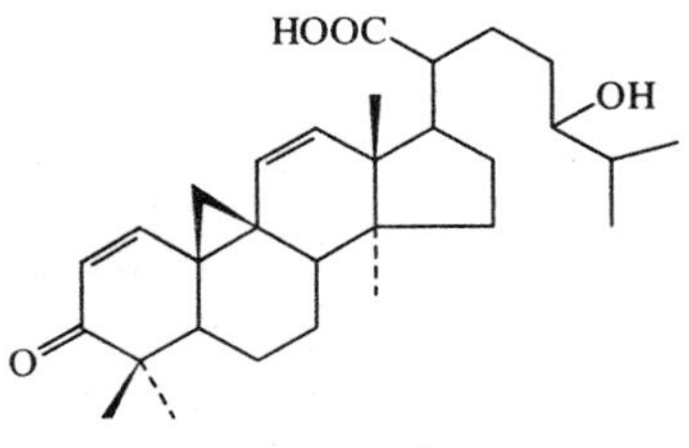

HOOC OH

(**2**)

HOOC OH

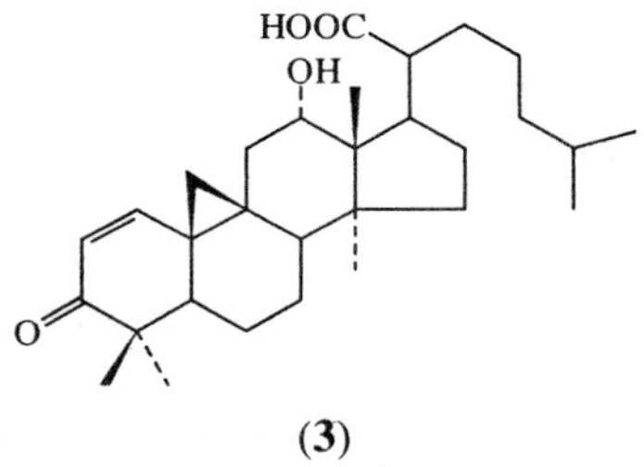

(**3**)

Metabolites isolated from the Combreaceae

The isolation of metabolites can be divided roughly into five significant stages starting in chronological order with the investigation of the extractives from *C. micranthum* in 1942 (Paris 1942) followed by the isolation of the substituted phenanthrenes (Letcher and Nhamo 1971); the gum exudates (Anderson and Bell. 1974); the acidic triterpenoid glycosides (Rogers *et al.* 1976) and the stilbenes and dihydrostilbenes (Pettit *et al.* 1988a).

Aminoacids and other nitrogen compounds

The activity of the drug "Kinkaleba" from the leaves of *C. micranthum* has been attributed to simple nitrogen compounds such as choline, betaine and combretins A and B (Paris 1942; Bassene 1989), whereas the presence of the amino acids *N*-methyl-L-tyrosine and its 4'-O-β-D-glucoside, 3-aminomethyl-L-phenylalanine, and 3-(3'-hydroxymethyl-L-phenylalanine) and its 3'-O-β-D-glucoside isolated from the fruit of *C. zeyheri* (Mwauluka *et al.* 1975a, 1975b; Panzini *et al.* 1993; Perosa 1992) has been attributed to fungal intrusion. With the exception of the simple indole alkaloids harman and eleagnine isolated from the roots of *G. senegalensis* (Combier 1977), there have been no other reports of alkaloids from the Combretaceae.

Substituted phenanthrenes and dihydrophenanthrenes

An investigation of the borer and termite resistance of *C.apiculatum* heartwood resulted in the isolation of seventeen substituted phenanthrenes and 9,10-dihydrophenanthrenes (**4-20**) from this tree and the heartwood of *C. molle, C. psidioides, C. hereroense* (Letcher *et al.* 1971, 1972 and 1973; Malan and Swinny 1993) and *C. caffrum* (Pettit *et al.* 1982). Certain of these compounds (**6, 17**) totally inhibited the growth of *Penicillium expansum* in antifungal tests (Malan and Swinny 1993) and three of the 9,10-dihydrophenanthrenes (**16, 18, 20**) isolated from *C. caffrum* showed reasonable antileukemic activity in the P-388 murine system (Pettit *et al.* 1988b).

Combretaceae gum exudates

Impetus for the study of these gums was supplied by the extreme Sahelian droughts of 1972-74, which devastated the *Acacia* population that supplied the bulk of the world's gum arabic requirements. The drought did not affect the *Combretum* forests further south and since Combretaceae gums are used extensively as adhesives and in other technological applications (*Anogeissus latifolia* is the main component of gum ghatti), they were investigated as possible replacements for the *Acacia* gums (Anderson and Bell 1977).

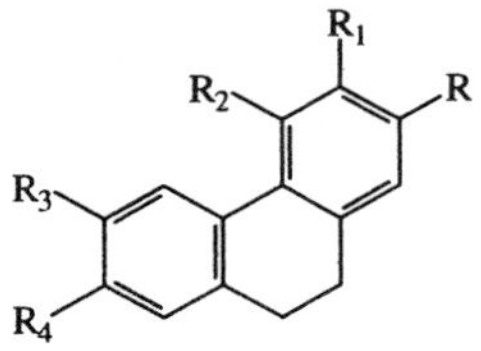

(4)-(11)

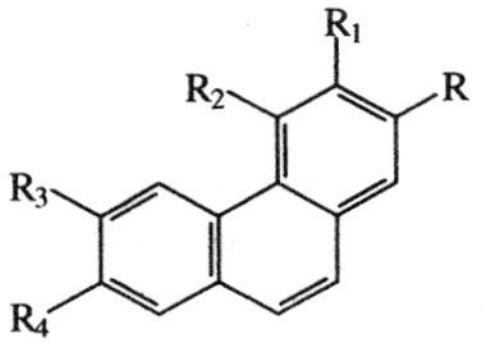

(12)-(20)

	R_1	R_2	R_3	R_4	R_5	PLANT
4	OCH3	OCH3	OH	OH	OH	*C. apiculatum*
5	OH	OCH3	OCH3	OH	OCH3	*C. apiculatum*
6	OCH3	OCH3	OH	OCH3	OH	*C. molle*, *C. apiculatum*
7	OCH3	OCH3	OCH3	OCH3	OH	*C. caffrum*
8	OH	OH	OCH3	OCH3	OH	*C. apiculatum*
9	OCH3	OH	OCH3	OH	OH	*C. apiculatum*
10	OCH3	OH	OCH3	OCH3	OH	*C. apiculatum*
11	OCH3	OH	OCH3	OCH3	OCH3	*C. apiculatum*
12	OCH3	H	OH	OCH3	OH	*C. apiculatum*
13	OCH3	OCH3	OH	OCH3	OH	*C. apiculatum*, *C. molle*
14	OH	OCH3	OCH3	OH	OCH3	*C. apiculatum*, *C. molle*
15	OCH3	OCH3	OH	OH	OH	*C. molle*, *C. apiculatum*
16	OCH3	OCH3	OCH3	OH	OH	*C. caffrum*, *C. apiculatum*, *C. hereroense*
17	OH	OCH3	OCH3	OCH3	OCH3	*C. caffrum*
18	OH	OCH3	OCH3	OCH3	OH	*C. caffrum*, *C. apiculatum*, *C. psidioides*
19	OCH3	OCH3	OCH3	OCH3	OH	*C. apiculatum*, *C. psidioides* *C. caffrum*
20	OCH3	OH	OCH3	OCH3	OCH3	*C. apiculatum*

Gum exudates from eleven *Combretum,* two *Terminalia* and two *Anogeissus* species have been examined and classified as arabinogalactan proteins (Anderson and Bell 1974 and 1977; Anderson *et al.* 1990 and 1991) and "gum *Combretum*" has been characterised as shown in Table 6.2.

Because of their high acidity and tannin content,"*Combretum* gums" are not permitted in foodstuffs, although they have been found to be the most frequent adulterant of commercial gum arabic (Anderson *et al.* 1991). The overall similarity of the gums makes this deception difficult to detect.

Table 6.2. Characteristics of "Gum *Combretum*"

- Dextrorotatory
- Low in Nitrogen; high in rhamnose
- High methoxy content
- High zinc content
- High in aspartic acid and glycine, low in hydroxyproline
- High in tannin
- Markedly hygroscopic - tend to "block" in storage
- High molecular mass, high acidity

Stilbenes and dihydrostilbenes (Combretastatins)

In 1979 an NIC screening programme found extracts from the Cape bushwillow, *C. caffrum*, to be cytotoxic and combretastatin (**21**) [*R*(-)-1-(3,4,5-trimethoxy-phenyl)-2-(3-hydroxy-4-methoxyphenyl) ethanol], the first of a series of unique stilbenes highly active against the murine P-388 lymphocytic leukemia cell line, was isolated from the methylene chloride-methanol extract of the whole plant (Pettit *et al.* 1982).

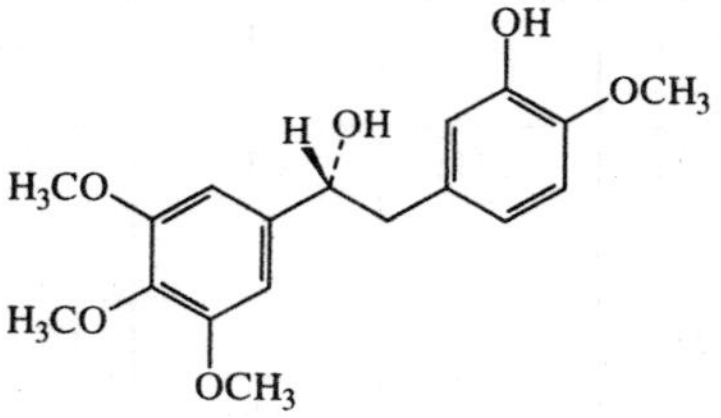

Combretastatin (**21**)

Large scale extraction of *C. caffrum* stemwood yielded further combretastatins designated A (**22-28**), B (**29-46**), C (**47**) and D (**48, 49**) according to the skeleton type (Pettit *et al.* 1987; Pettit and Singh 1987) plus a series of bibenzyls (Pettit *et al.* 1988c; Malan and Swinny 1993). In addition the fruit of *C. kraussii* yielded combretastatins A-1 (**22**) and B-1 (**29**) and their corresponding 2'-O-β-D-glucosides (**28,44**) in considerable yields compared to the yields from *C. caffrum* (Pelizzoni *et al.* 1993). For reasons that are probably connected to recent uncharacteristic annual climatic variations in South Africa, attempts to isolate the glucosides from fruit produced in the last few seasons has proved unsuccessful!

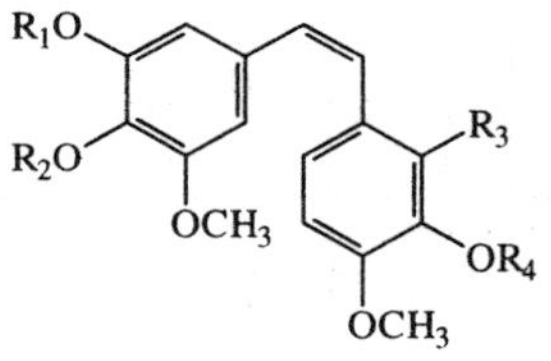

	R_1	R_2	R_3	R_4		PLANT
22	CH_3	CH_3	OH	H	combretastatin A-1	*C. caffrum,* *C. kraussii*
23	$-CH_2-$		OH	H	combretastatin A-2	*C. caffrum*
24	H	CH_3	H	OH	combretastatin A-3	*C. caffrum*
25	CH_3	CH_3	H	OH	combretastatin A-4	*C. caffrum*
26	H	CH_3	H	CH_3	combretastatin A-5	*C. caffrum*
27	H	CH_3	H	CH_3	combretastatin A-6 = (*E*)-combretastatin A-5	*C. caffrum*
28	CH_3	CH_3	OH	OGlc		*C. kraussii*

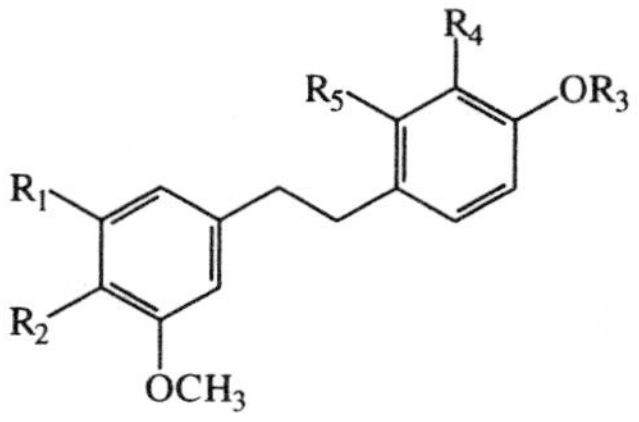

	R_1	R_2	R_3	R_4	R_5		PLANT
29	OCH_3	OCH_3	CH_3	OH	OH	combretastatin B-1	*C. caffrum,* *C. kraussi*
30	$O-CH_2-O$		CH_3	OH	OH	combretastatin B-2	*C. caffrum*
31	OCH_3	OCH_3	H	OH	H	combretastatin B-3	*C. caffrum*
32	OCH_3	H	H	OH	H	combretastatin B-4	*C. caffrum*
33	OCH_3	H	CH_3	OH	H		*C. caffrum*

34	OCH_3	H	H	H	H		*C. caffrum*
35	OCH_3	OCH_3	H	H	H		*C. psidioides, C. caffrum*
36	OH	H	H	H			*C. apiculatum*
37	OH	H	CH_3	OH	H		*C. apiculatum*
38	OH	OCH_3	H	OCH_3	H		*C. apiculatum*
39	OH	OH	CH_3	OCH_3	H		*C. apiculatum*
40	OH	OCH_3	H	OH	H		*C. apiculatum*
41	OH	OH	CH_3	H	H		*C. apiculatum*
42	OH	OCH_3	H	H	H		*C. apiculatum, C. molle*
43	OCH_3	OH	H	H	H		*C. apiculatum, C. psidioides*
44	OCH_3	OCH_3	CH_3	OH	O-Glc		*C. kraussii*
45	OCH_3	OH	CH_3	OH	OH	combretastatin B-5	*C. kraussii*
46	OCH_3	OH	CH_3	OH	O-Glc		*C. kraussii*

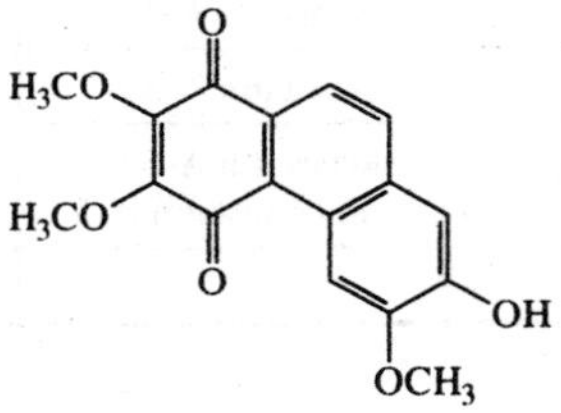

combretastatin C-1 (**47**)

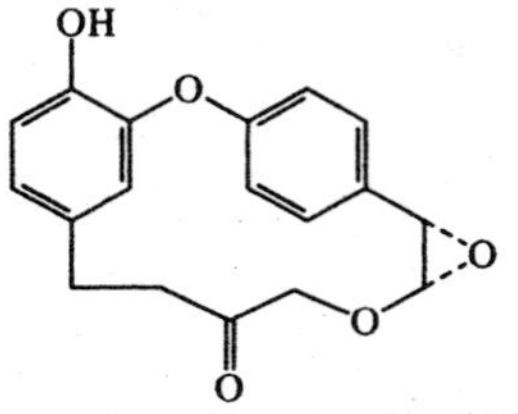

combretastatin D-1 (**48**)

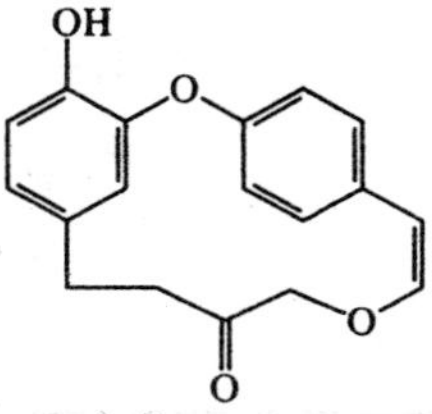

combretastatin D-2 (**49**)

Whereas all the combretastatins show some cytotoxic activity, certain of these compounds are especially effective. In particular, combretastatins A-1 (**22**) and B-1 (**29**) are potent inhibitors of microtubule assembly *in vitro* and were, at the time, among the most potent inhibitors of the binding of colchicine to tubulin (Pettit *et al.* 1987). In addition, in tests on mammalian sensory neurons, combretastatin B-1 (**29**) had rapid and completely reversible effects on a variety of potassium ion channels without any marked effects on calcium or sodium channels (Verotta *et al.* 1994).

However, the most potent cancer cell growth inhibitor of the series was found to be combretastatin A-4 (**25**) (Pettit *et al.* 1995). It is the most potent inhibitor of colchicine binding to tubulin yet discovered and inhibits tubulin polymerisation, retards strongly the growth of the murine lymphocytic leukemia L1210 and P388 cell lines as well as human colon cancer lines and is a potent antimitotic agent. As a result of this activity, combretastatins A-4 (**25**) and B-1 (**29**) are the subjects of patents (Pettit and Singh 1991, US Patent and Pelizzoni *et al.* 1994, IT Patent).

Bibenzyls and combretastatins have been found in trace quantities in fruit extracts of *C. bracteosum* and *C. zeyheri* (Lanfossi, M., University of Milan, unpublished data, 1993) and in *C. erythrophyllum* root bark (Brookes, B., Mangosuthu Technicon, Durban, personal communication, 1995).

Acidic triterpenoids and their glycosides

The serendipitous discovery that the triterpenoid molluscicide mollic acid (**50**) and its glycosides (**51-53**) are secreted as a surface coating on *C. molle* leaves and fruit through glandular, scale-like trichomes present on the leaf and fruit epidermis, has resulted in the isolation of a rich variety of acidic triterpenoids and their glycosides. Since these trichomes appear in great profusion on the epidermis of leaves and fruit of all species belonging to the subgenus *Combretum,* all other species of this subgenus were tested and found to have similar secretions. Both the composition of the mixture of triterpenoid acids and their saponins that are secreted and the anatomy of the trichomes responsible for the secretions have been found to be species specific. Thus trichome anatomy and trichome secretions provide an important combination of taxonomically useful characters for this subgenus and will be discussed later.

Since nearly all the acidic triterpenoids produce water soluble salts, the harvesting of the triterpenoids and their glycosides is achieved by simply washing fresh leaves with warm, 1% bicarbonate solution and acidifying the resultant solution (Lawton *et al.* 1991). Subjected to TLC analysis using a relatively polar solvent (ethyl acetate:chloroform:formic acid; 5:4:1) and the spray reagent *p*-anisaldehyde/c.H_2SO_4 in ethanol (5:5:90), each species gives a fingerprint of different coloured spots that is a powerful chemotaxonomic tool. In nearly all species studied so far, geographic and seasonal variation have little or no effect on the composition of the extracts (Carr and Rogers 1987).

Thus far the triterpenoids isolated from the *Combretum* belong almost exclusively to two distinct groups: *viz* 30-carboxy-1α-hydroxycycloartanes (**1-3, 50-56**) and 29-carboxy-1α-hydroxyoleanes (**57-69**) (Pegel and Rogers 1976, 1985; Rogers and Subramony 1988; Panzini *et al.* 1993; Rogers 1988, 1989a, 1898b; Osborne and Pegel 1985; Jossang *et al.* 1996). With the exception of the *C. molle* fruit extract, which contains both oleane and cycloartane triterpenoids and their mono- and bidesmosides, trichome secretions contain either one group or the other

(Panzini *et al.* 1993). There also appears to be a correlation between the size and morphology of the trichomes and group that is secreted. This will be discussed further-see Fig. 6.6.

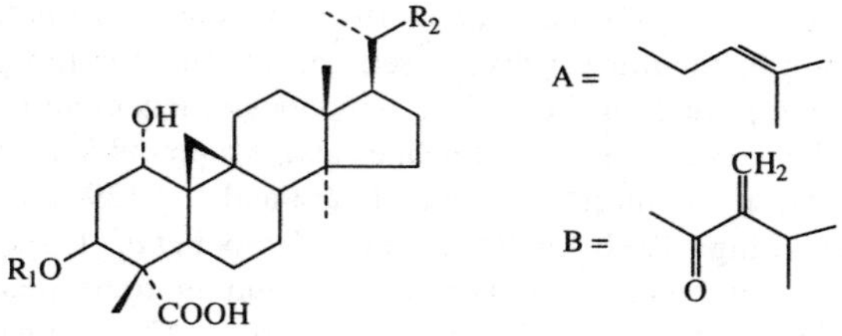

	R_1	R_2		PLANT
50	H	A	mollic acid	*C. molle*, *C. leprosum*
51	-D-glucosyl	A	molic acid glucoside	*C. molle*
52	-D-xylosyl	A	mollic acid xyloside	*C. molle*
53	-L-arabinosyl	A	mollic acid arabinoside	*C.molle*, *C. edwardsii*
54	H	B	jessic acid and its methyl ester	*C. elaeagnoides*
55	-L-arabinosyl	B	jessic acid arabinoside	*C. elaeagnoides*
56	-D-xyloside	B	jessic acid xyloside	*C. molle* (fruit)

Molluscicidal activity of the triterpenoids

The sodium salts of mollic acid glucoside (**51**), toxic to *Biomphalaria glabrata* snails at a concentration of 12 ppm (Rogers 1989), is the major 30-carboxy-1α-hydroxycycloartane glycoside trichome secretion washed from the leaf surface of *C. molle* by a hot, 1% $NaHCO_3$ solution (yield ± 0.8% dried and ± 0.25% fresh whole leaf). This salt qualifies as an ideal, economically viable, plant-derived molluscicide for use in Third World environs in Africa according to World Health Organisation criteria (Table 6.3).

Of the other triterpenoids tested, the sodium salt of imberbic acid (**57**) is toxic to *B. glabrata* snails at concentrations of 20 ppm whereas its bidesmosides (**58**) and (**60**) are inactive (Hostettmann, K., University of Lausanne, personal communication, 1987)

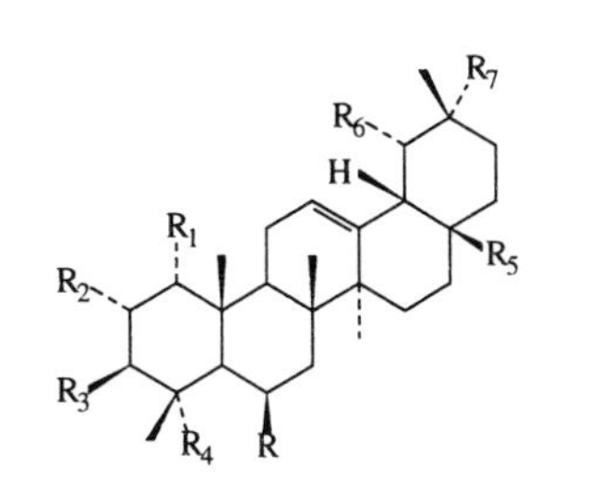

	R	**R_1**	**R_2**	**R_3**	**R_4**	**R_5**	**R_6**	**R_7**		**PLANT**
57	H	OH	H	OH	CH_3	CH_3	H	COOH	imberbic acid	*C. imberbe*
58	H	OH	H	OH	CH_2-O-^{4}AcRha	CH_3	H	COOH		*C. imberbe*, *C. padoides*
59	H	OH	H	O-Rha	CH_2-O-4AcRha	CH_3	H	COOH		*C. imberbe*, *C. padoides*
60	H	OH	H	O-Rha	CH_2O-Rha	CH_3	H	COOH		*C. padoides*
61	H	OH	H	O-Rha	CH_2O-Rha	CH_3	H	CH_2OH		*C. padoides*
62	H	OH	H	O-Rha	CH_2O-^{4}AcRha	CH_3	H	CH_2OH		*C. padoides*
63	H	OH	H	OH	CH_2O-Rha	CH_3	H	COOH		*C. imberbe*
64	H	OAc	H	OH	CH_2O-Rha	CH_3	H	COOH		*C. imberbe*
65	H	H	OH	OH	CH_2OH	COOH	OH	CH_3	arjungenin	*C. nigricans*
66	H	H	OH	OH	CH_2OH	COOGlc	OH	CH_3	arjunglucoside I	*C. molle* *C. nigricans*
67	OH	H	OH	OH	CH_3	COOH	OH	CH_3	arjunetin	*C. molle*
68	OH	H	OH	OH	CH_2OH	COOH	OH	CH_3	combregenin	*C. nigricans*
69	OH	H	OH	OH	CH_2OH	COOGlc	OH	CH_3	combre-glucoside	*C. nigricans*

Table 6.3. Advantages of mollic acid glucoside (**51**) as a molluscicide in the control of schistosomiasis in Africa

- *C. molle* is the most widespread *Combretum* in Africa occurring from the Transkei in the south and as far north as Arabia.
- A medium sized tree, it produces a high mass of leaves each season.
- The leaves are a renewable resource and are easier to harvest than berries or roots.
- The extraction process is cheap; a 200 litre drum, firewood and inexpensive $NaHCO_3$ are needed to prepare solutions.
- Unsophisticated labour would be able to carry out the process successfully.
- Mollic acid glucose is stable and has a low toxicity to mammals.

Allelopathic activity of epidermal leaf secretions from *Combretum* species.

It was observed that, although planted in thick coastal forest with ample rainfall, the area under the leaf canopy of a thicket bushwillow, *C. padoides*, remained barren. Since this tree and most other *Combretum* are deciduous, the area at the base of the tree is covered with a thick layer of leaves every autumn. To check whether the triterpenoid secretions on the leaves were responsible for this inhibition of undergrowth, 1% bicarbonate extracts from *C. padoides, C. molle, C. moggii, C. apiculatum, C. zeyheri, C. imberbe, C. celestroides, C. erythrophyllum* and *C. oxystachyum* were tested against a variety of commercially available seeds and seeds of *C. molle* (Rogers, 1991). Whereas seeds germinated readily in a 1% bicarbonate solution, germination was totally inhibited in almost ALL the solutions from the *Combretum*; the few seeds that did germinate did not grow. Furthermore, germination of *C. molle* seeds was significantly retarded by its own leaf extract although a solution of mollic acid glucoside (**51**), the main constituent of this leaf extract, did not have the same drastic inhibitory effects. As a control, a similar extract from *Albizia adianthifolia,* a local tree, enhanced germination and stimulated growth of the seedlings.

Intercontinental Chemotaxonomic links between African and South American *Combretum*

The isolation from the South American species, *C. leprosum,* of the cycloartenoid mollic acid (**50**) previously found only in several African *Combretum* species (Facundo 1993) establishes a direct "intercontinental" chemotaxonomic link between species on the two continents and presents intriguing phytochemical and phytogeographic questions related to the ancestry and development of this genus (Table 6.4).

Table 6.4. "Intercontinental" *Combretum* relationships

	Approximate Number of Species			
SUBGENUS	Africa	Madagascar	Asia	S.America
Combretum	98	1	18	28
Cacoucia	60	5	8	5

Treatment of leaves from South American and Indian species with 1% bicarbonate solution has yielded mixtures of acidic triterpenoids similar to those found in South African species (Rogers, C.B., University of Durban-Westville, unpublished data) and the extract from the Brazilian liana, *C. rotundifolium,* yielded two acidic dammarane arabinofuranosides, (**70**, **71**) (Rogers 1995). This shows that the epidermal trichomes function in the same way on all three continents.

HO, COOH, O, Arabinose

(**70**)

HO, COOH, O, Arabinose

(**71**)

These results give some idea of the age of the genus and its origin. Since the continents had separated by a significant amount approximately 120 million years ago and the separation was complete 64 million years later (Fig. 6.2), the *Combretum* must have been part of the flora of Gondwanaland ± 200 million years ago. The distribution of species in Table 6.3 supports the evidence that Africa formed the centre of Gondwanaland.

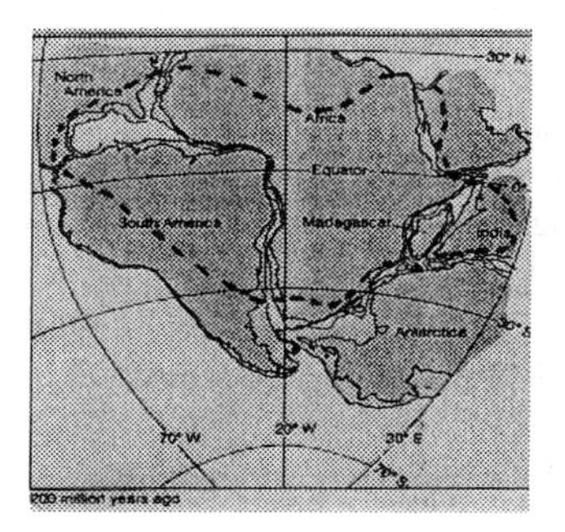

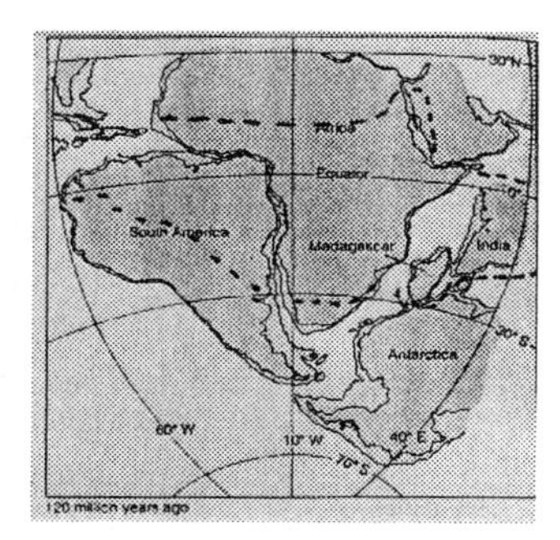

Fig. 6.2. Gondwanaland [----- Present range of the *Combretum*].

The anatomy of the epidermal trichomes and their mode of action in the *Combretum*

The discovery by electron microscopy that the often copious, sticky secretions present on young *Combretum* leaves were produced by glandular, scale-like trichomes that crowd the epidermis of all species of the subgenus *Combretum* (Fig. 6.3) (Stace 1969), coincided with the discovery that the secretions consisted of acidic triterpenoids and their glycosides (Lawton and Rogers 1991). Each trichome consists of a basal cell embedded in the epidermis and a two celled stalk subtending a cap (Fig. 6.4); it is the species specificity of the morphology of the cells in this one-cell-thick cap that has established this as such a valuable taxonomic character (Stace 1981).

Fig. 6.3. Surface of the leaf of *Combretum caffrum* showing trichomes in various stages of development.

In young trichomes, secretions accumulate within thewalls of the cap cells until the outer layer of the wall and cuticle break away (Fig. 6.5.) releasing the triterpenoid secretion as a sticky layer on the surface of the leaf. Evidence thus far suggests that there is a correlation between the trichome morphology and the class of triterpenoid secreted. Trichomes that are larger and have a complicated arrangement of cap cells secrete the oleane group of triterpenoids, whereas trichomes with smaller caps and less complicated cell patterns produce triterpenoids of the cycloartane group. This bifurcation in triterpenoid synthesis is illustrated in Fig. 6.6.

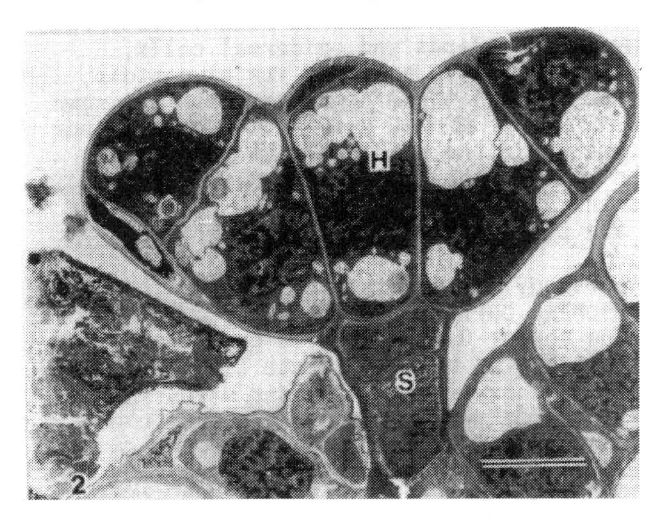

Fig. 6.4. Tangential longitudinal section through a trichome on a young leaf of *C. molle* showing a 2 celled stalk (S) subtending a head/cap (H). (Bar = 0.5mm).

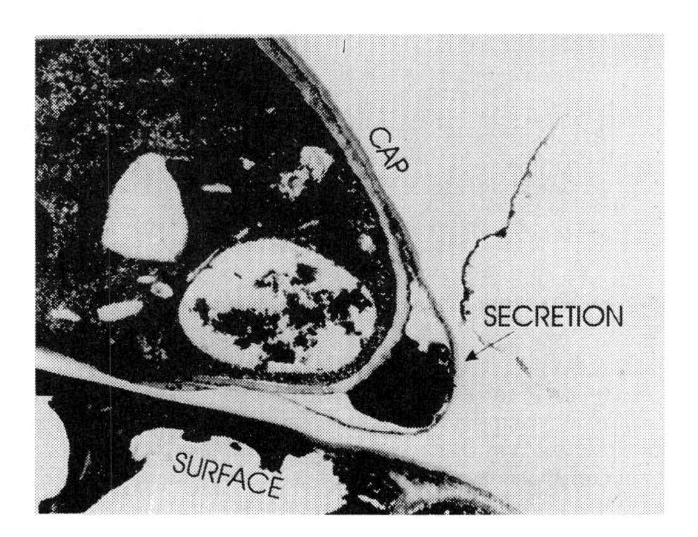

Fig. 6.5. Section through a trichome showing the outer layers of the wall and the cuticle about to rupture.

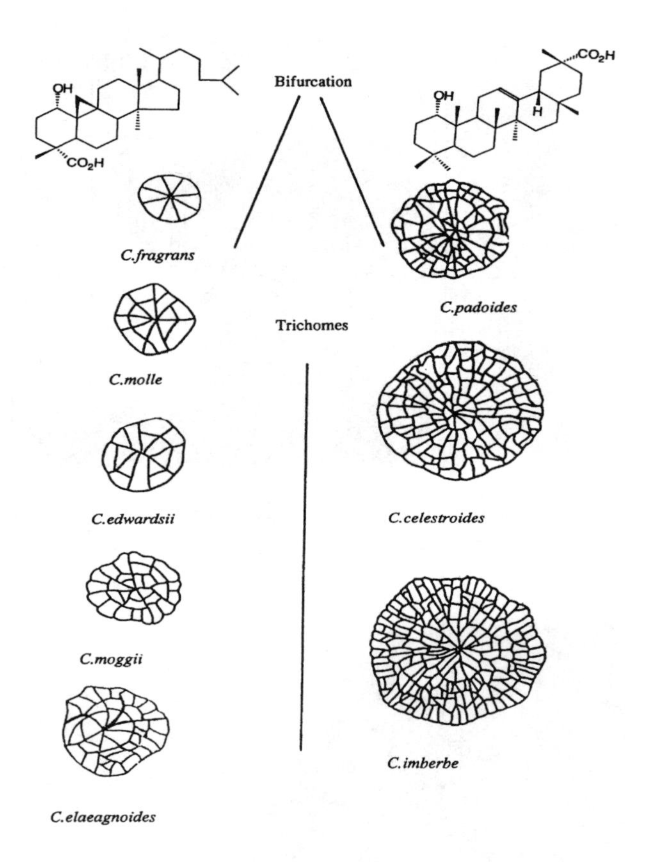

Fig. 6.6. Correlation of trichome anatomy with triterpenoid group secreted.

Since the composition of the acidic triterpenoid mixtures is also species specific, such relationships are to be expected. In particular, the triterpenoid mixtures and scale anatomy of *C. edwardsii, C. collinum, C. molle* and *C. moggii* are almost identical; the latter two species are so similar that this chemical evidence has convinced some botanists to review the distinction (Hennessy, E.F., University of Natal, personal communication, 1990).

Of less importance chemotaxonomically and of little taxonomic importance are the epidermal stalked glands that cover the epidermis of *Combretum* subgenus *cacoucia* species. The triterpenoid mixtures are much simpler and the stalked glands have no distinctive features.

Conclusion

The reported uses of the Combretaceae in traditional medicine has lead to a burgeoning interest in this family with an increasing number of genera other than the *Combretum* now under investigation. The isolation and activity of the combretastatins represents a significant advance in medicinal plant chemistry and the high yields and variety of still to be evaluated acidic triterpenoids from the leaf epidermis of the *Combretum* show promise as molluscicidal and allelopathic agents. The acidic triterpenoid "fingerprint" given by each *Combretum* species is now well established as a chemotaxonomic tool in the biology of this genus (yields are so high and the extraction procedure so simple, that *Combretum* leaves could prove a commercial source of these triterpenoids). Recent work shows that *Terminalia* species give similar "fingerprints" (Rogers, C.B., University of Durban-Westville, unpublished data).

Investigations of the genera *Anogeissus* and *Guiera* have yielded promising metabolites and initial investigations of the *Terminalia* (Rogers, C.B., University of Durban-Westville, unpublished data) are encouraging. No studies on the underground parts of any of the genera have been reported but, given the richness of the leaf and fruit extracts, these should provide a wealth of interesting metabolites.

References

Almagboul, A.Z., Bashir, A.K., Salih, A. Karim, M., Farouk, A., and Khalid, S.H. (1988). Antimicrobial activity of certain Sudanese plants used in folklorc medicine. Screening for antibacterial activity V. *Fitoterapia* **59,** 57-62.

Anderson, D.M.W. and Bell, P.C. (1974). Uronic acid materials XLIV. Composition and properties of thé gum from *Terminalia sericea* and *T. superba. Phytochemistry* **13,** 1871-1874.

Anderson, D.M.W. and Bell, P.C. (1977). The composition of the gum exudates from some *Combretum* species. *Carbohydrate Research* **57**, 215-221.

Anderson, D.M.W. and Wang Weiping. (1990). Composition of the Gum from *C. paniculatum* and four other gums which are not permitted food additives. *Phytochemistry* **29,** 1193-1195.

Anderson, D.M.W., Millar, J.R.A., and Wang Weiping. (1991). The gum exudates from *C. nigricans* gum, the major source of the West African 'gum *Combretum'. Food Additives Contamination* **8,** 423-436.

Bassene, L., Olschwang, D., and Pousset, J.L. (1989). Nonsaponifiables from the leaves of *Combretum micranthum* G.Don (Kinkeliaba). *Herba Hungarica* **28,** 75-80.

BSA Police Rhodesia (1970). Gwanda Docket SDD 19/70 and Poison File, E/3 28901, National Herbarium, Harare, Zimbabwe.

Carr, J.D. and Rogers, C.B. (1987). Chemosystematic studies of the genus *Combretum* (Combretaceae). 1. A convenient method of identifying species of this genus by a comparison of the polar constituents extracted from leaf material. *South African Journal of Botany* **53,** 173-176.

Combier, H., Becchi, M., and Cave, A. (1977). Alkaloids of *Guiera senegalensis. Plantes Médicinales et Phytothérapie* **11,** 251-253.

Facundo, V.A., Andrada, C.H.S., Silveira, E.R., Braz-Filho, R., and Hufford, C.D. (1993) Triterpenoids and flavonoids from *Combretum leprosum. Phytochemistry* **32,** 411.

Gelfand, M., Mavi, S., Drummond, R.B., and Ndemera, B. (1985). *The Traditional Medical Practitioner in Zimbabwe.* Mambo Press. Harare.

Jossang, A., Seuleiman, M., Maidou, E., and Bodo, B. (1996). Pentacyclic triterpenes from *Combretum nigricans. Phytochemistry* **41,** 591.

Kokwaro, O. (1976). *Medicinal Plants of East Africa.* East African Literature, Nairobi.

Lawton, J.R. and Rogers, C.B. (1991). Localisation of triterpenoids in young leaves of *Combretum molle. Electron Microscopy Society of South Africa* **21,** 27-28.

Lawton, J.R., Govender, H., and Rogers, C.B. (1991). Mollic acid glucoside: A possible (Third World) answer to the control of schistosomiasis in South and Central Africa. *Planta Medica* **57,** A711.

Letcher, R.M. and Nhamo, L.R.M. (1971). Chemical constituents from the Combretaceae. Part 1. Substituted phenanthrenes and 9,10-dihydrophenanthrenes from the heartwood of *Combretum apiculatum. Journal of the Chemical Society (C),* 3070-3076.

Letcher, R.M. and Nhamo, L.R.M. (1973). Chemical constituents from the Combretaceae. Part IV. Phenanthrene derivatives from the heartwood of *Combretum hereroense. Journal of the Chemical Society, Perkin Transactions 1,* 1179-1181 and references cited therein for Parts II (1971) and Parts III (1972).

Mahmood, N., Moore, P.S., and DeTommasi N. (1993). Inhibition of HIV infection by caffeoylquinic acid derivatives. *Antiviral Chemistry and Chemotherapy* **4,** 235-240.

Malan, E. and Swinny, E. (1993). Substituted bibenzyls, phenanthrenes from the heartwood of *Combretum apiculatum. Phytochemistry* **34,** 1139-1142.

Mwauluka, K., Charlwood, B.V., Briggs, J.M., and Bell, E.A. (1975a). N-Methyl-L-thyrosine from seeds of *Combretum zeyheri. Phytochemistry* **14,** 1657.

Mwauluka, K., Charlwood, B.V., Briggs, J.M., and Bell, E.A. (1975b). L-3(3-aminomethylphenyl)-alanine, a new aminoacid from seeds of *Combretum zeyheri. Biocheme und Physiologie der Pflanzen* **168,** 15.

Osborne, O. and Pegel, K.H. (1985). Methyl jessate 1a, 11α-oxide, a further novel triterpenoid ester from *Combretum elaeagnoides. South African Journal of Chemistry* **38,** 83.

Panzini, I., Pelizzoni, F., Verotta, L., and Rogers, C.B. (1993). Constituents of the fruit of South African *Combretum* species: Part 1. *South African Journal of Science* **89,** 324.

Paris, R (1942). "Kinkeliba", a west-African drug. *Bulletin of Science and Pharmacology* **49**, 181-186.

Pegel, K.H. and Rogers, C.B. (1985). The characterisation of mollic acid 3-β-D-xyloside and its genuine aglycone mollic acid, two novel 1α-hydroxycycloartenoids from *Combretum molle. Journal of the Chemical Society, Perkin Transactions 1,* 1711-1715.

Perosa, A. (1992) *Thesis Dissertation,* University of Milan.

Pettit, G.R. and Singh, S.B. (1987). Isolation, structure and synthesis of combretastatins A-2, A-3 and B-2. *Canadian Journal of Chemistry* **65,** 2390-2396.

Pettit, G.R. and Singh, S.B. (1991). Combretastatin A-4. *United States Patent,* Patent Number 4, 996, 237.

Pettit, G.R., Cragg, G.M., Herald, D.L., Schmidt, J.M., and Lohavanijaya, P. (1982). Isolation and structure of combretastatin. *Canadian Journal of Chemistry* **60,** 1374-1376.

Pettit, G.R., Singh, S.B., Niven, M.L., and Schmidt, J.M. (1987). Isolation, structure and synthesis of combretastatins A-1, B-1, potent new inhibitors of microtubule assembly, derived from *Combretum caffrum. Journal of Natural Products* **50,** 119-131.

Pettit, G.R., Singh, S.B., Niven, M.L., and Schmidt, J.M. (1988a). Cell growth inhibitory dihydrophenanthrene and phenanthrene constituents of the African tree *Combretum caffrum. Canadian Journal of Chemistry* **66,** 406-413.

Pettit, G.R., Singh, S.B., Niven, M.L., Schmidt, J.M., Hamel, E., and Lin, C.M. (1988b). Isolation, structure, synthesis, and antimitotic properties of combretastatins B-3 and B-4 from *Combretum caffrum. Journal of Natural Products* **51,** 517-527.

Pettit, G.R., Singh, S.B., and Niven, M.L. (1988c). Isolation and structure of combretastatin D-1: A cell growzh inhibitory macrocyclic lactone from *Combretum caffrum. Journal American of the Chemical Society* **110,** 8539.

Pettit, G.R., Hamel, E., Lin, C.M., Alberts, D.S., and Garcia-Kendall, D. (1989). Isolation and structure of the strong cell growth and tubulin inhibitor combretastatin A-4. *Experentia* **45,** 209-11.

Pettit, G.R., Singh, S.B., Boyd, M.R., Hamel, E., Pettit, R.K., Schmidt, J.M., and Hogan, F. (1995). Antineoplastic agents. 291. Isolation and synthesis of combretastatins A-4, A-5, and A-6. *Journal of Medicinal Chemistry* **38,** 1666.

Pelizzoni, F., Verotta, L., Rogers, C.B., Colombo, R., Pedrotti, B., Balconi, G., Erba, E., and D'incalci, M. (1993). *Natural Products Letters* **1,** 273.

Pelizzoni, F., Verotta, L., Colombo, R., and D'incalci, M. (1994) *Italian Patent* Combretastatin derivatives with antitumour activity and process for the preparation thereof. *PCT Int. Appl WO* 9, 405, 682.

Pousset, J.L., Rey, J.P., Levesque, J., Coursaget, P., and Galen, F.X. (1993). Hepatitis B surface antigen (HBsAg) inactivation and angiotensin-converting enzyme (ACE) inhibition *in vitro* by *Combretum glutinosum* Perr. (Combretaceae) extract. *Phytotherapy Research* **7,** 101-102.

Rimondo, A.M., Pezzuto, J.M., Farnsworth, N.R., Santisuk, T., and Rentrakul, V. (1994). New lignans from *A.acuminata*. *Journal of Natural Products* **57,** 896-904.

Rogers, C.B. (1988). Pentacyclic triterpenoid rhamnosides from *Combretum imberbe* leaves. *Phytochemistry* **27,** 3217-20.

Rogers, C.B. (1989a). Isolation of the 1α-hydroxycycloartenoid mollic acid α-L-arabinoside from *Combretum edwardsii* leaves. *Phytochemistry* **28,** 279-81.

Rogers, C.B. (1989b). New mono- and bi-desmosidic triterpenoids isolated from *Combretum padoides* leaves. *Journal Natural Products* **52,** 528-533.

Rogers, C.B. (1996). Acidic dammarane arabinofuranosides from *C. rotundifolium. Phytochemistry* **40,** 833-36.

Rogers, C.B. and Pegel, K.H. (1976). Mollic acid 3-β-D-glucoside, a novel 1α-hydroxycycloartane saponin from *Combretum molle* (Combretaceae). *Tetrahedron Letters* **47,** 4299-4302.

Rogers, C.B. and Subramony, G. (1988). The Structure of imberbic Acid, a 1α-hydroxy pentacyclic triterpenoid from *Combretum imberbe. Phytochemistry* **27,** 531-533.

Rogers, E. (1991). Natal Schools Science Expo Project.

Sanui, S.B. and Okor, D.I. (1983). The fluoride content of Nigerian chewing sticks. *Medicinal Science Library Compendium* **11**, 604.

Singh, S.B. and Pettit, G.R. (1989). Isolation, structure and synthesis of combretastatins C-1. *Journal of Organic Chemistry* **54,** 4105-4114.

Singh, S.B. and Pettit, G.R. (1990). Antineoplastic agents. 206. Structure of the cytostatic macrocyclic lactone combretastatin D-2. *Journal of Organic Chemistry* **55,** 2797-2800.

Stace, C.A. (1969). The significance of the leaf epidermis in the taxonomy of the Combretaceae. II. The genus *Combretum* subgenus *Combretum* in Africa. *Botanical Journal of the Linnean Society* **62,** 131-168.

Stace, C.A. (1981). The significance of the leaf epidermis in the taxonomy of the Combretaceae: conclusions *Botanical Journal of the Linnean Society* **81,** 327-339.

Verotta, L., Guatteo, E., Wanke, E., and Rogers, C.B. (1994). The search for the "hiccup" nut toxin. *International Research Congress on Natural Products* Halifax (Canada).

Watt, J.M. and Breyer-Brandwijk, M.G. (1962). *Medicinal and Poisonous Plants of Southern and Eastern Africa* 2nd Ed. E. and S. Livingstone Ltd., London.

7. Overview of the chemistry of aloes of Africa

E. DAGNE

Addis Ababa University, Department of Chemistry, P.O. Box 30270, Addis Ababa, Ethiopia

---

Introduction

The genus *Aloe* consists of more than 360 species distributed mainly in Tropical Africa, Madagascar and Southern Arabia (Mabberley 1987) and includes herbs, shrubs and trees. The leaves are fleshy, strongly cuticularized and are usually prickly at the margins. The flowers are of various colours, white, yellow, pink, greenish and red.

The bitter leaf exudates of some *Aloe* species are commercially important sources of the laxative aloe drug and are also used in the cosmetics industry as additives in shampoos, shaving and skin care creams (Leung 1970) and in the treatment of skin disorder and in particular as topical medication for the treatment of burns (Rowe *et al.* 1941). The exudate has also been used as bittering agent in alcoholic beverages. The term *aloe* is derived from the Arabic word *alloeh* which means a shining bitter substance (Tyler *et al.* 1976). Medicinally, the gel and dried leaf exudates of *Aloe* species have been used since ancient civilizations of the Egyptians, and Mediterranean peoples (Trease and Evans 1976). Egyptian copts used aloe to treat eye diseases, swellings and digestive disorders. Aloe was known to the Greeks for Alexander the Great is said to have been advised to conquer the island of Socotra near the East African shores to get aloe drug to treat his wounded soldiers (Maniche 1989). *Aloe* species still enjoy a very wide folkloric usage in many parts of the world and are also used in modern medicine. In commercial circles "Cape aloe" means the dried latex of the leaves of *Aloe ferox* Miller while "Curaçao aloe" is the latex from *Aloe vera* Miller (U.S. Pharmacopeia 1979).

Except for a few species which have been made cosmopolitan most *Aloe* species are confined to Africa and Arabia. It is therefore important for Africa that these species which have proved to be a store of diverse and interesting natural products be studied from many aspects. This paper highlights the chemistry of *Aloe* species and supplements the review by Reynolds (1985a).

The two most important analytical methods in the study of leaves and roots of *Aloe* are TLC and HPLC. Reynolds and Herring (1991) recommended for TLC of leaf components a mixture of di-isopropyl ether/*n*PrOH and water (7:5:1) followed by use of the lower layer of $CHCl_3$/EtOH/H_2O (7:3:1) to develop the spots in a direction at right angle to the first. For root constituents we have found (Dagne *et al.* 1994) petrol/$CHCl_3$ (1:1), $CHCl_3$-EtOAc (7:3) and for analysis of leaf components EtOAc-MeOH-H_2O (77:13:10) to be quite suitable solvent systems. Developed TLC plates may be viewed under UV_{254} and UV_{366}, or sprayed with 0.5% aqueous solution of Fast Blue Salt B followed by spraying with caustic soda solution (Jork *et al.* 1990). 5-Hydroxyaloin gives characteristic violet-brown color when sprayed with 5% aqueous sodium metaperiodate (Rauwald and Beil 1993).

Reversed phase HPLC of the methanol extract of leaf exudate has proved to be one of the best methods for establishing chemical profile in *Aloe*. Whereas it is very difficult to distinguish between aloin A and B by TLC, these epimers can be easily distinguished by HPLC (Reynolds and Herring 1991). The roots of several species of *Aloe* and *Lomatophyllum* were analysed by Van Wyk *et al.* (1995b, 1995c) by TLC and HPLC for the presence of nine anthraquinones and pre-anthraquinones which are known to be characteristic constituents of roots of *Aloe*. The results support the inclusion of the genus *Lomatophyllum* in *Aloe*. Furthermore comparative studies of roots of 46 species belonging to the genera *Bulbine*, *Bulbinella* and *Kniphofia* (family Asphodelaceae) (Van Wyk *et al.* 1995a) revealed the relationships as well as differences of the two families Asphodelaceae and Aloaceae, which until recently were kept together in one family.

Constituents of *Aloe* species

The leaves and roots of *Aloe* species elaborate many interesting secondary metabolites belonging to different classes of compounds including alkaloids, anthraquinones, pre-anthraquinones, anthrones, bianthraquinoids, chromones, coumarins and pyrones. The three most important constituents of commercial aloe drug are the anthrones aloin A and B (**42**), and the chromones aloesin (**63**) and aloeresin A (**64**). We discuss below briefly each of the major classes of *Aloe* compounds.

Alkaloids

The piperidine alkaloid γ-coniceine and other related hemlock alkaloids were reported to occur in seven *Aloe* species with very restricted distribution (Dring *et al.* 1984). By screening 224 *Aloe* species for alkaloids Nash *et al.* (1992) found 21% of the species positive and identified tyramine and its derivatives in 18 species. In view of the potential toxicity of many alkaloids, the authors pointed out the importance of screening for alkaloid prior to the use of *Aloe* plants as medicines.

Anthraquinones and pre-anthraquinones

Several free anthraquinones occur in roots and leaves of *Aloe* species. Aloe-emodin (**1**) is a typical leaf constituent and is wide spread in the genus. Chrysophanol (**10**) occurs both in roots (Yagi *et al.* 1977a) and leaves (Dagne and Alemu 1991) while nataloe-emodin (**6**) has so far been reported only from leaves (Conner *et al.* 1987). The anthraquinones in leaves may be present as O-glycosides as is the case in compounds **2** and **7**. The anthraquinones, physcion and emodin which are oxygenated at the 6 position are not found in *Aloe*. Aloesaponarin I (**8**), aloesaponarin II (**9**), desoxyerythrolaccin (**12**), helminthosporin (**3**), isoxanthorin (**5**) and laccaic acid D methyl ester (**14**) were isolated first from roots of *A. saponaria* (Yagi *et al.* 1974) but have recently been shown to occur in roots of many other *Aloe* species (Dagne *et al.* 1994).

Thus two main types of anthraquinones are present in the roots of *Aloe*, these are 1,8-dihydroxyanthraquinone (*e.g.* chrysophanol, aloe-emodin) and 1-hydroxy-8-methylanthra-quinone (*e.g.* aloesaponarin I). Whereas anthraquinones of the former type are known to occur both in leaves and roots, those that belong to the latter type are confined only to roots. In a recent study of the roots of 172 species of *Aloe*, Van Wyk *et al.* (1995a) detected 1,8-dihydroxy-anthraquinones in almost all and 1-hydroxy-8-methylanthraquinones in 129 *Aloe* species. As shown in Figure 7.1. these two types of anthraquinones appear to be derived through two parallel biogenetic routes of the polyketide pathway, differing by the way the octaketide chain folds (Leistner 1973).

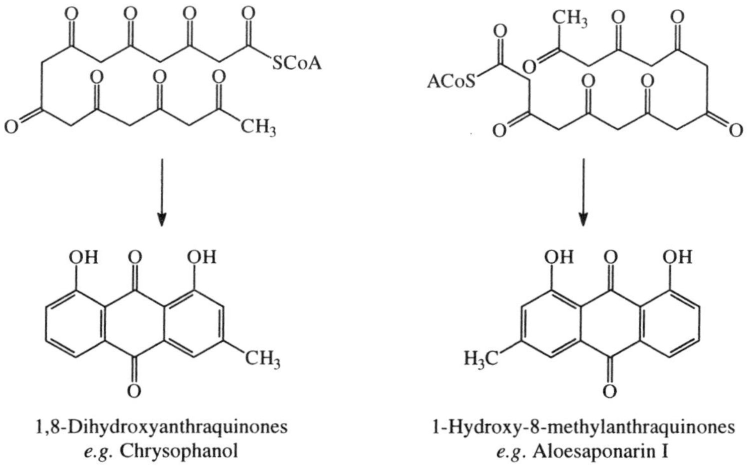

1,8-Dihydroxyanthraquinones
e.g. Chrysophanol

1-Hydroxy-8-methylanthraquinones
e.g. Aloesaponarin I

Fig. 7.1. Two folding mechanisms of octaketides leading to 1,-dihydroxy- and 1-hydroxy-8-methyl-anthraquinones.

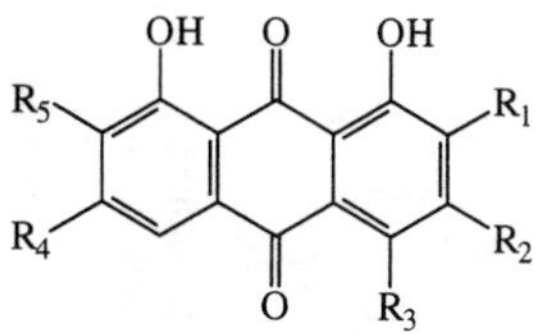

	R_1	R_2	R_3	R_4	R_5	
(**1**)	H	CH_2OH	H	H	H	Aloe-emodin
(**2**)	H	CH_2O-Rha	H	H	H	Aloe-emodin-11-O-rhamnoside
(**3**)	H	H	OH	CH_3	H	Helminthosporin
(**4**)	H	CH_2OH	H	H	OH	7-Hydroxyaloe-emodin
(**5**)	OCH_3	H	OH	CH_3	H	Isoxanthorin
(**6**)	H	CH_3	H	H	OH	Nataloe-emodin
(**7**)	H	CH_3	H	H	O-Glc	Nataloe-emodin-2-O-glucoside

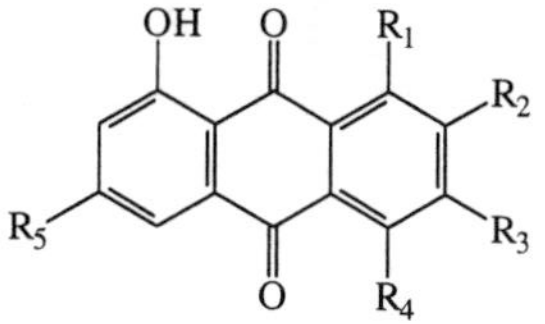

	R_1	R_2	R_3	R_4	R_5	
(**8**)	CH_3	$COOCH_3$	OH	H	H	Aloesaponarin I
(**9**)	CH_3	H	OH	H	H	Aloesaponarin II
(**10**)	H	H	H	H	CH_3	Chrysophanol
(**11**)	CH_3	H	H	H	CH_3	Chrysophanol-8-methyl ether
(**12**)	CH_3	H	OH	H	OH	Desoxyerythrolaccin
(**13**)	H	H	H	OH	CH_2OH	1,5-Dihydroxy-3-hydroxymethyl-anthraquinone
(**14**)	CH_3	$COOCH_3$	OH	H	OH	Laccaic acid D-methyl ether
(**15**)	OCH_3	OH	H	H	CH_3	Nataloe-emodin-8-methyl ether

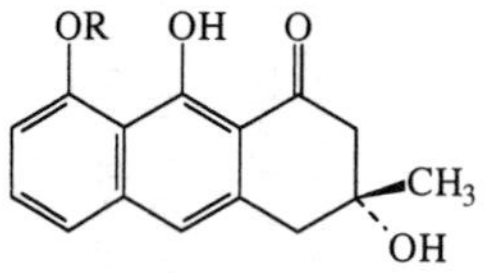

(**16**) R=CH_3 Aloechrysone
(**17**) R=H Prechrysophanol

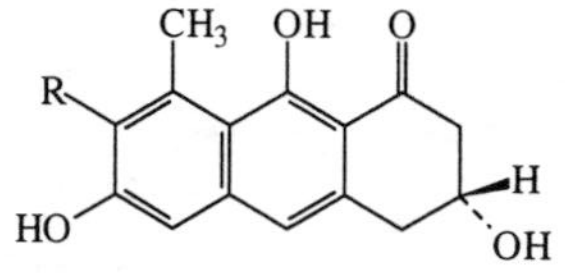

(**18**) R=$COOCH_3$ Aloesaponol I
(**19**) R=H Aloesaponol II

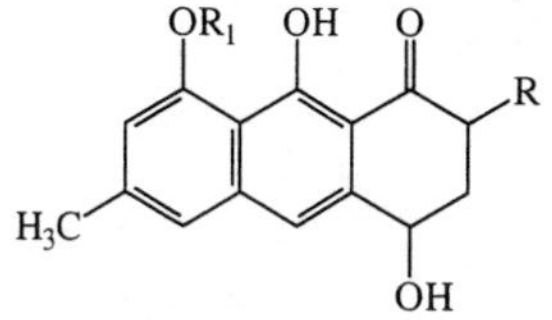

(**20**) R=R_1=H Aloesaponol III
(**21**) R=OCH_3, R_1=H Aloesaponol IV
(**22**) R=H, R_1=Glc Aloesaponol III-8-O-glc
(**23**) R=OCH_3, R_1=Glc Aloesaponol IV-8-O-glucoside

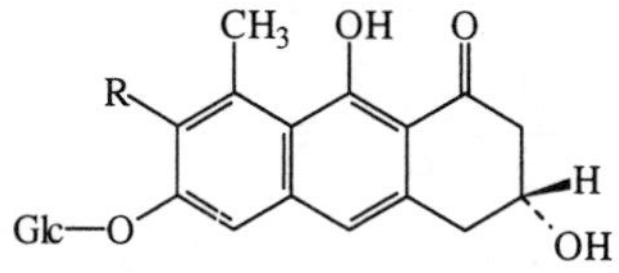

(**24**) R=$COOCH_3$ Aloesaponol I-6-O-glucoside
(**25**) R = H Aloesaponol II-6-O-glucoside

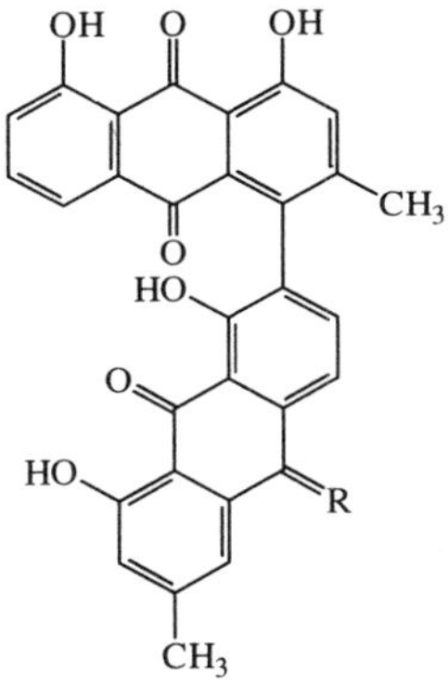

(**26**) R=O Asphodelin
(**27**) R=H_2 Bianthracene III

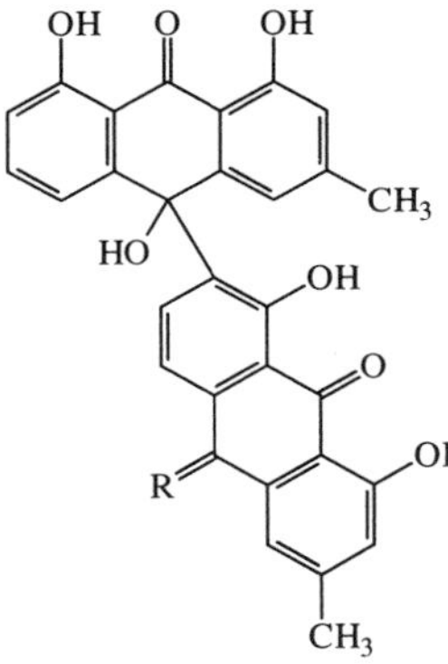

(**28**) R=O Bianthracene II
(**29**) R=H_2 Bianthracene IV

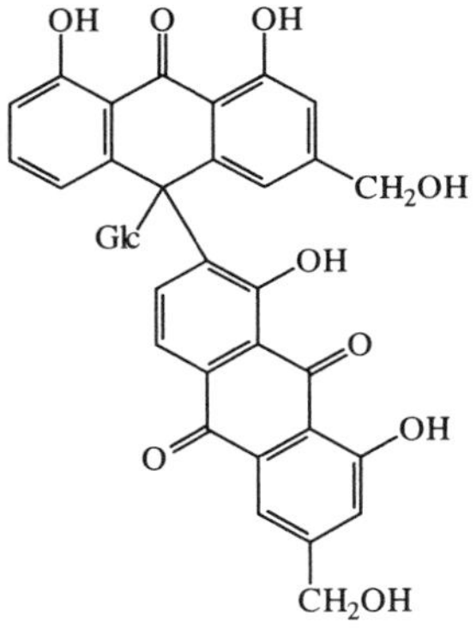

(**30**) Elgonicardine

Several pre-anthraquinones, which could be considered as progenitors of the above two types of anthraquinones, have been isolated and characterized mainly from subterranean parts of *Aloe*. However, the pre-anthraquinone aloechrysone (**16**) was detected both in roots and leaves of four *Aloe* species from Ethiopia (Dagne and Alemu 1991). It is interesting to note that the related genus *Gasteria* (Aloaceae), elaborates *Aloe* type pre-anthraquinones both in the leaves and roots (Dagne *et al.* 1996). The pre-anthraquinones could be readily converted to the corresponding anthraquinones by treatment with base (Yenesew *et al.* 1993).

Anthrones

Anthrones are by far the most important of all the classes of compounds present in *Aloe* species. The most outstanding members of this class are aloin A and B (**42a, 42b**), which are collectively known as barbaloin because they were first isolated from Barbados aloe. Aloin A and B are two diastereomeric C-glucosides that differ in the configuration at C-10 of the aloe-emodin anthrone moiety. These compounds are believed to be mainly responsible for the bitter and purgative properties of the well known commercial aloe drug, which is principally made up of the leaf exudates of *A. ferox* and *A. vera*. The leaf exudate of *A. ferox* may contain up to 10% barbaloin (Groom and Reynolds 1987). However not all *Aloe* species are found to contain barbaloin. In a screening of 240 *Aloe* species, barbaloin was found to occur in exudates of 85 of the species examined (Reynolds 1985b) usually in 10-20% concentration. Although *A. littoralis* Baker is reported to be positive for presence of barbaloin by Reynolds(1985) and also reported to be present to the extent of 18.2 % by Groom and Reynolds (1987), our analysis by TLC and HPLC of the exudate of *A. littoralis* did not show the presence even of a trace of barbaloin (Dagne *et al.*, *Phytochemistry,* in press). Instead we isolated

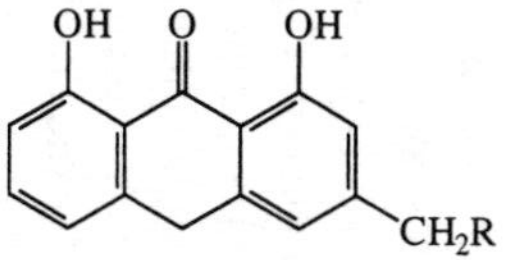

(**31**) R=OH Aloe-emodinanthrone
(**32**) R=H Chrysophanolanthrone

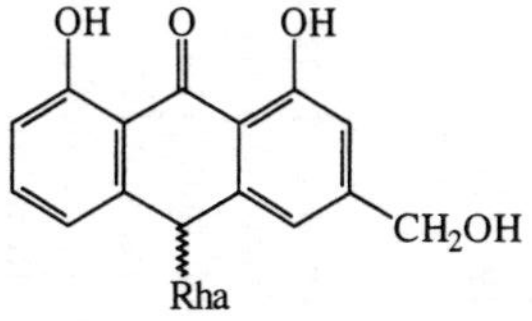

(**33**) Aloe-emodinanthrone-10-C-Rhamnoside

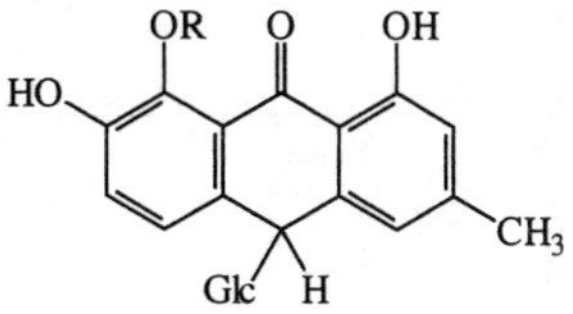

(**34**) R=CH_3 Homonataloin
(**35**) R=H Nataloin

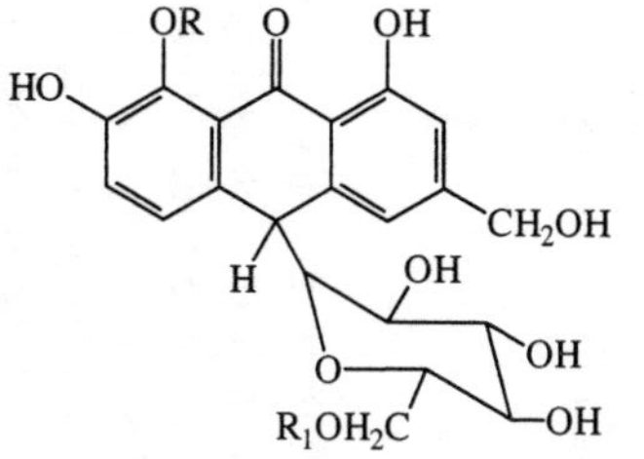

(**36**) R=R_1=H (7-Hydroxyaloin)
(**37**) R=H, R_1= *p*-coumaroyl (7-Hydroxyaloin-6'-O-*p*-coumaroyl)
(**38**) R=CH_3, R_1=H (8-O-Methyl-7-hydroxyaloin)
(**39**) R=CH_3, R_1= cinnamoyl (6'-O-Cinnamoyl-8-O-methyl-7-hydroxyaloin)

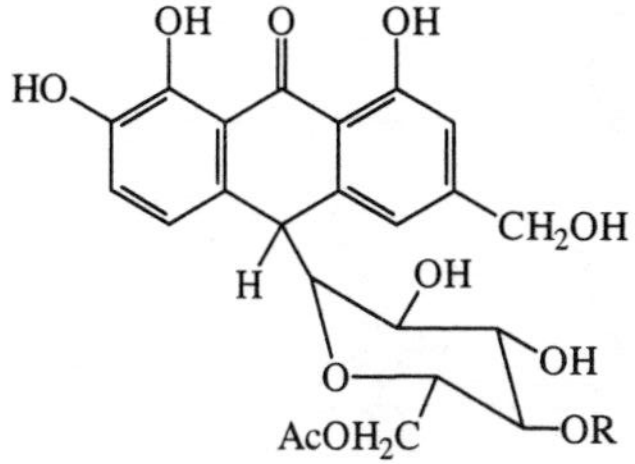

(**40**) R=H (7-Hydroxyaloin-6'-O-acetate)
(**41**) R=Ac (7-Hydroxyaloin-4',6'-diacetate)

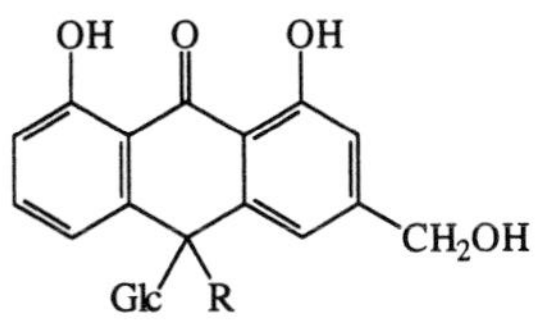

(**42a**) R=α-H Aloin A
(**42b**) R=β-H Aloin B

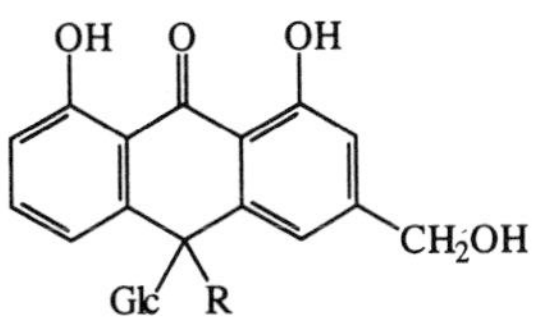

(**43a**) R=α-OH Hydroxyaloin A
(**43b**) R=β-OH Hydroxyaloin B

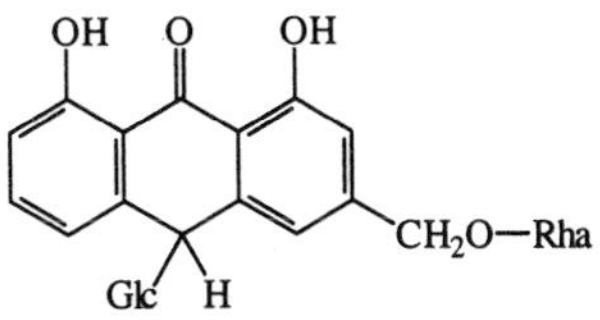

(**44**) Aloinoside

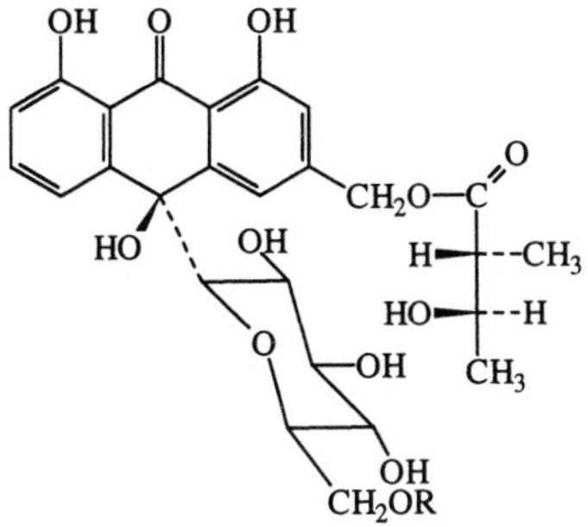

(**45**) R=H Deacetyllitoraloin
(**46**) R=Ac Littoraloin

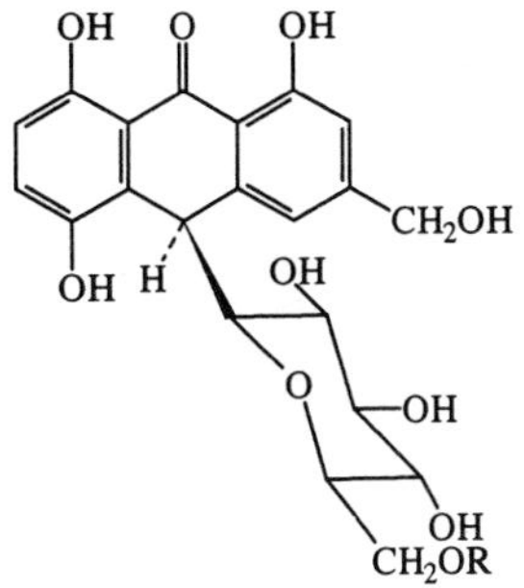

(**47**) R=H 5-Hydroxyaloin
(**48**) R=caffeoyl Microstigmin A

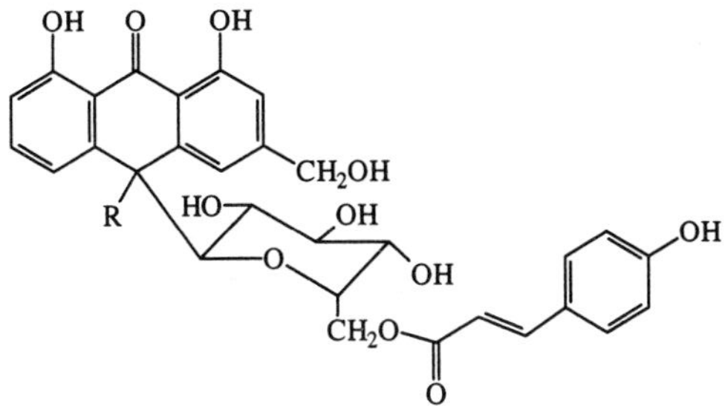

(**49a**) R=α-H Microdontin A
(**49b**) R=β-H Microdontin B

10-hydroxyaloin B (**43b**) and its two novel nilate ester derivatives littoraloin (**46**) and deacetyllittoraloin (**45**). The chemotaxonomic significance of these compounds in *Aloe* has been recently published (Viljoen *et al.* 1996).

Interestingly barbaloin and homonataloin seem to be mutually exclusive, with the notable exception of *A. mutabilis* (Reynolds 1990). It should be pointed out that aloin should not be regarded to be confined only to *Aloe* species as it has also been found in the extracts of cascara bark (*Rhamnus purshiana* D.C.) (Manitto *et al.* 1990). Rauwald and Lohse (1992) have also reported occurrence of 10-hydroxy-aloin B (**43b**) in *Rhamnus* sp.

The determination of the absolute configuration of aloin A and B (**42a**, **42b**) has engaged several workers (Rauwald *et al.* 1989; Manitto *et al.*1990). In the true natural product, *i.e,* aloin B, the glucose moiety attached to C-10 has the α orientation (*i.e.*, 10*R*,1'*S*) and the β orientation (*i.e.*, 10*S*,1'*S*) follows for aloin A.

Biosynthetic study of Grün and Franz (1980) has shown that aloin B is the true natural product but is gradually converted to aloin A. That study also established that aloin B is formed by attachment of glucose to aloe-emodinanthrone (**31**), a compound detected so far in flowers but not in leaves of *Aloe* (Sigler and Rauwald 1994). In 10-hydroxyaloin B (**43b**) and its two novel nilate ester derivatives **45** and **46** obtained from *A. littoralis*, the glucose also has the α configuration indicating that hydroxylation at C-10 occurs prior to epimerization of the natural aloin B (Dagne *et al., Phytochemistry,* in press). On the other hand, 5-hydroxyaloin A (**47**) is known only in the A form (Rauwald and Beil 1993) *i.e.* with the β orientation for the glucose moiety at C-10, an observation which is also the case for its natural derivative microstigmin A (**48**) a novel compound that we recently found in *A. microstigma* (Dagne *et al., s*ubmitted to *Phytochemistry*).

Roots of *Aloe* spp. elaborate in the main anthraquinones and pre-anthraquinones. It has recently been shown that inflorescence of *Aloe* also produce anthrones (Sigler and Rauwald).

Benzene/naphthalene derivatives

Several naphthalene and benzene based secondary metabolites have been reported from *Aloe* species. One of the first such compounds to be reported is the naphthalene derivative of isoeleutherol-5-O-glucoside (**58**) isolated by Yagi *et al.* (Yagi *et al.* 1977b) from the subterranean stems of *Aloe saponaria*. It is interesting to note that such a glycoside is present in the subterranean part of *Aloe.* The aglycone isoeleutherol (**57**) was reported for the first time as a natural product by our group (Dagne *et al.* 1994) from roots of more than a dozen *Aloe* species belonging to the series Saponariae. Isoeleutherol was conspicuously absent from other series investigated, indicating its chemotaxonomic significance in delineating members of the Saponariae series from other series. The insecticidal compound pluridone (**52**) isolated from roots of the South African *A. pluridens* is the only example of a sulfur containing compound ever isolated from *Aloe*. The recently reported 1,1-diphenylethane (**50**) from Cape aloe (Speranza *et al.* 1994) and plicataloside (**62**) from *A. plicatilis* (Wessels *et al.* 1996) have added more variety to benzene- and naphthalene-derived compounds found in *Aloe* species. Furthermore the discovery of the tetrahydronaphthalenes feroxidin (**54**), feroxin A (**55**) and B (**56**) in Cape aloe by Speranza *et al.* (1990, 1992) is a further testimony of the diversity of the constituents of this aloes of commerce.

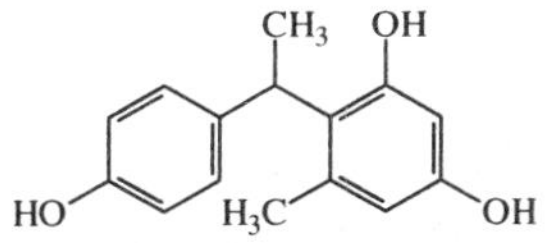

(**50**) 1,1-Diphenylethane

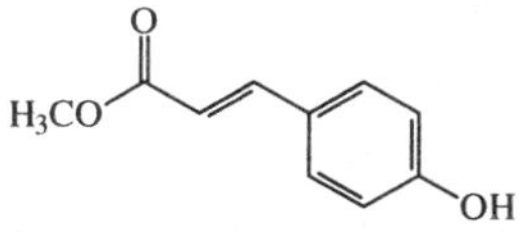

(**51**) Methyl-*p*-coumarate

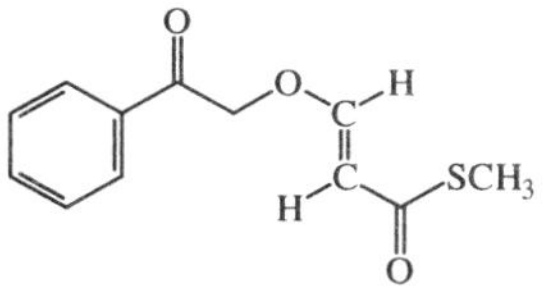

(**52**) Pluridone

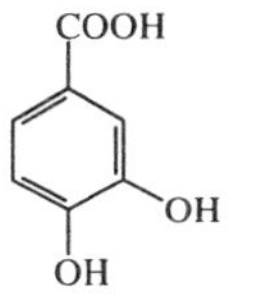

(**53**) Protocatechuic acid

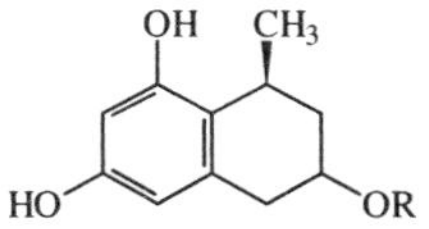

(**54**) R=H Feroxidin
(**55**) R=Glc Feroxin A
(**56**) R=4,6-Dicoumaroyl-Glc = Feroxin B

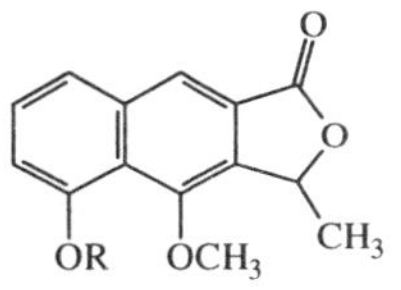

(**57**) R=H Isoeleutherol
(**58**) R=Glc Isoelutherol-5-O-glucoside

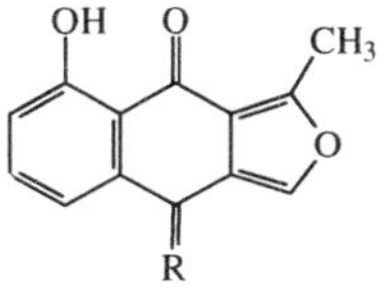

(**59**) R=H_2
(**60**) R=O

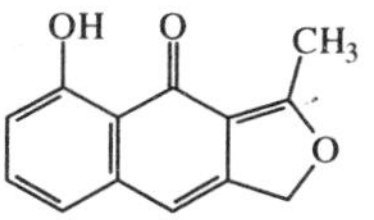

(**61**)

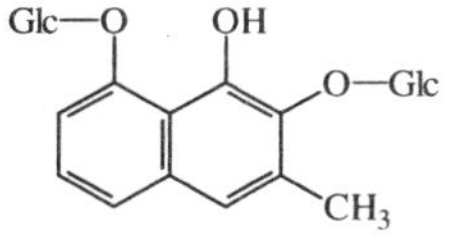

(**62**) Plicataloside

Chromones

Aloesin (**63**) formerly called aloeresin B, is one of the three most significant constituents of aloe drug, the other two being barbaloin and aloeresin A. Random screening of *Aloe* species indicated its presence in leaves of at least 30% of the species examined (Reynolds 1985a). Its structure was established as **63** in by Haynes *et al.* (1970) and subsequently in 1972 its aglycone named as aloesone (**77**) was recognized as an *Aloe* leaf constituent by Holdsworth (1972). The structure of aloeresin A, first proposed incorrectly as a *p*-coumarate ester of aloesin esterified on C_6 of the sugar moiety, was later on revised to structure **64** in which the ester was placed on C_2 of the sugar (Gramatica *et al.* 1982). Mebe (1987) reported the aloeresin derivative (**68**) from *A. excelsa* but incorrectly named it as 2'-*p*-methoxycoumaroylaloeresin when it should have been called either 2'-*p*-O-methylcoumaroylaloesin or 2'-*p*-methoxycinnamoylaloesin.

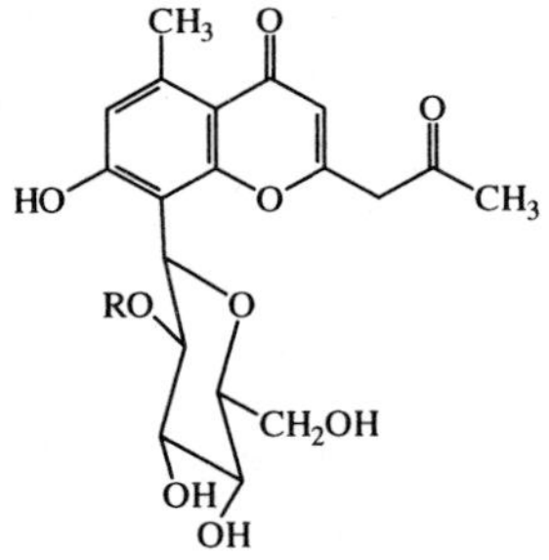

(**63**) R=H	Aloesin or Aloeresin B
(**64**) R=*p*-coumaroyl	Aloeresin A
(**65**) R=cinnamoyl	Aloeresin F
(**66**) R=feruloyl	2'-O-Feruloylaloesin
(**67**) R=*cis*-*p*-coumaroyl	Iso-aloeresin A
(**68**) R=*p*-O-methylcoumaroyl	2'-O-Methylcoumaroylaloesin
(**69**) R=tigloyl	2'-O-Tigloylaloesin

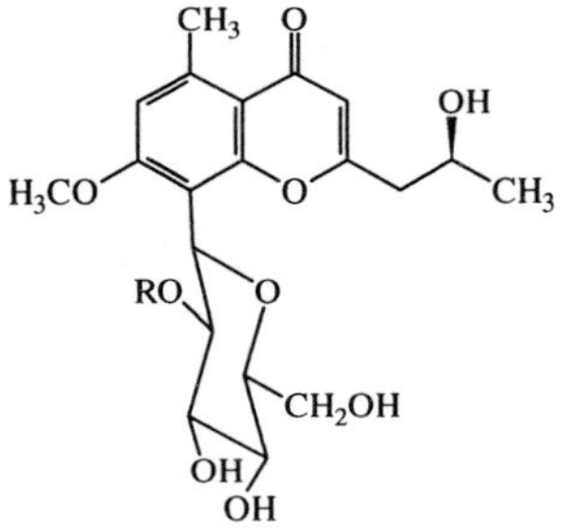

(**70**) R=*p*-coumaroyl Aloeresin D
(**71**) R=caffeoyl Rabaichromone

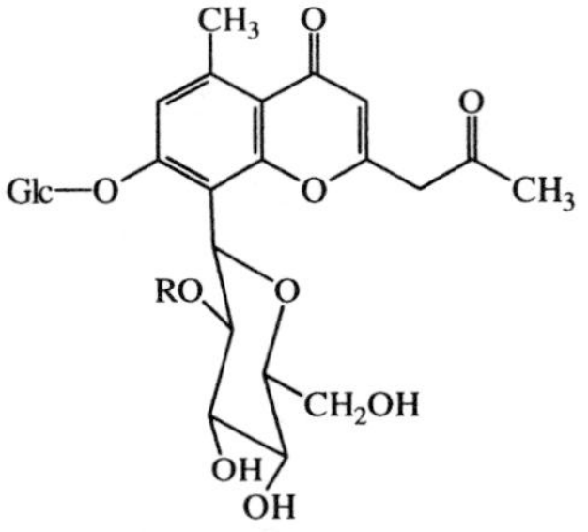

(**72**) R=*p*-coumaroyl Aloeresin C
(**73**) R=cinnamoyl Aloeresin E

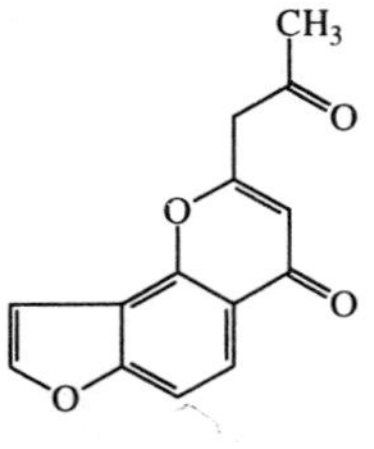

(**74**) Furoalesone

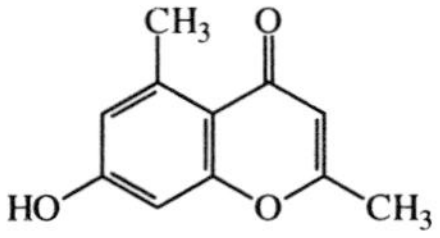

(**75**)

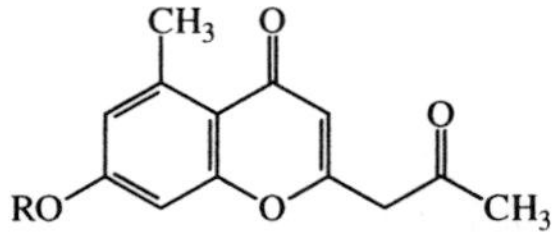

(**76**) Aloesol

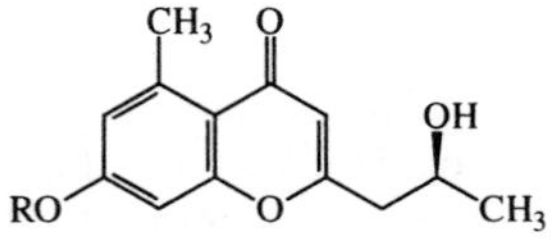

(**77**) Aloesone

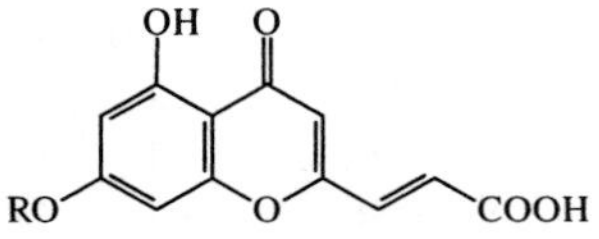

(**78**)

TLC and HPLC examination of *Aloe rupestris* Bak. leaf exudate showed the total absence of aloin and homonataloin as well as aloeresin A, C and D but instead revealed the presence of two major and several other minor constituents. Isolation of the major components resulted in their characterization as aloesin (**63**) and the new natural product 7-O-methylaloesin (Dagne *et al.*, submitted to *Biochemical Systematics and Ecology*). The latter compound is of considerable chemotaxonomic value since it is present in most species of *Aloe* series Asperifoliae Berger.

Pyrones

Aloenin (**81**), a phenylpyrone derivative, is a relatively infrequently encountered bitter component of *Aloe* leaf exudate, whose revised structure was reported in 1974 (Suga *et al.*1974). Aloenin aglycone (**82**) and the coumaroyl ester **85** were

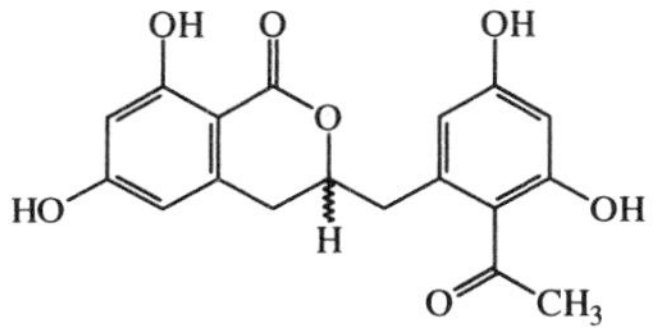

(**79**)

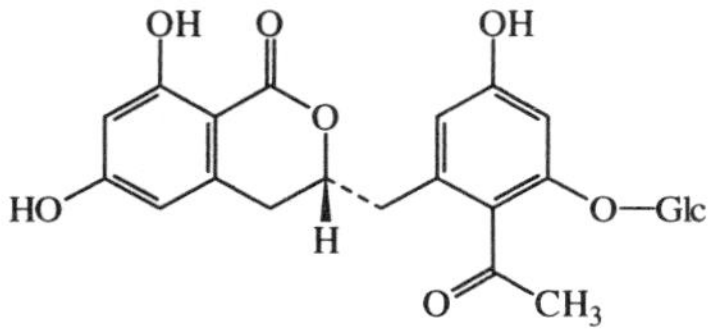

(**80**) Feralolide

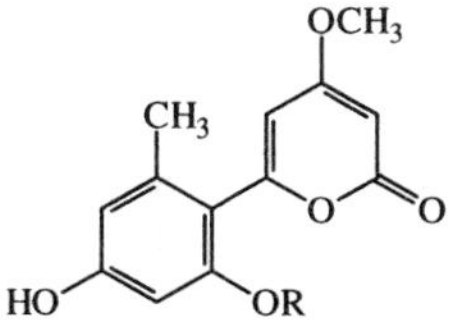

(81) R=Glc Aloenin
(82) R=H Aloenin aglycone

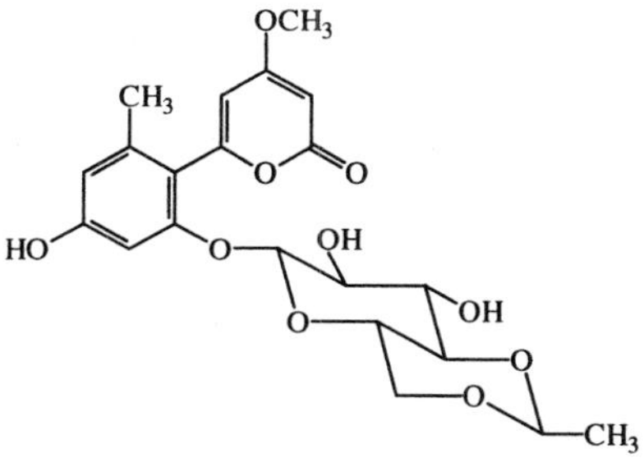

(**83**) Aloenin acetal

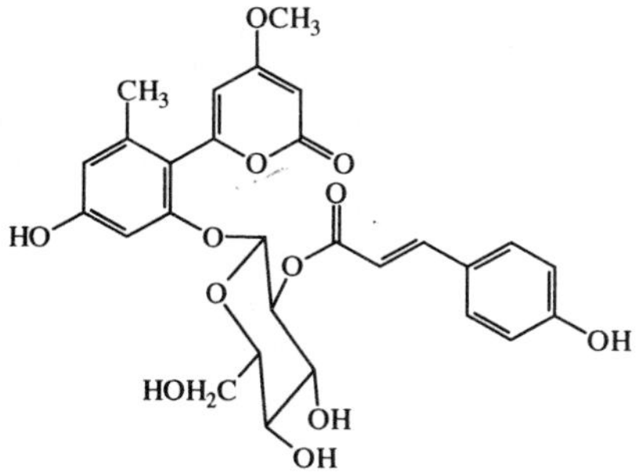

(**84**) R=Glc Aloenin B
(**85**) R=H Aloenin-2'-*p*-coumaroyl ester

reported recently (Conner *et al.* 1987). Aloenin B (**84**) is one of the major (13.5%) constituents of commercial Kenya aloe (Speranza *et al.* 1986).

Contribution to the study of the chemistry of aloes of Ethiopia

There are nearly 36 species of *Aloe* in Ethiopia with 18 endemics. The ethanol extract of the leaves of the endemic Ethiopian species *A. berhana* now renamed as *A. debrana* (Demissew and Gilbert, in *Flora of Ethiopia and Erythrea*, Addis Ababa, in press). yielded the interesting pre-anthraquinone, aloechrysone (**16**) (Dagne and Alemu 1991) in addition to chrysophanol, aloe-emodin, barbaloin and β-sitosterol. Aloechrysone fluoresces strongly under 366 nm UV light, a property that helps to identify it during separation using silica gel columns and plates. Aloechrysone was first detected in the roots of *A. berhana* (Dagne *et al.* 1992) and was later found also in roots of other *Aloe* species (Dagne *et al.* 1994). It is the most likely precursor of chrysophanol-8-methyl ether (**6**), an anthraquinone that also occurs in roots of *A. berhana* and other *Aloe* species. The isolation of the simple, biologically important, aromatic acid, 3,4-dihydroxybenzoic acid, also known as protocatechuic acid, from *A. berhana* leaves is noteworthy (Dagne and Alemu 1991). Although this acid is known to occur in other plants such as in coffee pulp (Dagne and Alemu 1991), this was the first report of its occurrence in *Aloe* species. Other constituents of roots of *A. berhana* include: aloesaponol I, laccaic acid D methyl ester, aloesaponol III, aloesaponarin I, chrysophanol-8-methyl ether and chrysophanol (Dagne *et al.* 1994).

Likewise study of the leaves of *A. megalacantha* yielded, barbaloin, chrysophanol, aloinoside and β-sitosterol (Dagne and Alemu 1991), while its roots (Dagne *et al.* 1994) afforded aloechrysone, aloesaponarin I, aloesaponol I, aloesaponarin II, aloesaponol II, aloesaponol III, asphodelin, chrysophanol, chrysophanol-8-methyl ether, helminthosporin and laccaic acid D methyl ester.

A recent study (Yenesew *et al.* 1993) of the subterranean stem of this species, *i.e.*, *A. lateritia* Engl. var. *graminicola* (formerly *A. graminicola*) resulted in the isolation and characterization of prechrysophanol (**25**), which can be considered as the direct progenitor of chrysophanol. This plant has also been shown (Dagne *et al.* 1994) to be among the few species of *Aloe* that elaborate in their roots the chemotaxonomicaly important 2-naphtoic acid derivative, isoeleutherol (**57**), which has been indicated as a chemotaxonomic marker for a group of *Aloe* lumped in the series Saponariae. The presence of isoeleutherol in the roots of *A. kefaensis* Gilbert & Sebsebe and *A. macrocarpa* Tod. also confirms their placement in the above mentioned series Saponariae.

A. pulcherrima is one of the most beautiful of the *Aloe* taxa found in Ethiopia. It is known as "Sete-Eret" and is mainly used to accelerate wound healing. Chemical analysis of its leaves revealed the conspicuous absence of barbaloin but showed instead nataloin and 7-hydroxy-barbaloin as its major constituents (Dagne and Alemu 1991).

The root chemistry of *A. calidophila, A. camperi, A. gilbertii, A. pulcherrima, A. rivae, A. schelpei, A. sinana* is similar to that of *A. megalacantha*. However, *A. secundiflora* differs because it does not contain aloechrysone, a compound also absent from all members of the Saponariae series (Dagne *et al.* 1994).

References

Confalone, P.N., Huie, E.M., and Patel, N.G. (1983). The isolation, structure determination and synthesis of pluridone, a novel insecticide from *Aloe pluridens*. *Tetrahedron Letters* **24**, 5563-5566.

Conner, J.M., Gray, A.I., Reynolds, T., and Waterman, P.G. (1987). Anthraquinone, anthrone and phenylpyrone components of *Aloe nyeriensis* var. kedongensis leaf exudate. *Phytochemistry* **26**, 2995-2997.

Council of Europe (ed.) (1981). *Flavoring substances and natural sources of flavorings*, p. 61. Maisonneuve, Moulins-les-Metz (France)

Dagne, E. and Alemu, M. (1991). Constituents of the leaves of four *Aloe* species from Ethiopia. *Bulletin of the Chemical Society of Ethiopia* **5**, 87-91.

Dagne, E., Casser, I., and Steglich, W. (1992). Aloechrysone, a dihydroanthracenone from *Aloe berhana*. *Phytochemistry* **31**, 1791-1793.

Dagne, E., Yenesew, A., Asmellash, S., Demissew, S., and Mavi, S. (1994). Anthraquinones, pre-anthraquinones and isoeleutherol in the roots of *Aloe* species. *Phytochemistry* **35**, 401-406.

Dagne, E., Van Wyk, B.-E., Mueller, M., and Steglich, W. (1996).Three dihydro-anthracenones from *Gasteria bicolor*. *Phytochemistry* **41**, 795-799.

Dring, J.V., Nash, R.J., Roberts, M.F., and Reynolds, T. (1984). Hemlock alkaloids in Aloes. Occurrence and distribution of γ-coniceine. *Planta Medica* **50**, 442-443.

Gramatica, P., Monti, D., Speranza, G., and Manitto, P. (1982). Aloe revisited the structure of Aloeresin A. *Tetrahedron Letters* **23**, 2423-2429.

Groom, Q.J. and Reynolds, T (1987). Barbaloin in *Aloe* species. *Planta Medica* **53**, 345-348.

Grün, M. and Franz, G. (1980). Studies on the biosynthesis of aloin in *Aloe arborescens*. *Planta Medica* **39**, 288.

Haynes, L.J. and Hodlsworth, D.K. (1970). C-glucosyl compounds. Part VI. Aloesin, a C-glucosyl chromone from *Aloe* sp. *Journal of the Chemical Society (C)*, 2581-2586.

Holdsworth, D.K. (1972). Chromones in *Aloe* species. Part II. Aloesone. *Planta Medica* **22**, 54-58.

Jork, H., Funk, W., Fischer, W., and Wimmer, H. (1990). *Thin-Layer Chromatography: reagents and detection methods*, p. 288. VCH, Weinheim.

Leistner, E. (1973). Quinonoid Pigments. In *Phytochemical Methods* (ed J.B. Harborne). Chapman and Hall, London.

Mabberley, D.J. (1987). *The plant Book: a portable dictionary of Higher Plants*. Cambridge University Press, Cambridge.

Manniche, L. (1989). *An ancient Egyptian Herbal*. British Museum Publication Ltd., London.

Manitto, P., Monti, D., and Speranza, G. (1990). Studies on Aloe. Part 6. Conformation and absolute configuration of aloins A and B and related 10-*C*-glucosyl-9-anthrones. *Journal of the Chemical Society, Perkin Transactions* **1**, 1297-1300.

Mebe, P.P. (1987). 2'-*p*-Methoxycoumaroylaloesin, a *C*-glucoside from *Aloe excelsa*. *Phytochemistry* **26,** 2646-2647.

Nash, R. J., Beaumont, J.,Veitch, N.C., Reynolds, T., Benner, J., Hughes,C.N.G., Dring, J.V., Bennett, R.N., and Dellar, J.E. (1992). Phenylethylamine and piperidine alkaloids in *Aloe* species. *Planta Medica* **58**, 84-87.

Rauwald, H. W. and Beil, A. (1993). 5-Hydroxyaloin A in the genus *Aloe*, Thin layer chromatographic screening and high performance liquid chromatographic determination. *Zeitschrift für Naturforschung* **48c**, 1-4.

Rauwald, H.W. and Lohse, K. (1992). Structure revision of 4-hydroxyaloin: 10-hydroxyaloins A and B as main in *vitro*-oxidation products of the diastereomeric aloins. *Planta Medica* **58**, 259-262.

Rauwald, H.W., Lohse, K., and Bats, J.W. (1989). Configurations of aloin A and B, two diastereomeric C-glucosylanthrones from *Aloe* species. *Angewandte Chemie, International English Edition* **28**, 1528-1529.

Reynolds, T. (1985a). The compounds in *Aloe* leaf exudates a review. *Botanical Journal of the Linnean Society* **90**, 157- 177.

Reynolds, T. (1985b). Observations on the phytochemistry of the *Aloe* leaf-exudate compounds. *Botanical Journal of the Linnean Society* **90**, 179-199.

Reynolds, T. (1990). Comparative chromatographic patterns of leaf exudate components from shrubby aloes. *Botanical Journal of the Linnean Society* **102**, 273-285.

Reynolds, T. and Herring, C. (1991). Chromatographic evidence of the geographical origin of *Aloe arborescens* introduced into Gibraltar. *British Cacti and Succulents Journal* **9**, 77-79.

Rowe, T. D., Lovell, B K., and Parks, L. (1941). Further observations on the use of *Aloe vera* leaf in the treatment of third degree x-ray reactions. *Journal of the American Pharmaceutical Association* **30**, 266-268.

Sigler, A. and Rauwald, H. W. (1994). *Aloe* plants accumulate anthrone-type anthranoids in inflorescence and leaves, and tetrahydroanthracenes in roots. *Zeitschrift für Naturforschung* **49c**, 286-292.

Speranza, G., Dada, G., Lunazzi, L. Gramatica, P., and Manitto, P. (1986). Aloenin B, a new diglucosylated 6-phenyl-2-pyrone from Kenya aloe. *Journal of Natural Products* **49**, 800-805.

Speranza, G., Manitto, P., Monti, D., and Lianza, F. (1990). Feroxidin, a novel 1-methyltetralin isolated from Cape aloe (*Aloe ferox*). *Tetrahedron Letters* **31**, 3077-3080.

Speranza, G., Manitto, P., Monti, D., and Pezzuto, D. (1992). Studies on Aloe, Part 10. Feroxins A and B, two *O*-glucosylated 1-methyltetralins from Cape aloe. *Journal of Natural Products* **55**, 723-729.

Speranza, G., Corti, S., and Manitto, P. (1994). Isolation and chemical characterization of a new constituent of Cape aloe having the 1,1-diphenylethane skeleton. *Journal of Agricultural and Food Chemistry* **42**, 2002-2006.

Suga, T., Hirata, T., and Tori, K. (1974). Structure of aloenin, a bitter glucoside from *Aloe* species. *Chemistry Letters:* 715-718.

Tyler, V.E., Brady, L.R., and Robbers, J.E. (1976). *Pharmacognosy*, 7th ed. Lea and Febiger, Philadelphia.

Trease, G. E. and Evans, W. C. (1976). *Pharmacognosy*, 12th ed., p. 404. Bailliere Tindall, London.

U.S. Pharmacopeia: XX-The National Formulary XV (1979). p. 21. Marck, Easton.

Van Wyk, B.E., Yenesew, A., and Dagne, E. (1995a). Chemotaxonomic survey of anthraquinones and pre-anthraquinones in roots of *Aloe* species. *Biochemical Systematics and Ecology* **23**, 267-275.

Van Wyk, B.E., Yenesew, A., and Dagne, E. (1995b), Chemotaxonomic significance of anthraquinones in the roots of Asphodeloideae (Asphodelaceae) *Biochemical Systematics and Ecology* **23**, 277-281.

Van Wyk, B.-E, Yenesew, A., and Dagne, E. (1995c). The chemotaxonomic significance of root anthraquinones and pre-anthraquinones in the genus *Lomatophylum* (Asphodelaceae). *Biochemical Systematics and Ecology* **23**, 805-808.

Viljoen, A. M., Van Wyk, B.-E., and Dagne, E. (1996). The chemotaxonomic value of 10-hydroxyaloin B and its derivatives in *Aloe* series Asperifoliae Berger. *Kew Bulletin* **51**, 159-168.

Wessels, P. L., Holzapfel, C. W., Van Wyk, B.-E., and Marais, W. (1996). Plicataloside, an *O,O*-di-glycosylated naphthalene derivative from *Aloe plicatilis*. *Phytochemistry* **41**, 1547-1551.

Yagi, A., Makino, K., and Nishioka, I. (1974). Studies on the constituents of *Aloe saponaria* HAW. I. The structures of tetrahydroanthracene derivatives and the related *anthraquinones Chemical and Pharmaceutical Bulletin* **22**, 1159-1166.

Yagi, A., Makino, K., and Nishioka, I. (1977a). Studies on the constituents of *Aloe saponaria* HAW. II. The structures of tetrahydroanthracene derivatives, aloesaponol III and -IV. *Chemical and Pharmaceutical Bulletin* **25**, 1764-1770.

Yagi, A., Makino, K., and Nishioka, I. (1977b). Studies on the constituents of *Aloe saponaria* HAW. III. The structures of phenol glucosides. *Chemical and Pharmaceutical Bulletin* **25**, 1771-1776.

Yenesew, A., Ogur, J.A., and Duddeck, H. (1993). (*R*)-Prechrysophanol from *Aloe graminicola*. *Phytochemistry* **34**, 1442-1444.

8. Quinones and other phenolic compounds from marketed African plants

B.M. ABEGAZ[1], G. ALEMAYEHU[2], T. KEBEDE[2], D. MAHAJAN[1] AND M. M. NINDI[1]

[1]*Department of Chemistry, University of Botswana, Private Bag 0022, Gaborone, Botswana and* [2]*Department of Chemistry, Addis Ababa University, P.O.Box 1176, Addis Ababa, Ethiopia*

Introduction

Almost all traditional markets in Africa have sections where plants are sold for a variety of uses. A closer look at even modern shopping centers will reveal that there are thriving businesses of native plants. These uses include medicinal, culinary, fragrance, majico-medical, etc. In each region one finds indigenous plants that have emerged from the local culture and tradition as established items of commerce for that particular community. It is worth noting that many clients have established the utility and efficacy of these plants out of personal previous experiences and so simply proceed to buy them in very much the same way as one would buy common over-the-counter drugs and other personal hygiene aids.

We have been studying plants that are sold in African markets. We have conducted surveys in such markets in several countries in Africa, especially in Ethiopia, Kenya, Uganda, Tanzania (Abegaz and Demissew 1992) and Botswana. This report will deal with our recent findings in which we have identified novel anthraquinone and naphthalene glucosides from *Rhamnus prinoides* (Rhamnaceae) and bianthraquinone pigments from *Senna* (Fabaceae) species.

The genus Rhamnus

The genus *Rhamnus* has been widely investigated. The most comprehensive review of the family being that of Hegnauer (1973). As many as 24 species have been described in the chemical literature. The genus has been a source of a variety of flavonoids which are based on quercetin and kaempferol and their glycosides. Several anthrones, anthraquinones and their glycosides have also been reported. The well known cathartic drug, Cascara bark, is derived from the Asian *R. purshiana* DC Bark (Tyler *et al.* 1988). This plant has been investigated extensively and several C-10 glycosides of emodin (cascarosides A-D), O,Cdiglucosylanthrones (cascarosides E and F) (Mannito *et al.* 1993, 1995) as well as the diastereomeric 10-hydroxyaloins A and B (Rauwald *et al.* 1991) have been reported. There are also several phytochemical reports on Asian *R. formosana* (Kalidhar 1992; Lin and Wei 1993), *R. wightii* (Peppalla *et al.* 1991) and *R. nakahari* (Lin and Wei 1994) revealing the presence of many anthraquinones and naphthalenic derivatives. 6-O-Rhamnosides of emodin (known as frangulin A, B, and glucofrangulin) have also been found in *R. fallax* (Kinget 1967).

Rhamnus prinoides

In Africa, the genus *Rhamnus* is represented by only two taxa, namely - *R. prinoides* and *R. staddo. R. prinoides,* known in Ethiopia by the *Amharic* name: *Gesho*, is a plant which grows up to 6 meters. It is also known to occur in Cameroon, Sudan, throughout East Africa to South Africa, Angola and in Arabia (Thulin 1989). It is cultivated in Ethiopia, specially in Tigrai, in North Shoa around Kara Kori and Sebeta, just west of Addis Ababa. *Gesho* is an important commodity and is sold in almost every traditional market in Ethiopia.

The leaves and stems of *Gesho* are indispensable ingredients in the making of the traditional fermented beverages *Tella* and *Tej*. In doing so, care is always taken to remove the fruits of the plant from the leaves and stems. The fruits are, however, used for the treatment of ring worm infections. *Tella* is a malt beverage, like beer. *Tej* is also a fermented beverage based on honey. Over 5 million people consume these beverages everyday in Ethiopia. Although it is generally known that *Gesho* imparts the characteristic bitterness of these beverages, more precise understanding of the scientific role of this plant in this traditional brewing process is emerging only very slowly. The first scientific report on *R. prinoides* is that of Salgues (1962) who described the presence of inorganic cations, organic acids and the flavonoid derivative rhamnetin. He also claimed that the leaf extract was toxic to rabbits. The role of *Gesho* in the fermentation process has been investigated (Kleyn and Hough 1971; Sahle and Gashe 1991) and it is claimed that the plant regulates the microflora responsible for the fermentation process. These reports indicate further that the bitterness of the brew is directly related to the amount of *Gesho* added. It has also been reported that extracts from *Gesho* can be used as a commercial hopping agent for beer (Tessema 1994).

Secondary metabolites of the fruits of R. prinoides

Emodin, physcion, emodinanthrone, emodinbianthrone, rhamnazin and prinoidin (**1**), a novel anthrone rhamnoside diacetate were reported by Abegaz and Dagne (1988). Subsequently the isolation of minor pigments of the fruits, other mono- (**6**), di- (**3, 4, 5**), and triacetates (**2**) of emodin were reported (Abegaz and Peter 1995). It is not clear if these isomeric acetates are true natural products or artefacts formed during preparative thin-layer chromatography on silica gel. However, solutions of these compounds in chloroform are indefinitely stable in the NMR tube. In earlier work we had isolated a dimer of prinoidin which we initially thought was a natural product. We have now made the observation that the amount of this dimer increases with prolonged contact of the natural products to silica gel during flash chromatography. We are, therefore, inclined to conclude that this dimer is an artefact. The structure of the dimer (**7**), was nevertheless characterized. FABMS clearly indicated the molecular ion at 970, which is consistent with a dimer of prinoidin. The 600 MHz ^{1}H-NMR showed the presence of four chelated hydroxyl signals and eight aromatic proton resonances at δ6.93, 6.84, 6.79, 6.71, 6.61, 5.48, 5.35 and 5.27 (See Table 8.1). The striking feature of the spectrum is the highly shielded aromatic protons at 5.35 and 5.27 ppm. Further more, these two signals and two others at 6.93, 6.84 are significantly broadened as shown in Fig. 8.1. The two shielded signals are assigned to the 5 and 5' protons which are most probably forced out of the plane of the ring by the adjacent sugar substituent on each of the anthraquinone moiety. The reason for the broadening of the four signals is not entirely clear and we are not sure if they are caused by a scissoring type dynamic behaviour. The proton signals of one of the rhamnose moiety is also more shielded. The type of conformation which would be consistent with the observed spectrum is at the present time unclear.

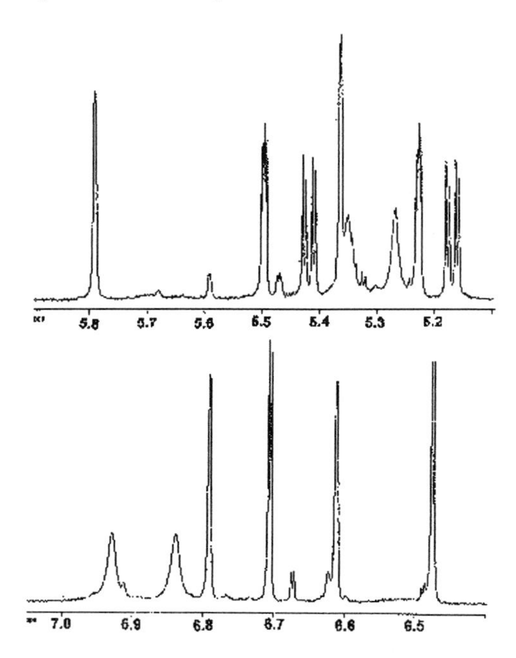

Fig. 8.1. Partial ^{1}H-NMR Spectrum of dimer **7** (600 MHz) showing aromatic proton signals.

Table 8.1. ^{1}H-NMR spectral data for dimer **7** (600 MHz, $CDCl_3$, δ ppm)

Protons	δ(ppm); J (Hz)	Sugar protons	δ(ppm); J (Hz)
H-2	6.79 *s*	H-1"	5.79 *d* (1.40)
H-4	6.93 *brs*	H-2"	5.50 *dd* (1.79, 3.49)
H-5	5.35 *brs*	H-3"	5.42 *dd* (3.51, 10.04)
H-7	6.71 *d* (2.30)	H-4"	3.65 *m*
H-10	4.48 *d* (3.89)	H-5"	3.95 *m*
C3-Me	2.48 *s*	C5"-Me	1.43 *d* (6.06)
H-2'	6.61 *s*	H-1"'	5.36 *d* (1.78)
H-4'	6.84 *brs*	H-2"'	5.23 *dd* (1.96, 3.41)
H-5'	5.27 *brs*	H-3"'	5.17 *dd* (3.46, 9.87)
H-7'	6.45 *d* (2.30)	H-4"'	3.65 m
H-10'	4.45 *d* (3.88)	H-5"'	3.75 *m*
C3'-Me	2.12 *s*	C5"'-Me	1.33 *d* (6.20)
-OH	12.43 *s*	CO-Me	2.22 *s*
-OH	11.90 *s*	CO-Me	2.15 *s*
-OH	11.81 *s*	-COMe	2.14 *s*
-OH	11.64 *s*	-CO-Me	2.13 *s*

Secondary metabolites of the leaves of R. prinoides

We have identified 11 secondary metabolites from the leaves. These include the known anthracene derivatives: chrysophanol, physcion and emodin; the flavonoids rhamnocitrin, rhamnezin, quercetin and 3-O-methylquercetin; and the naphthalenic derivatives sorigenin (**9**), musizin (**10**) and the previously unknown β-sorigenin-8-O-β-D-glucoside, (geshoidin, **8**). An organoleptic evaluation on the above compounds revealed that geshoidin is the most significant bitter substance of the leaves. 3-O-Methylquercetin also displayed bitterness. Further organoleptic evaluation was made by five volunteers who independently confirmed that geshoidin possesses bitter properties. It is interesting to note that geshoidin is bitter despite the presence of a glucose moiety in the structure. The aglycone, sorigenin (**9**) is in fact not bitter at all. The toxicity of geshoidin to brine shrimp (*Artemia salina)* was evaluated at seven different concentrations over a range of 64 folds (15 to 1000 μg/ml)[*]. No lethality was observed. Preliminary assay for possible cytotoxicity of geshoidin has also been negative[*]. Although rigorous toxicity tests should be conducted on geshoidin, the results obtained so far suggest that this compound may have commercial potential. The critical question in determining the structure of geshoidin was to provide evidence to show that the

[*] We are grateful to Ato Mesfin Bogale of the Faculty of Science, Addis Ababa University for the brine shrimp assay and to Dr. R. Becker of the University of the North for the cytotoxicity assays.

sugar moiety was attached to the 8-position and not at the alternative C-9 position. In fact the alternative structure is known and had been reported from *R. wightii* (Peppalla 1991).The identification of geshoidin was based on spectroscopic and chemical evidence (Abegaz and Kebede 1995). The EI as well as CIMS of geshoidin failed to show a molecular ion. But ESIMS yielded $[M+23]^+$ ion at 401.

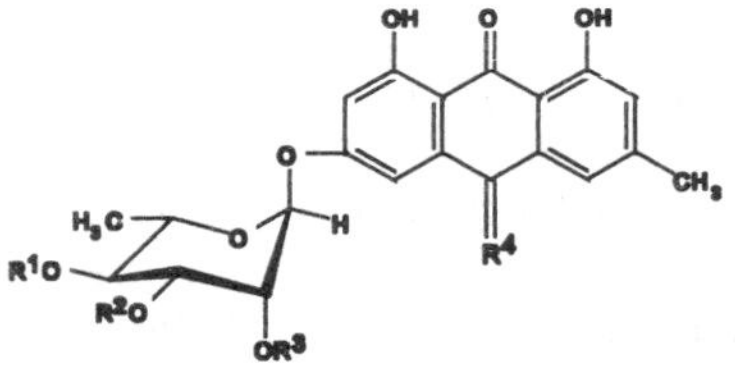

	R^1	R^2	R^3	R^4
1	H	Ac	Ac	H, H
2	Ac	Ac	Ac	H,H
3	H	Ac	Ac	O
4	Ac	Ac	H	O
5	Ac	H	Ac	O
6	H	H	Ac	O

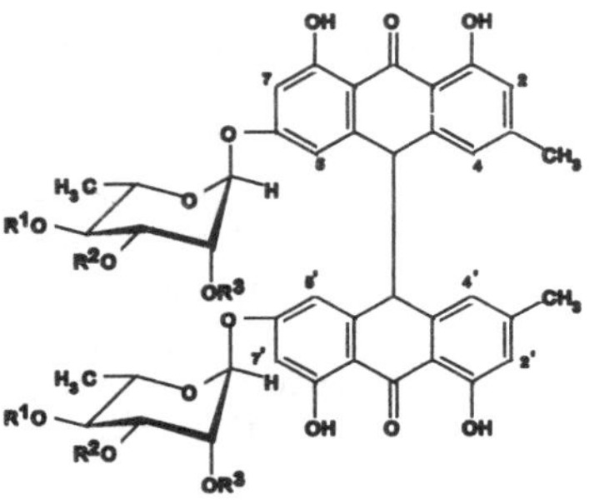

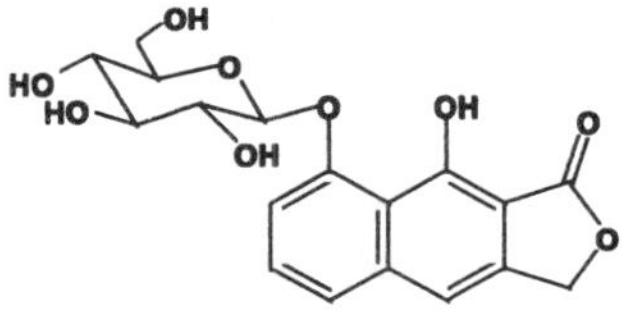

Collision induced dissociation of this parent ion by application of an offset voltage of 35V resulted in the appearance of an ion at m/z of 239 which is believed to arise by loss of the glucose unit from the $[M+23]^+$ ion. The co-occurrence of the two naphthalenic compounds **9** and **10** in the leaves suggests a biogenetic relationship of these two compounds and geshoidin (**8**). It seems very probable that musizin (**10**) undergoes oxidative cyclization to sorigenin (**9**) and glucosylation at position 9 to yield geshoidin (**8**).

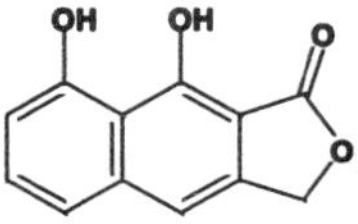

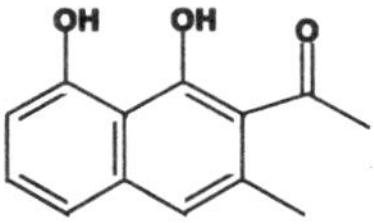

The genus *Senna*

Senna is an important genus which has yielded important purgative drugs. The most famous is *Cassia senna (C. acutifolia)*, known as Alexandrian senna or *Cassia angustifolia*, also known as Tinnevelly senna, or a mixture of the two species. The biologically active constituents of *Senna* are the hydroxyanthracene glycosides known as sennosides. Many members of *Senna* have for a long time been considered in a broader classification together with the now separate genera of *Cassia* and *Camechrista* (Thulin 1989). We have studied four taxa, namely: *S. didymobotrya, S. septemtrionalis, S. longiracemosa* and *S. multiglandulosa* and have identified several anthraquinones, anthrones, preanthraquinones, and novel bianthraquinones.

Senna septemtrionalis (synonyms: *Cassia laevigata, C. floribunda, S. floribunda)*

This legume has been shown to contain 8-mono- and digalactosides of physcion as well as chrysophanol and emodin (Singh *et al.* 1980). Our studies on the leaves yielded the common anthraquinones, emodin and physcion, in addition to two novel bianthraquinone pigments for which the names floribundone-1 (**12**) and floribundone-2 (**13**) were given (Alemayehu *et al.* 1988). These compounds represented the second set of examples of an anthraquinone dimer with a 5-7'-bianthracene linkage. The mixture of floribundone-1 and 2 was difficult to separate and it was observed that a solution of floribundone-2 (**13**) was easily oxidized to floribundone-1 (**12**).

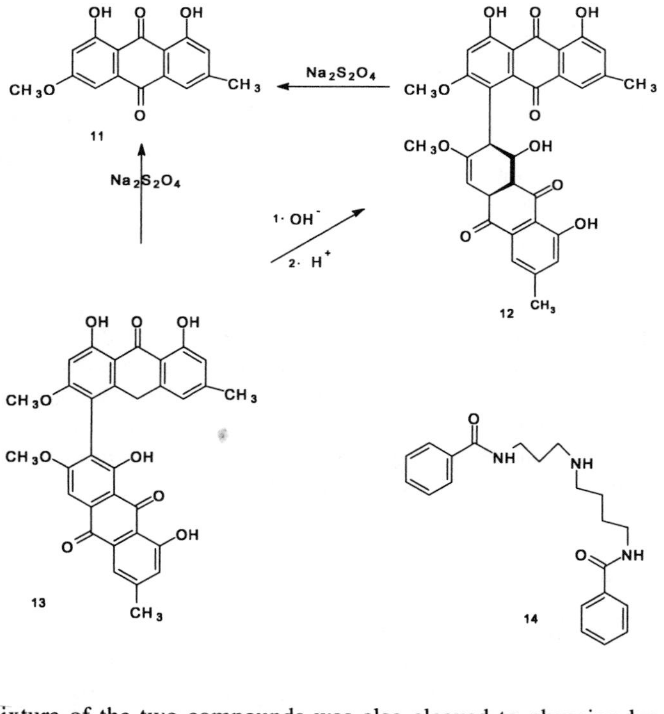

The mixture of the two compounds was also cleaved to physcion by reaction with sodium dithionite. The reductive cleavage of floribundone-2 (**13**) presumably yields physcion anthrone which would be oxidized to physcion (**11**) during work up. Alternatively and most probably, the oxidation of floribundone-2 (**13**) to floribundone-1 (**12**) may take place faster than the cleavage to yield physcion anthrone directly. Also isolated from the leaves was a N^1,N^8-dibenzoyl-spermidine (**14**) and other traces of pigments which at present are unidentified. Floribundone-1 (**12**) has since then been reported from Mexican *Senna* species (Barba *et al.* 1993) and its atropi-isomer from *Cassia torosa* of Japanese origin (Kitanaka and Takido 1995).

Senna longiracemosa (synonym: *C. longiracemosa*)

The leaves yielded chrysophanol, physcion, torachrysone (**15**), rubrofusarin (**16**), nataloe-emodin (**17**), 10,10'-bichrysophanol (**19**), 10,10'-chrysophanol-physcion (**20**), 10,10'-chrysophanol-isophyscion (**21**), 10,10'-biisophyscion (**22**) and 10-hydroxy-10,7'-(chrysophanol-anthrone)-chrysophanol (**23**) (Alemayehu *et al.* 1993). Compounds **17**, **20**, **21**, and **22** were reported for the first time and **23** had not been reported from the genus *Senna* previously. The root bark also yielded 2-methoxy-stypandrone (**18**) in addition to chrysophanol, physcion, emodin, **19, 20**, **21**, and **22** (Alemayehu 1989).

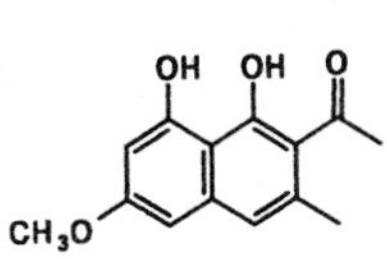

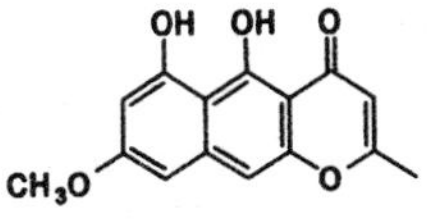

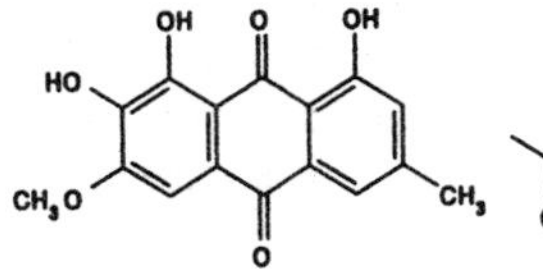

Senna multiglandulosa (synonyms: *Cassia multiglandulosa, S. tomentosa, C. tomentosa)*

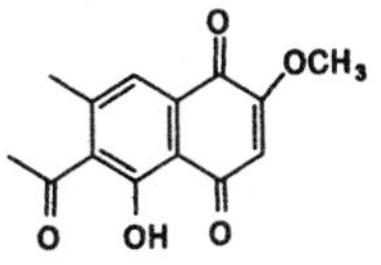

Our earlier investigation of the leaves and stems had yielded chrysophanol, emodin, physcion and four bianthraquinones: floribundone-1 (**12**), torosanin-9,10-quinone (**25**), anhydrophlegmacin-9,10-quinone (**26**) and the novel 1,4-quinone: 9-(physcion-7-yl)-5,10-dihydroxy-2-methoxy-7-methyl-1,4-anthraquinone (**27**) (Abegaz *et al.* 1994). Torosanin-9,10-quinone had been reported as an oxidation product of torosanin obtained from the unripe seeds of *Cassia torosa* (Kitanaka and Takido 1982). The most interesting compound obtained from this plant is the 1,4-quinone **27** for which the name sengulone is proposed. This is the first example of an anthraquinone dimer containing a 1,4-quinone moiety. We were initially intrigued by the possibility of another structure (**28**) which closely fitted the spectroscopic data for compound **27**. Difnoe experiments unequivocally established that the compound isolated from the leaves was **27** and not **28.** Thus,

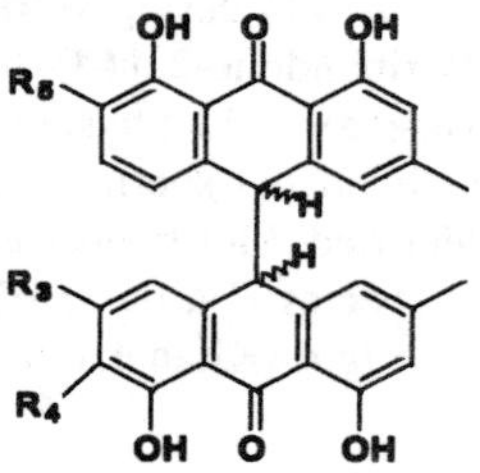

	R_3	R_4	R_5
19	H	H	H
20	OCH_3	H	H
21	H	OCH_3	H
22	H	OCH_3	OCH_3

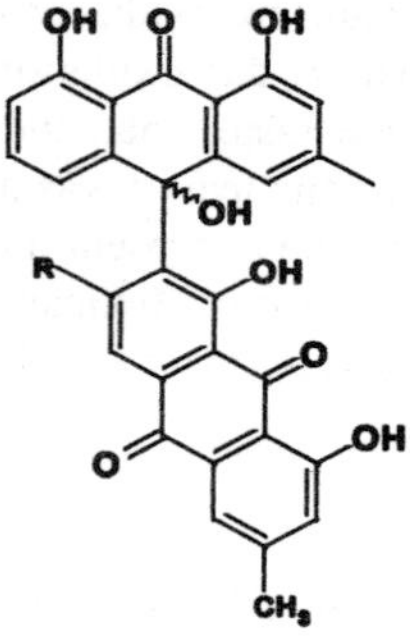

23 R = H
24 R = OCH_3

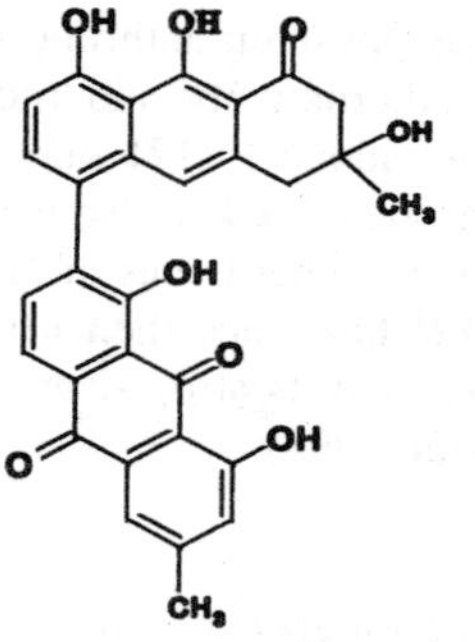

25

irradiation of the Me signal at δ 2.49 led to enhancement of the Ar-H signals at 7.70 and 7.11, and irradiating the other Me signal at δ 2.32 led to the enhancement of two Ar-H signals at 6.95 and 6.79. On the other hand, irradiation of the methoxy signals at 3.82 and 3.88 each led to a corresponding increase in the signals of only one quinonoid proton at 6.20 and 7.59, respectively. These noe data enabled us to reject structure **28** in favour of **27** for the structure of sengulone. Survey of the literature indicates that 1,4-anthraquinones have not been reported from higher plants. There are, however, a few 1,4-anthraquinones isolated from *Aspergillus cristatus* (Laatsch and Anke 1982).

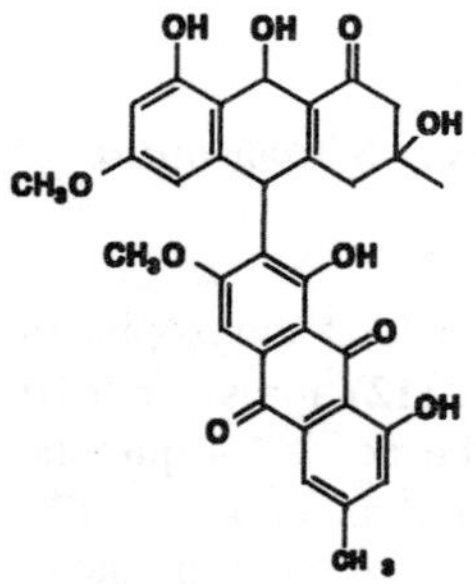

26

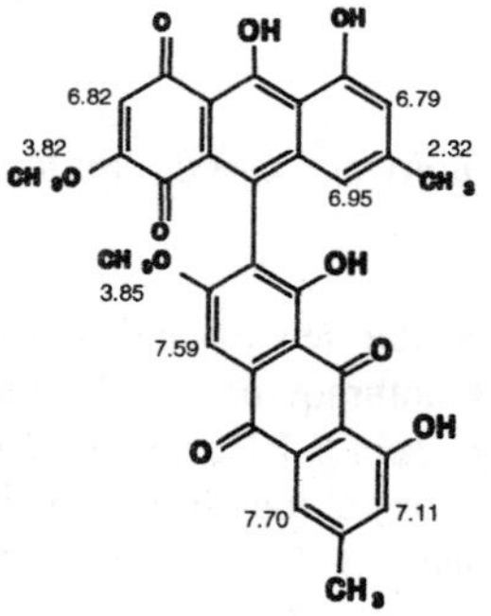

27

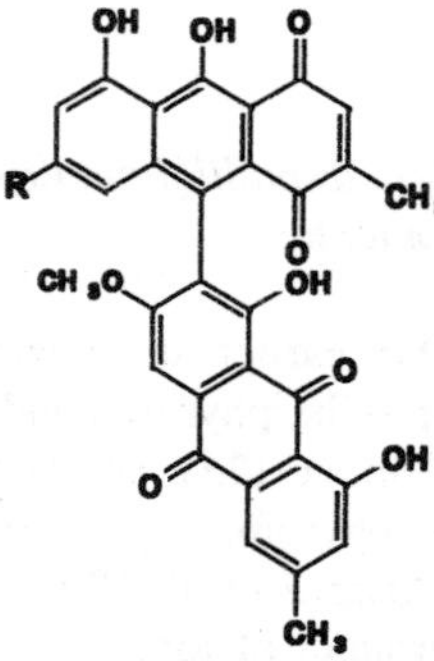

28 R = OCH3
29 R = H

In a subsequent study we undertook to examine the chemical constituents of the seeds of *S. multiglandulosa* and we were able to identify an isomer of sengulone (named iso-sengulone) which fits the rejected alternative structure described above (**28**) in addition to other pigments, namely physcion, torosachrysone (**30**), the bianthraquinones floribundone-1 (**12**) and anhydrophlegmacin-9,10-quinone (**26**). Difnoe data were consistent with the structure assigned for **28** as were homonuclear 2D-NMR measurements (COSY45) (Alemayehu and Abegaz 1996).

Senna didymobotrya (synonym: *Cassia didymobotrya*)

The leaves yielded the monoanthracene derivatives, chrysophanol, physcion, aloe-emodin (**31**), fallacinol (**32**), rhein (**33**), parietinic acid (**34**) and the preanthraquinone, torosachrysone (**30**). The isolation of fallacinol (**32**) and parietinic acid (**34**) constituted the first report on the occurrence of these two substances from higher plants (Alemayehu *et al.* 1989). Previously they had been reported from lichens and cultures of *Eurotium echinolatum* (Thomson, 1971).

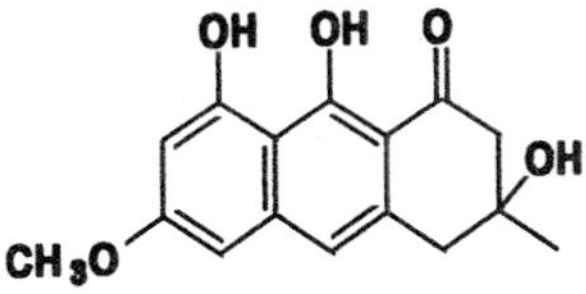

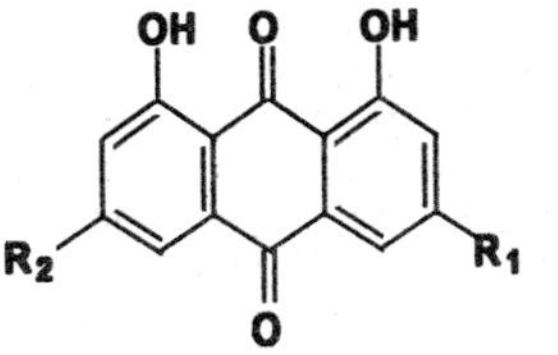

	R_1	R_2
31	CH_2OH	H
32	CH_2OH	OCH_3
33	COOH	H
34	COOH	OCH_3

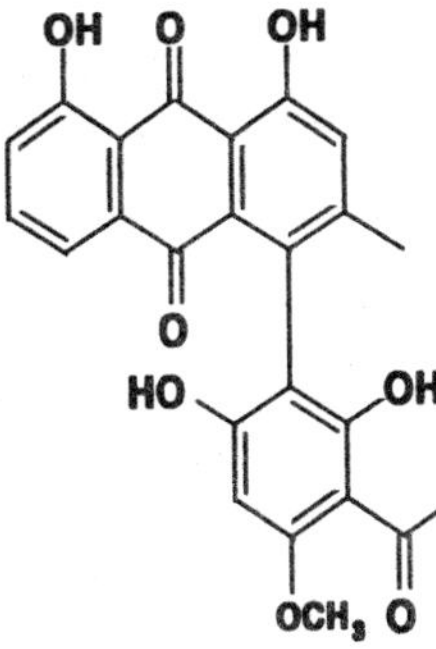

Chrysophanol, aloe-emodin and rhein had been reported earlier from this plant by Egyptian workers (El-Sayyad and Ross 1983). Further examination of the pods of this plant has resulted in the isolation of the common anthraquinones, chrysophanol, emodin and physcion and the known but novel compound knipholone (**35**). Knipholone which contains an acetylphloroglucinol methyl ether moiety attached to C-4 of chrysophanol was first isolated by Dagne and Steglich (1984) from *Kniphofia foliosa.* Although it has been reported in a number of *Kniphofia* and *Bulbine* species since then (Van Staden and Drewes 1994), this is the first report of this unique anthraquinone in the family Fabaceae. In addition the pods also yielded two new bianthraquinones (**24** and **29**). Compound **29** constitutes the third example of 1,4-quionone in the genus *Senna* (Alemayehu *et al.* 1996).

References

Abegaz, B. M. and Dagne, E. (1988). Anthracene derivatives of *Rhamnus prinoides*. *Bulletin of the Chemical Society of Ethiopia* **2**, 15-20.

Abegaz, B. M. and Demissew, S. (1992). *Evaluation of the status of medicinal plants and other useful plants*. A report for the United Nations University, Addis Ababa.

Abegaz, B.M. and Kebede, T. (1995). Geshoidin: A bitter principle of *Rhamnus prinoides* and other constituents of the leaves. *Bulletin of the Chemical Society of Ethiopia* **9**, 107-114.

Abegaz, B.M. and Peter. M.G. (1995). Emodin and emodinanthrone rhamnoside diacetates from fruits of *Rhamnus prinoides*. *Phytochemistry* **39**, 1411-1414.

Abegaz, B.M., Bezabeh, M., Alemayehu, G., and Duddeck, H. (1994). Anthraquinones from *Senna multiglandulosa*. *Phytochemistry* **35**, 465-468.

Alemayehu, G. (1989). Chemical investigation of some Ethiopian *Senna* species. PhD dissertation, Addis Ababa University, Addis Ababa.

Alemayehu, G. and Abegaz, B.M., (1996). Bianthraquinones from *Senna multiglandulosa*. *Phytochemistry* **41**, 919-921.

Alemayehu, G., Abegaz, B.M., Snatzke, G., and Duddeck, H. (1988). Bianthraquinones and a spermidine alkaloid from *Cassia floribunda*. *Phytochemistry* **27**, 3255-3258.

Alemayehu, G., Abegaz, B., Snatzke, G., and Duddeck, H. (1989). Quinones of *Senna didymobotrya*. *Bulletin of the Chemical Society of Ethiopia* **3**, 37-40.

Alemayehu, G., Abegaz, B.M., Snatzke, G., and Duddeck, H. (1993). Bianthraquinones from *Senna longiracemosa*. *Phytochemistry* **32**, 1273-1277.

Alemayehu, G., Hailu, A., and Abegaz., B.M. (1996). Bianthraquinones from *Senna didymobotrya*. *Phytochemistry* **42**, 1423-1425.

Barba, B., Dîaz, J.G.and, Herz, W. (1992). Anthraquinones and other constituents of two *Senna* species. *Phytochemistry* **31**, 4374-4375.

El-Sayyad, S.M. and Ross, S. A. (1983). A phytochemical study of some *Cassia* species cultivated in Egypt. *Journal of Natural Products* **46**, 431-432.

Dagne, E. and Steglich, W. (1984). Knipholone: a unique anthraquinone derivative from *Kniphofia foliosa*. *Phytochemistry* **23**, 1729-1731.

Hegnauer, R. (1973). *Chemotaxonomie der Pflanzen*, Band VI, p. 64. Birkhauser Verlag, Basel.

Kalidhar, S. B. (1992). Reassessment of the structures of an anthraquinone glycoside from *Rhamnus formosana*. *Phytochemistry* **31**, 2905-2906.

Kinget, P.R. (1967). Recherches sur les drogues a principes anthraquinoniques. *Planta Medica* **15**, 233-237.

Kitanaka, S. and Takido, M. (1982). Dimeric hydroanthracenes from the unripe seeds of *Cassia torosa*. *Phytochemistry* **21**, 2103-2106.

Kitanaka, S. and Takido, M. (1995). (S)-5,7'-(Biphyscion-8-glucoside from *Cassia torosa*. *Phytochemistry* **39**, 717-718.

Kleyn, J. and Hough, J. (1971). The Microbiology of Brewing. *Annual Review of Microbiology* **25**, 583.

Laatsch, H. and Anke, H. (1982). Viocristin, isoviocristin und hydroxyviocristin - Struktur und Synthese. *Liebigs Annalen der Chemie*, 2189-2215.

Lin, C-N. and Wei, B-L (1993). Anthraquinone and naphthalene glycosides from *Rhamnus nakahari*. *Phytochemistry* **33**, 905-908.

Lin, C-N. and Wei, B-L (1994). Flavonol and naphthalene diglycosides from *Rhamnus nakahari*. *Journal of Natural Products* **57**, 294-297.

Mannito, P., Monti, D., Speranza, G., Mulinacci, N., Vincieri, F., Griffini, A., and Pifferi, G. (1993). Conformational studies of natural product. Part 4. Conformation and absolute configuration of cascarosides A,B,C,D. *Journal of the Chemical Society, Perkin Transactions 1*,1577-1580.

Mannito, P., Monti, D., Speranza, G., Mulinacci, N., Vincieri, F.F, Griffini, A., and Pifferi, G. (1995). Studies on Cascara. Part 2. Structure of cascarosides E and F. *Journal of Natural Products* **58**, 419-423.

Peppalla, S.B., Jammula, S.R., Telikepalli, H., Bhattiprolu, K.V. and Rao, K.V.J. (1991). A naphthalene glucoside from *Rhamnus wtghtii. Phytochemistry* **30**, 4193-4194.

Rauwald, H.W., Lohse, K., and Bats, J.W. (1991). Neue Untersuchung über Inhaltstoffe aus Aloe- und Rhamnus-Arten, XIII. Konfigurations- und Konformationsbestimmung der diastereomeren Oxanthron-C-glucosyle 10-Hydroxyaloin A und B. *Zeitschrift für Natürforschung* **46**, 551-557.

Sahle, S. and Gashe, B.A. (1991). The microbiology of Tella fermentation. *SINET: Ethiopian Journal of Science* **14**, 93.

Salgues, R. (1962). New chemical and toxicological studies on the genus *Rhamnus. Qualitas Plantarum et Materiae Vegetabiles* **9**, 15-32.

Singh, J., Tiwari, A.R., and Tiwari, R.D. (1980) Anthraquinones and flavonoids of *Cassia laevigata* roots. *Phytochemistry* **19**, 1253-1254.

Tessema, A.D. (1994). PhD dissertation, Moscow State Academy of Food Products, Moscow.

Thomson, R. H. (1971). *Naturally occurring quinones,* 2nd edn. Academic press, London.

Thulin, M. (1989) in *Flora of Ethiopia* Vol.3 (eds. I. Hedberg and S. Edwards) Addis Ababa and Asmara, Ethiopia and Uppsala, Sweden.

Tyler, V.E., Brady, L.R. and Robebers, J.E. (1988). *Pharmacognosy*, pp. 60-62. Lea and Febiger, Philadelphia.

Van Staden, L.F. and Drewes, S.E. (1994). Knipholone from *Bulbine latifolia* and *Bulbine frutescens. Phytochemistry* **35**, 685-686.

9. Phytochemical studies of medicinal plants from Malawi

J. D. MSONTHI[1], K. HOSTETTMANN[2] AND M. MAILLARD[2]

[1]Chemistry Department, University of Swaziland, Private Bag 4, Kwaluseni, Swaziland and [2]Institut de Pharmacognosie et Phytochimie, BEP Université de Lausanne, CH-1015 Lausanne, Switzerland

---

Introduction

Medicinal plants from Malawi have been analyzed phytochemically and biologically largely through collaboration with the Universities of Lausanne, Switzerland; of Rome, Italy and the Technical University of Berlin, Germany.

The selection of plants for investigation has been based on interviews with traditional healers of the Herbalists Association of Malawi under the Chairmanship of Mr. James Gangire Phiri.

A large percentage of the plants selected by the traditional healers gave positive leads to the activity claimed from their medicinal uses. We also used random selection based on literature reviews and chematoxonomic relationships. The studies on these plants included collection, extraction, purification, *in vitro* activity-guided fractionation, isolation of active principles, derivatization and further *in vitro* bioassays.

Included here are antitumoral, antifungal, antibacterial, molluscicidal, hypoglycemic, antifeedant and, to some extent, immunostimulant activities.

Depending on the activity of the compounds, some of them could be used directly after further studies on toxicity, biodegradation and efficacy through standardization.

This chapter covers some of the reports on phytochemistry of medicinal plants from Malawi, with emphasis on biologically active compounds. In order to clarify the organization of this chapter, the results have been separated into three main groups: compounds exhibiting molluscicidal activity, fungitoxic natural products and miscellaneous structures

Results

Molluscicidal activity

Schistosomiasis (bilharzia) is a parasitic disease affecting millions of people in Africa, as well as in South America and Asia. It is caused by nematodes (*Schistosoma* sp.) that colonize the bladder or intestines. The parasite life cycle needs contact with water sources where the parasite's eggs can hatch into miracidia and enter freshwater snails, such as *Biomphalaria glabrata, Bulinus globosus*, etc. Once in the snail, thousands of cercaria are produced which can eventually penetrate the intact skin of humans in contact with the water source in question.

One way to prevent the transmission of the disease is to destroy the intermediate host of the parasite, by the use of molluscicides.

Plants have been shown (Hostettmann and Marston 1987) to be an interesting source of new natural molluscicides, and those originating in Malawi are listed below.

Talinum tenuissimum Dinter (Portulacaceae)

The tubers of *Talinum tenuissimum* are used in Malawi, according to traditional healers, for the treatment of schistosomiasis. In the course of systematic screening studies on compounds with molluscicidal activity, Gafner *et al.* (1985) noticed that the aqueous extract of the tubers of *T. tenuissimum* also killed *B. glabrata* snails, at a concentration as low as 25 ppm within 24 hours. This observation led to the isolation of new triterpenoid saponins (**1-3**from the crude plant extract.

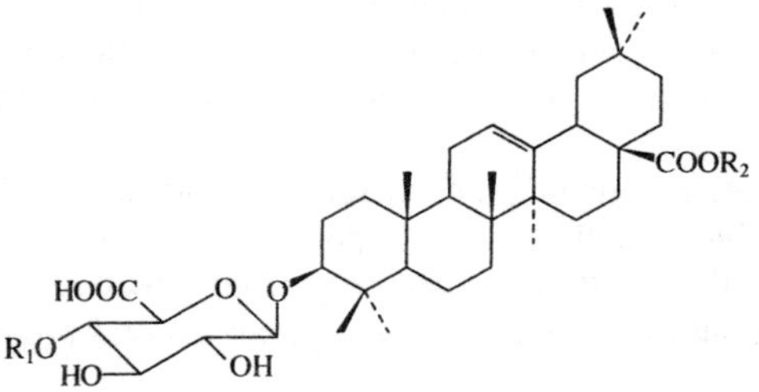

	R_1	R_2
(**1**)	xylose	glucose
(**2**)	xylose	H
(**3**)	H	H

The major bidesmosidic saponin **1** isolated from the methanolic extract was inactive against *B. glabrata*, whereas the monodesmosidic saponins **2** and **3** killed snails at a concentration of 1.5 ppm within 24 hours.

Cussonia spicata Thumb. (Araliaceae)

The bark of *Cussonia spicata* and other species of the genus *Cussonia* are used in African traditional medicine against malaria. An infusion of the roots of *C. spicata* prevents skin irritation and is antifebrile.

The water extract of the stem bark of *Cussonia spicata* showed an activity of 400 ppm within 24 hours against *B. glabrata* snails. This activity was strong enough to undertake a phytochemical investigation of this plant and two molluscicidal saponins (**4**, **5**) were finally isolated from the stem bark of *C. spicata* using MPLC on a RP-8 support.

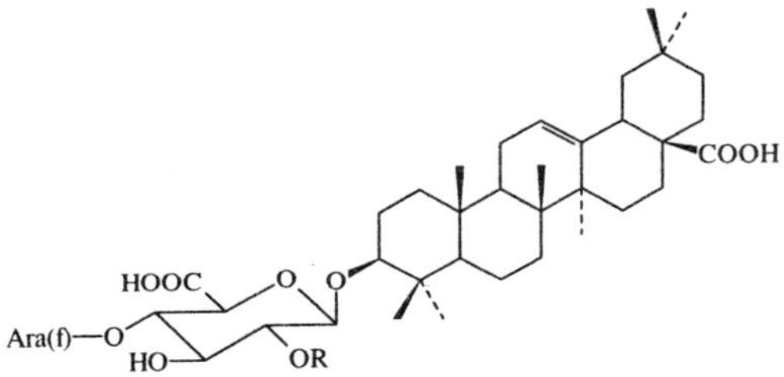

(**4**) R = H

(**5**) R = galactose

Saponin **4** was toxic to *B. glabrata* at 12.5 ppm and compound **5** at 100 ppm. Furthermore, a preliminary screening for spermicidal activity against human spermatozoids showed an activity at a concentration of 1 ppm and 3 ppm within 3 minutes for saponins **4** and **5**, respectively (Gunzinger *et al.* 1986)

Diospyros zombensis White (Ebenaceae)

Numerous *Diospyros* species are utilized in Africa as chewing sticks, but *D. zombensis* is especially used by the traditional healers of Malawi for the treatment of schistosomiasis. In addition to flavonoid glycosides and biologically active naphthoquinones (7-methyljuglone (**12**, see later), and isodiospyrin), molluscicidal saponins were isolated for the first time in the Ebenaceae family from the methanol extract of the root bark (Gafner *et al.* 1987).

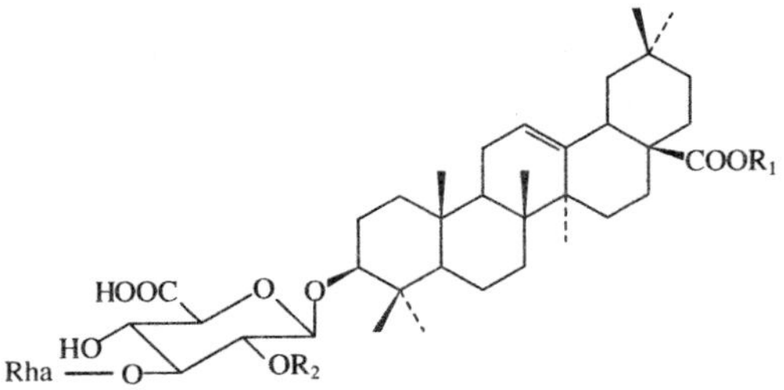

	R_1	R_2
(**6**)	H	H
(**7**)	glucose	H
(**8**)	glucose	xylose
(**9**)	H	xylose

As generally observed in nature, the bidesmosidic saponins (**7** and **8**) were not active against *B. glabrata*, whereas the monodesmosidic saponins **6** and **9** exhibited strong activity (3 ppm within 24 hours)

Clerodendrum wildii Moldenke (Verbenaceae)

Mi-saponin A (**10**), another molluscicidal triterpenoid saponin, has been isolated from the roots of *Clerodendrum wildii*, a medicinal plant from Malawi claimed to be active against intestinal parasites, or in the treatment of malaria. This bitter bidesmosidic saponin showed toxicity towards *B. glabrata* snails (25 ppm) (Toyota *et al.* 1990).

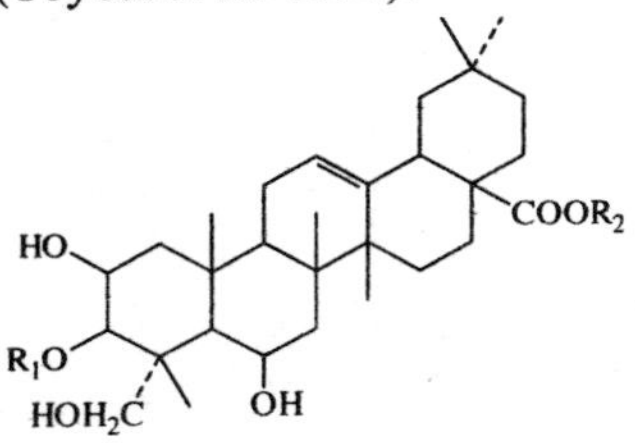

HO, R_1O, HOH_2C, OH, $COOR_2$

(**10**) R_1=Glc; R_2=-Ara^2-Rha^4-Xyl^3-Rha

(**11**) R_1=R_2=H Protobassic acid

Fungicidal activity

Due to the increasing incidence of opportunistic systemic mycoses associated with AIDS or treatment by immunosuppressive drugs, there is an urgent need to find new antifungal compounds. The study of medicinal plants from Malawi led to the isolation of compounds of diverse structures that showed toxicity activity against a phytopathogenic fungus *Cladosporium cucumerinum* in a TLC bioassay using the spores of this fungus as target organism (Homans and Fuchs 1970). For example, the triterpene glycoside Mi-saponin A (**10**) inhibited the growth of this fungus in the TLC bioassay at 1.5 μg. Its aglycone, protobassic acid **11** was also active in this test when 3.3 μg were spotted onto the TLC plate (Toyota *et al.* 1990). The naphthoquinone 7-methyl juglone (**12**) isolated from the twigs of *Diospyros usambarensis* (Marston *et al.* 1984) or from the root bark of *D. zombensis* (Gafner *et al.* 1987) is one of the most active compounds that have been so far encountered in the *C. cucumerinum* bioassay, being antifungal even at 0.025 μg.

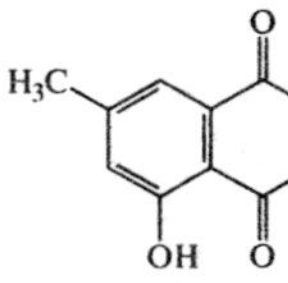

O, H_3C, OH, O

(**12**)

Synthetic antimycotic compounds like the imidazole and triazole derivatives (*i.e.* miconazole, propiconazole and sulconazole) have been tested in this bioassay as positive controls. They inhibited fungal growth at 1, 0.1 and 0.01 µg, respectively, when spotted onto the TLC plate. Azoles act through interference with ergosterol biosynthesis. Sulconazole is thought to perturb also the glucan-synthase or chitin-synthase and this might account for its efficacy against *C. cucumerinum* (Rahalison *et al.* 1994).

The fungicidal activities of different crude extracts of medicinal plants from Malawi, or of isolated compounds were also measured against human pathogenic microorganisms (Rahalison 1994).

Clerodendrum uncinatum Schinz. (Verbenaceae,

In Malawi, the powdered root bark of *Clerodendrum uncinatum* is supposed to have contraceptive activity. A decoction of these roots is also used as gargle for sore throat and this plant is claimed by the healers to cure schistosomiasis.

In a preliminary biological screening, it was found that the petroleum ether extract of these roots showed interesting fungicidal activity in the TLC bioassay using *C. cucumerinum* spores.

Isolation of the main active compound, the hydroquinone diterpene uncinatone (**13**) was achieved by column chromatography on silicagel and further crystallization. Uncinatone was shown to inhibit the growth of the fungus down to 0.5 µg in the TLC assay (Dorsaz *et al.* 1985).

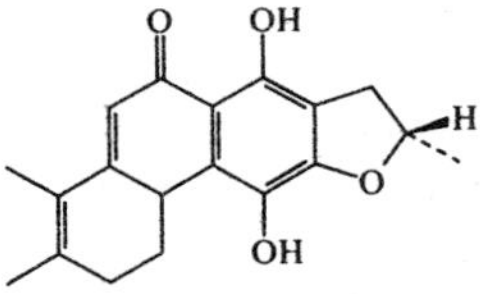

O OH H O OH

Uncinatone (**13**)

Hypericum revolutum Vahl. (Guttiferae)

Hypericum revolutum is a shrub native to South-East Africa, growing at high altitude in open mountain grassland at the margins of evergreen forest. The light petroleum ether extract of this plant was fungicidal in the TLC assay. Two new benzopyran ketones (**14, 15**) were responsible for this activity. In addition four pentacyclic dimers (**16-19**) of these products were also isolated from the lipophilic leaves and twigs extract of this plant. However, these compounds were devoid of any activity (Décosterd *et al.* 1987). From the same extract, cytotoxic products, hyperevolutin A and B (**20-21**), derivatives of the known active principle hyperforin, were also isolated (Décosterd *et al.* 1989).

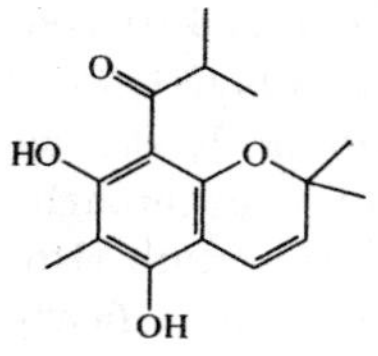

(14)

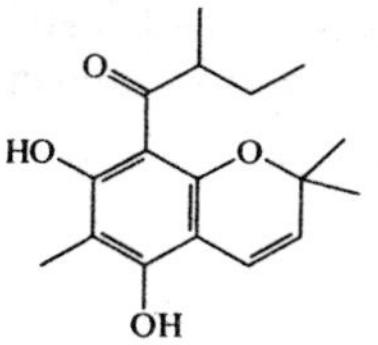

(15)

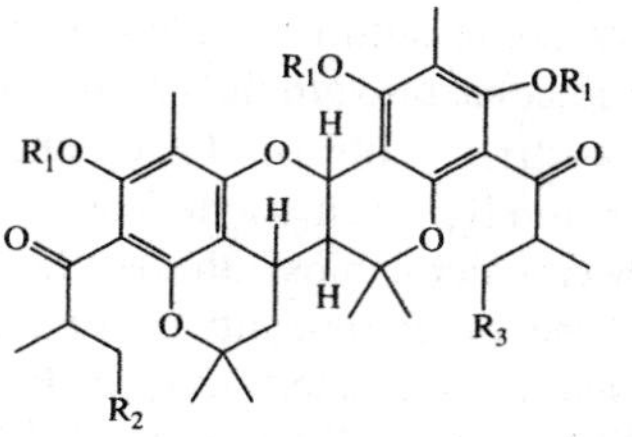

	R_1	R_2	R_3
(16)	H	H	H
(17)	H	H	CH_3
(18)	H	CH_3	H
(19)	H	CH_3	CH_3

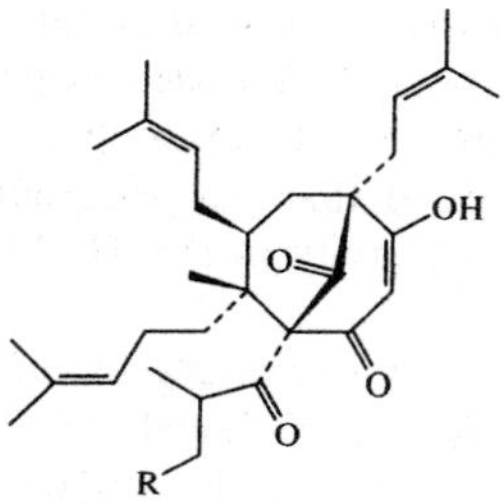

(20) R=H Hyperevolutin A

(21) R=CH_3 Hyperevolutin B

Dolichos marginata ssp. *erecta* E. Mey (Bak.) Verdc. (Leguminosae)

Some members of the genus *Dolichos* are used by the traditional healers to treat aches and pains. But the study of *Dolichos marginata* ssp. *erecta* (syn. *Sphenostylis erecta* E. Mey) was provoked by the observation that the lipophilic root extracts of this plant contained several components active in the TLC assay using the spores of *C. cucumerinum*. The major antifungal sphenostylins A-D (**22-25**) were isolated by a combination of medium and low pressure liquid chromatography.

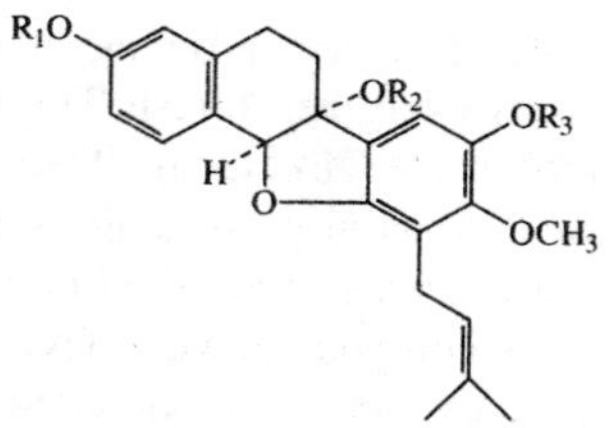

R_1=R_3=CH_3, R_2=H sphenostylin A (**22**)
R_1= R_2=R_3=H sphenostylin B (**23**)

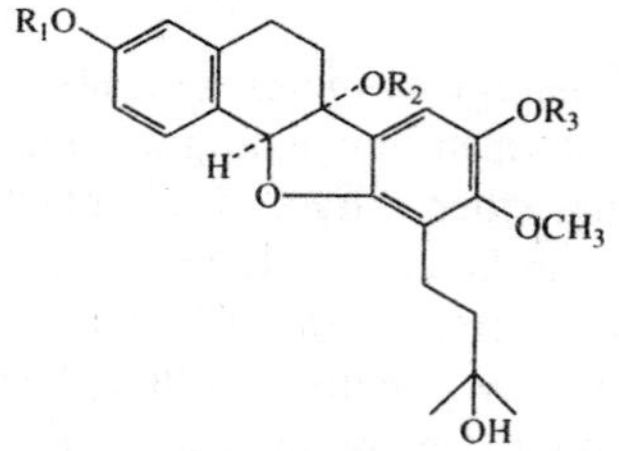

R_1= R_2=R_3=H sphenostylin C (**24**)
R_1=R_2=CH_3, R_3=H sphenostylin D (**25**)

The minimum amounts of sphenostylins A-D required to inhibit the growth of the fungus were respectively 6.25 μg, 10 μg, 50 μg and 20 μg (Gunzinger *et al.* 1988)

Helichrysum nitens Oliv. & Hiern (Asteraceae)

From the aerial parts of *Helichrysum nitens*, eight methoxylated flavonoids have been isolated (**26-33**), of which the majority exhibited strong activity against *C. cucumerinum* (Table 9.1). These compounds have been found externally deposited on the leaf and stem surfaces, suggesting that they should provide chemical barriers to the invasion of micro-organisms. Indeed, the methylated lipophilic flavonoids are especially suitable as protection against fungi and bacteria because of their ease in penetrating membranes. For this reason, the external accumulation of antifungal methylated flavones in *H. nitens* is of ecological significance (Tomas-Barberan *et al.* 1988).

Table 9.1. Antifungal epicuticular flavonoids from *H. nitens*

	R	R_1	R_2	R_3	R_4	Af.A [μg]
26	H	OCH_3		OCH_3		1
27	OCH_3	OCH_3		OCH_3		1
28	H	OCH_3	OCH_3	OCH_3		5
29	OCH_3	OCH_3	OCH_3	OCH_3		5
30	H	OCH_3	OCH_3	OCH_3	OCH_3	2
31	OCH_3	OCH_3	OCH_3	OCH_3	OCH_3	5
32	H	OH	OCH_3	OCH_3		50
33	H	OH	OCH_3	OCH_3	OCH_3	not active

Af.A: Antifungal activity: minimum quantities required to inhibit growth of spores of *C. cucumerinum* on TLC plate

This table also demonstrates how the antifungal activity shown by the fully methylated flavonoids decreases dramatically when the methyl group at position 5 is removed.

Diplolophium buchanani (Benth. ex Oliv.) Norman (Apiaceae)

Three phenylpropanoids and two furanocoumarins have been obtained from *Diplolophium buchanani* almost exclusively by centrifugal partition chromatography. Fractionation of this plant, endemic to the Zomba and Mulanje Plateaux of Malawi, afforded myristicin (**34**), elemicin (**35**), trans-isoelemicin (**36**), oxypeucedanin (**37**) and oxypeucedanin hydrate (**38**) (Table 9.2).

Table 9.2. Antifungal and larvicidal activities of compounds isolated from *Diplolophium buchanani*

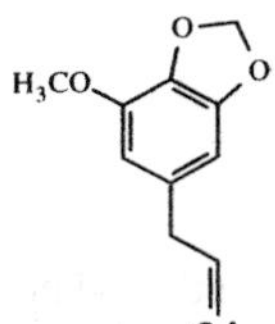

Compound	Antifungal[a] activity	Larvicidal[b] activity
34	20 µg	25 ppm
35 + 36	8 µg	100 ppm
37	1 µg	25 ppm
38	10 µg	inactive
Miconazole	1 µg	
β-Asarone		16 ppm

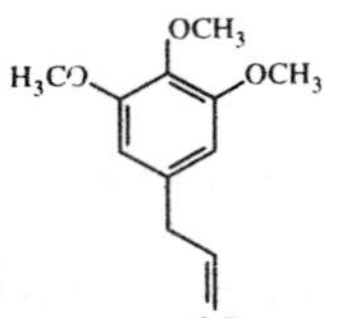

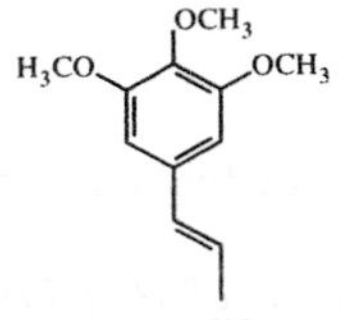

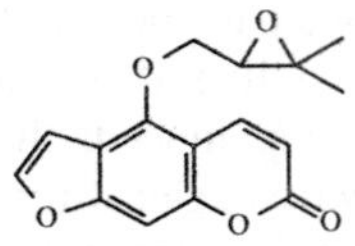

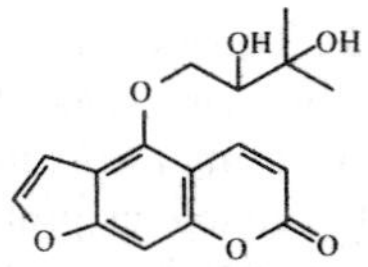

[a] minimum quantity required to inhibit growth of spores on TLC plate

[b] LD_{100} after 24 hours

Concerning the biological activities, the efficacy of the strongest antifungal compound, oxypeucedanin **37** was comparable to the commercially available fungicide miconazole. The other isolated compounds were also active but to a lesser degree. These five products were also tested in a larvicidal bioassay using *Aedes aegypti* larvae as target. Myristicin (**34**) and oxypeucedanin (**37**) were larvicidal at concentrations similar to that of the reference compound β-asarone (Marston *et al.* 1995)

Antifungal compounds isolated from medicinal plants from Malawi present various other structural features. For example, the antimalarially used leaves of *Heteromorpha trifoliata* (Umbelliferae) furnished the fungicidal polyacetylene falcarindiol (**39**) and sarisan (**40**), an isomer of myristicin after flash and low-pressure liquid chromatography (Villegas *et al.* 1988). Furthermore, a combination of different techniques of column liquid chromatography allowed the isolation of new naphthoxirene derivatives (**41-44**) and their glycosides from the fungitoxic dichloromethane extract of the root bark of *Sesamum angolense* (Pedaliaceae) (Potterat *et al.* 1987). From the same plant, Potterat *et al.* (1988) described the isolation of the inactive iridoid glucosides sesamoside (**45**), phlomiol (**46**), pulchelloside I (**47**) and 6β-hydroxypolamiide (**48**) together with the phenylpropanoid glycoside verbascoside.

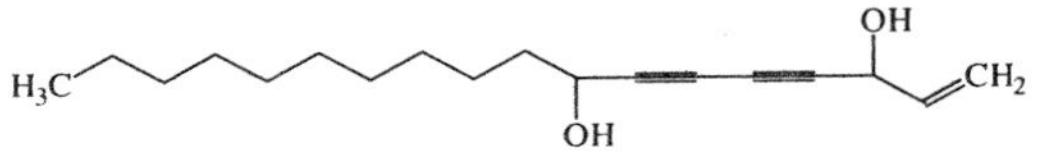

falcarindiol (**39**)

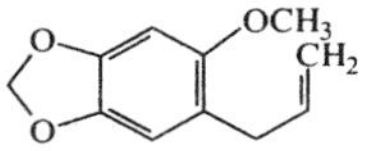

sarisan (**40**)

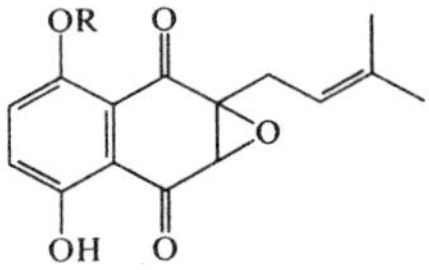

(**41**) R = H
(**42**) R = Glucose

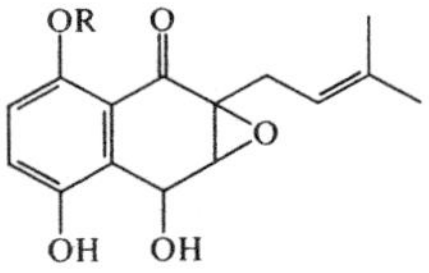

(**43**) R = H
(**44**) R = Glucose

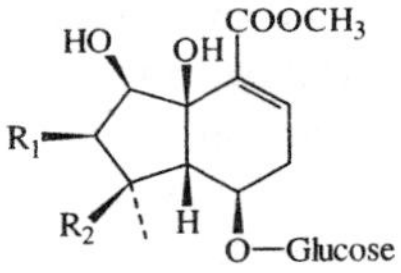

(**45**) R_1 = OH R_2 = OH
(**46**) R_1 = OH R_2 = OH
(**47**) R_1 = OH R_2 = OH

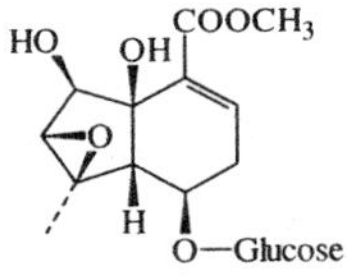

(**48**)

Finally, from the lipophilic crude extract of *Valeriana capense* (Valerianaceae), a series of antifungal valepotriates (including valtrate, isovaltrate, didrovaltrate, chlorovaltrate, valtrate hydrine B4, homovaltrate, dihomovaltrate, homodidrovaltrate, diavaltrate and isovaleroxyhydroxydidrovaltrate) was isolated by a combination of medium-pressure and semi-preparative high-pressure liquid chromatography. Although valepotriates were already isolated in the 1960's, their antifungal activity had not been described before this study (Fuzzati *et al.* 1996). In this work, valtrate was shown to possess a wide spectrum of antifungal activities, in particular against different phytopathogenic fungi (*C. cucumerinum*, *Erysiphe graminis*, etc.) where it gives similar toxicity to commercially used synthetic products.

Miscellaneous

In this section, an non-exaustive series of natural products isolated from medicinal plants from Malawi will be presented. Some of them, in particular the xanthones, are of crucial interest in the discovery new monoamine oxidase inhibitors that should eventually play a role in the management of depression. Others are interesting growth inhibitors of carcinoma cell lines and in some cases the isolation of the products was only been done from a phytochemical point of view.

Disturbances in monoamine oxidase (MAO) levels have been reported in a series of disorders (*i.e.* Parkinson's disease, Huntington's chorea, depression, anxiety, etc.). Thus, substances having modulating action on this enzyme should be of great pharmacological interest.

Xanthones from plant sources have been shown to be strong inhibitors of MAO (Suzuki *et al.* 1980, 1981; Schaufelberger and Hostettmann 1988). In order to find further active xanthones, some medicinal plants of Malawi have been investigated.

Polygala virgata Thumb. and *Polygala nyikensis* (Polygalaceae)

The genus *Polygala* (containing *ca* 500 species) is known to be a source of xanthones. *P. virgata* is a small shrub up to 1.5 m tall which is quite common on the high plateaux of Malawi. Fractionation of the dichloromethane extract of the root extract of this plant afforded different sinapoyl glycosides (Bashir *et al.* 1993), together with three new methoxylated isoflavones (**49-51**), and xanthones (**52, 53**) (Bashir *et al.* 1992).

The analysis of the lipophilic extract of the roots of *P. nyikensis*, an endemic species from the Nyika Plateau of northern Malawi, showed in preliminary screening the presence of at least two antifungal compounds. Isolation of these biologically active products afforded four simple xanthones (**54-57**), two of which were active against the plant pathogenic fungus *C. cucumerinum* at the minimal amounts of 0.6 and 0.4 μg, respectively (Marston *et al.* 1993).

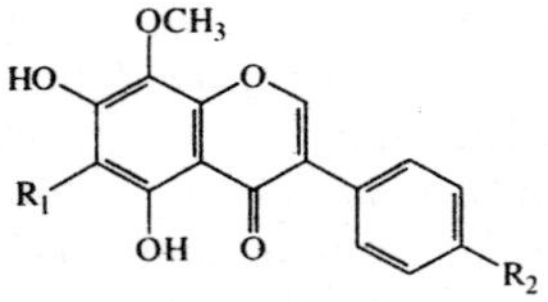

(**49**) $R_1=OCH_3$, $R_2=OH$
(**50**) $R_1=OCH_3$, $R_2=OCH_3$
(**51**) $R_1=H$, $R_2=OCH_3$

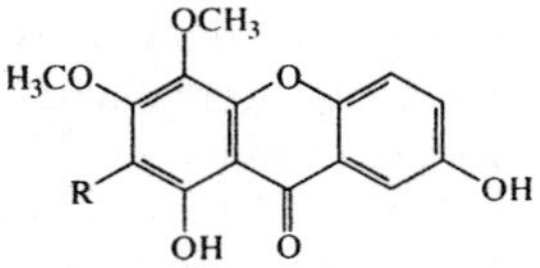

(**52**) $R=H$
(**53**) $R=OCH_3$

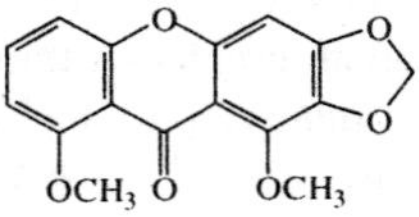

(**54**)

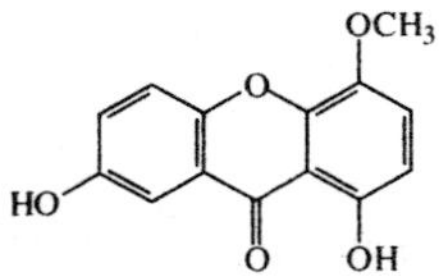

(**55**) 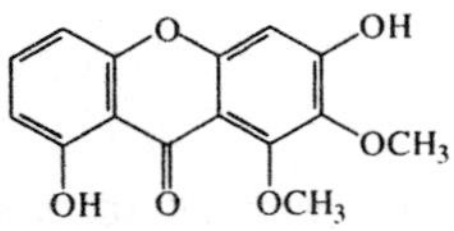 (**56**) 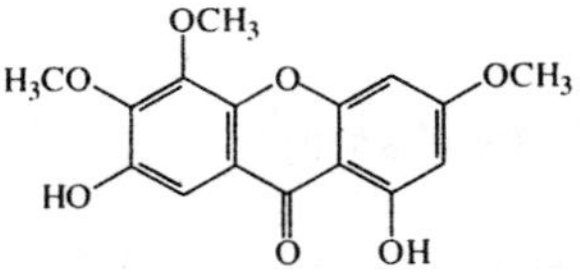 (**57**)

Ectiadiopsis oblongifolia (Meisn.) Schlecht. (Periplocaceae) and *Securidaca longipedunculata* Fresen. (Polygalaceae)

E. oblongifolia is a shrub of Western Africa, where it is used as a medicinal plant with various indications (aphrodisiac, stomach diseases, pains, etc.). Periplocaceae are classified by some authors as a distinct family, and by other as a subfamily of Asclepiadaceae.

The methanolic extract of the bark of this plant have been shown to contain two xanthone pigments (**58, 59**) (Galeffi *et al.* 1990). This discovery was of chemotaxonomic significance, owing to the absence of reports on the occurrence of this type of compound in Asclepiadaceae.

Securidaca longipedunculata is widely utilized in Africa as medicinal plant. Many healers have used this plant as a general remedy to treat different diseases and particularly to cure rheumatism, an indication which can be related to the large amount of methylsalicylate in the roots. These are quite toxic and some cases of death by introduction into the vagina have been reported. Also ingestion of the oil obtained from the seeds is reported to be fatal within a few hours.

Phytochemical investigation of the roots of *S. longipedunculata* collected in Malawi afforded one xanthone (**60**) with the rare oxygenation pattern 1,2,7 (Galeffi *et al.* 1990). Xanthones are known to be present in some representatives of the Polygalaceae family, but have only been found in the genus *Polygala.* Most of the isolated xanthones possess a typical 1,2,3 oxygenation pattern (Bashir 1993).

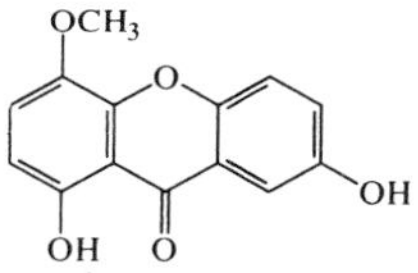

OCH_3 O OH OH O

(**58**)

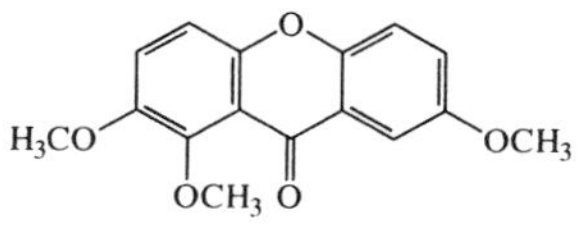

H_3CO O OCH_3 OCH_3 O

(**59**)

HO O OCH_3 OCH_3 O

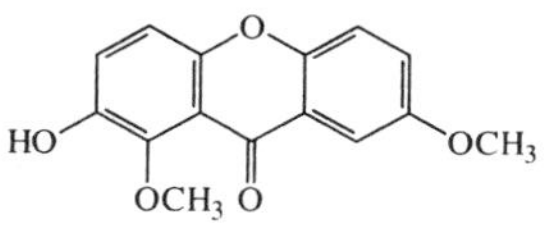

(**60**)

Chironia krebsii Griseb. (Gentianaceae)

Rpresentatives of the Gentianaceae family are also known to be natural sources of xanthones. The phytochemical investigation of the roots and aerial parts of the endemic species *Chironia krebsii* allowed the isolation and the characterization of a series of xanthone aglycones together with some glycosides (Wolfender *et al.* 1991). This work and further studies of the content of this plant, using new analytical methods including LC-MS techniques form a major part of chapter 2.

Hypoxis nyasica Bak. and *H. obtusa* Burch. (Hypoxidaceae)

The rhizomes of *Hypoxis* sp. are used in African traditional medicine for the treatment of urinary infections, prostatic hypertrophy and internal cancer (Watt and Breyer-Brandwijk 1962). Separation of the methanolic extract of the rhizomes of *H. nyasica*, using different chromatographic techniques, including liquid-liquid partition chromatography, afforded nyasoside (**62**) and the mononyasines A and B (**63**, **64**), three glucosides of nyasol (**65**), together with the known norlignan diglucoside hypoxoside (**61**) (Messana *et al.* 1989).

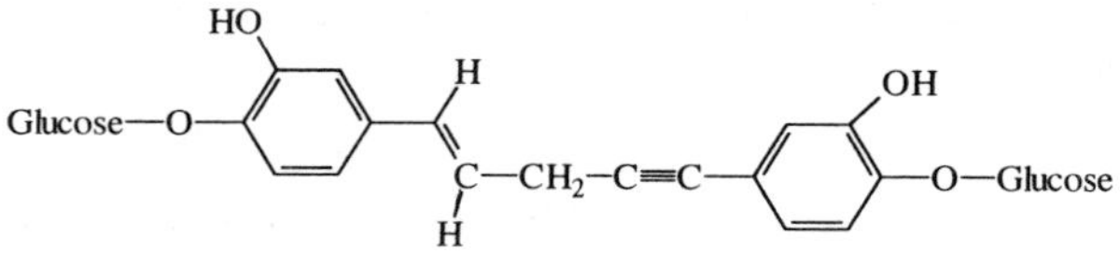

hypoxoside (**61**)

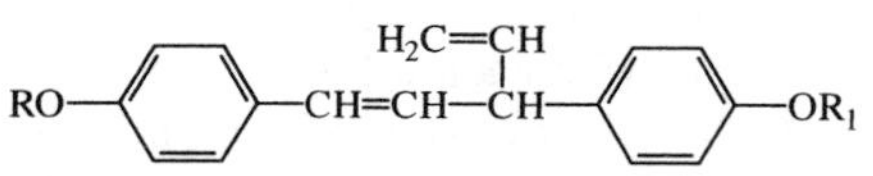

R=R$_1$=glucose: nyasoside (**62**)
R=H; R$_1$=glucose: mononyasine A (**63**)
R=H; R$_1$=glucose: mononyasine B (**64**)
R=R$_1$=H: nyasol (**65**)

The investigation of the methanolic extract of the whole fresh plant of *H. obtusa* has led to the isolation of a new phenolic glycoside named obtusaside (**66**) (Msonthi *et al.* 1990)

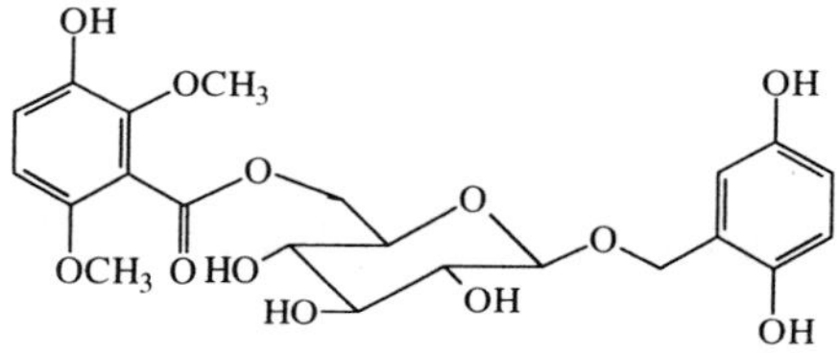

Obtusaside (**66**)

Psorospermum febrifugum Spach (Guttiferae)

Psorospermum febrifugum is a shrub which grows in many parts of Africa. The roots are used for treating wounds and the leaves and bark for skin diseases. It is supposedly a febrifuge and antileprous. Like many members of the Guttiferae, a yellow resin can be obtained from the bark and root bark.

A series of anthracene and anthraquinone derivatives (**67-71**), of which the new anthraquinone (**69**) and the new tetrahydroanthracene (**70**), were isolated from the root bark using a combination of flash-, low-pressure-chromatography on reversed phase and centrifugal-TLC. These compounds show *in vitro* cytoxic activity against the Co-115 human carcinoma cell line (Table 9.3) (Marston *et al.* 1986).

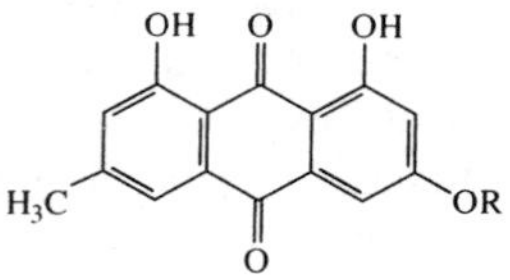

(**67**)

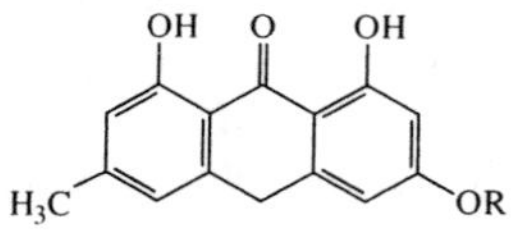

(**68**)

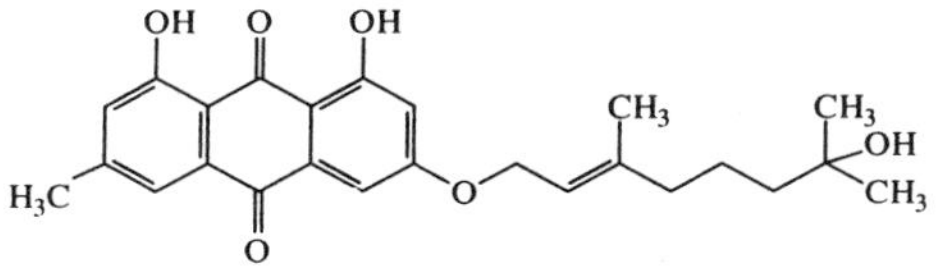

(**69**)

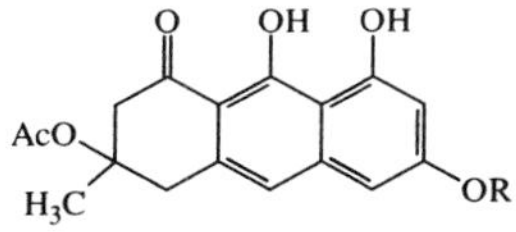

(**70**)

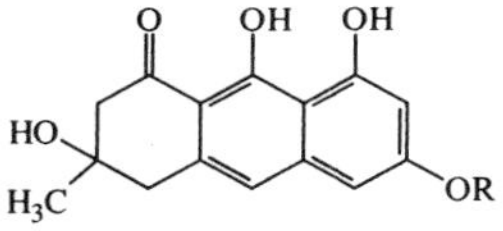

(**71**) Vismione D

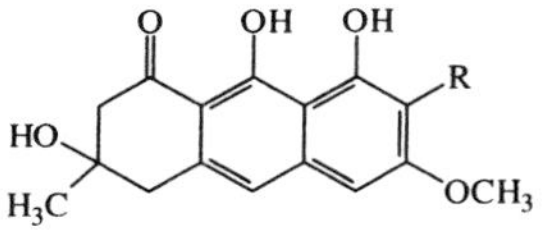

(**72**)

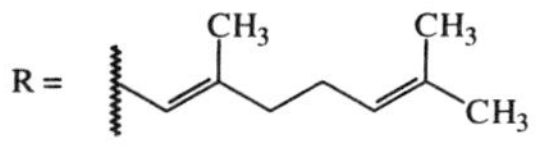

R = geranyl

Table 9.3. Cytotoxicities of *P. febrifugum* anthranoids to the human colon carcinoma cell line Co-115 after a 5 day incubation period

Compound	LD_{50} (μg/ml)
67	>10
69	>10
68	4.3
70	3.8×10^{-1}
71 (vismione D)	1.5×10^{-1}
5-Fluorouracil	6.5×10^{-2}
Vinblastine	5.5×10^{-3}

The isolated products gave varying degrees of cytotoxicity. While the two anthraquinones **67** and **69** were inactive, the two major components of the petroleum ether extract of the root bark of *P. febrifugum*, the tetrahydroanthracenes **70** and **71** exhibited major and reproducible toxicity to the Co-115 human colon carcinoma cell line. Their cytotoxicities approached the LD_{50} found for the clinically-important antitumor agent 5-fluorouracil, but they were less active than vinblastine, an antileukemic alkaloid isolated from *Catharanthus roseus* (Apocynaceae).

In a effort to isolate more important quantities of the biologically active constituents of *P. febrifugum*, a reinvestigation of the root bark has been performed. Since the separation of the anthranoid constituents by flash chromatography and low pressure reversed-phase chromatography resulted in considerable loss of material, centrifugal partition chromatography (CPC) was used. Thus, in a single CPC step (Sanki Cartridge system), three pure compounds **65**, **67** and **71** and a mixture of two anthranoid pigments (**71** and the minor component **72**) were obtained without any loss of material. A non-aqueous solvent system was used for their separation and a 100 mg sample of crude extract was separated within four hours using the upper phase of the solvent system nC_6H_{14}-MeCN-MeOH (40:25:10) as mobile phase (Marston *et al.* 1988).

Acanthospermum hispidum DC. (Asteraceae)

Traditional medicinal uses of *Acanthospermum hispidum* are numerous. It is used to treat stomach complaints, wounds, migraine. In Ivory Coast, a decoction is drunk as a purgative and counter-poison and an aqueous macerate is drunk and put into baths for arthritis and rheumatism. *Acanthospermum* species contain the characteristic canthospermolides, that show *in vitro* and *in vivo* anticancer activity. *Acanthospermum hispidum* has been investigated to look for more lactones of this type.

Phytochemical investigation of the aerial parts of *A. hispidum* led to the isolation of 10 sesquiterpene lactones (**73-82**), three of which are new naturally occurring compounds: 4*E*-acanthospermolide **73** and the 4*Z*-derivatives **76** and **82** (Jakupovic *et al.* 1986).

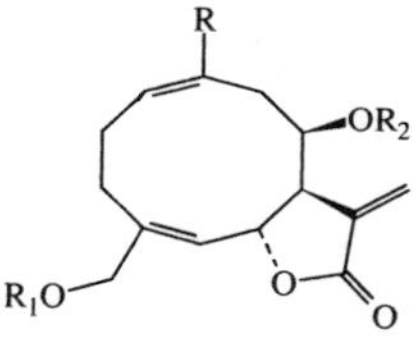

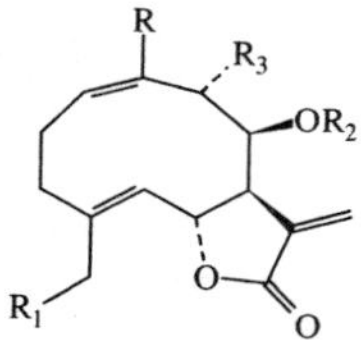

	R	R_1	R_2
73	CH_2OH	Ac	MeBu
74	CHO	H	MeBu
75	CHO	H	*i*Val

	R	R_1	R_2	R_3
76	CH_2OH	OH	*i*Bu	OAc
77	CH_2OH	OH	MeBu	OAc
78	CHO	H	MeBu	OAc
79	CHO	OH	MeBu	H
80	CHO	OH	*i*Val	H
81	CHO	OH	MeBu	OAc
82	CHO	OH	MeBu	OMe

Conclusion

As the above mentioned examples indicate, medicinal plants from Malawi can be considered as an important source of new biologically active compounds, with a wide range of pharmacological and therapeutical properties.

With the present socio-economic, socio-political and demographic constraints of the Third World, it is imperative that whatever research is performed must be converted from purely academic publications into a more practical utilization of the biologically active compounds so far isolated. It is indeed the role of governments, together with universities, pharmaceutical companies and related private industries to jointly promote collaborative development of such bioactive compounds for medical use both locally, nationally and for export.

Acknowledgments

One of the author (JD.M.) wishes to thank the IOCD chairman (Plant Chemistry) for inviting him to the symposium and all the members of the three institutions who have helped his research.

References

Bashir, A. (1993). *Isolation and Characterization of polyphenols from two species of the Polygalaceae family: Moninia sylvatica and Polygala virgata*. Ph.D. Thesis. University of Lausanne, Switzerland.

Bashir, A., Hamburger, M., Msonthi, J.D., and Hostettmann, K. (1992). Isoflavones and xanthones from *Polygala virgata*. *Phytochemistry* **31**, 309-311.

Bashir, A., Hamburger, M., Msonthi, J.D., and Hostettmann, K. (1993). Sinapic acid esters from *Polygala virgata*. *Phytochemistry* **32**, 741-745.

Décosterd, L.A., Stoeckli-Evans, H., Msonthi, J.D., and Hostettmann, K. (1987). New antifungal chromenyl ketones and their pentacyclic dimers from *Hypericum revolutum* Vahl. *Helvetica Chimica Acta* **70**, 1694-1702.

Décosterd, L.A., Stoeckli-Evans, H., Chapuis, J.-C., Msonthi, J.D., Sordat, B., and Hostettmann, K. (1989). New hyperforin derivatives from *Hypericum revolutum* Vahl. with growth-inhibitory activity against a human colon carcinoma cell line. *Helvetica Chimica Acta* **72**, 464-471

Dorsaz, A.-C., Marston, A., Stoeckli-Evans, H., Msonthi, J.D., and Hostettmann, K. (1985). Uncinatone, a new antifungal hydroquinone diterpenoid from *Clerodendrum uncinatum* Schinz. *Helvetica Chimica Acta* **68**, 1605-1610.

Fuzzati, N., Wolfender, J.L., Hostettmann, K., Msonthi, J.D., Mavi, S., and Molleyres, L.P. (1996). Isolation of antifungal valepotriates from *Valeriana capense* and the search for valepotriates in crude Valerianaceae extracts. *Phytochemical Analysis* **7**,76-85.

Gafner, F., Msonthi, J.D., and Hostettmann, K. (1985). Phytochemistry of African medicinal plants: Part. 3. Molluscicidal saponins from *Talinum tenuissimum* Dinter. *Helvetica Chimica Acta* **63**, 606-609.

Gafner, F., Chapuis, J.-C., Msonthi, J.D., and Hostettmann, K. (1987). Cytotoxic naphthoquinones, molluscicidal saponins and flavonols from *Diospyros zombensis*. *Phytochemistry* **26**, 2501-2503.

Galeffi, C., Federici, E., Msonthi, J.D., Marini-Bettolo, G.B., and Nicoletti, M. (1990). New xanthones from *Ectiadiopsis oblongifolia* and *Securidaca longipedunculata*. *Fitoterapia* **61**, 79-81.

Gunzinger, J., Msonthi, J.D., and Hostettmann, K. (1986). Molluscicidal saponins from *Cussonia spicata*. *Phytochemistry* **25**, 2501-2503.

Gunzinger, J., Msonthi, J.D., and Hostettmann, K. (1988). New pterocarpinoids from *Dolichos marginata* ssp. *erecta*. *Helvetica Chimica Acta* **71**, 72-76.

Homans, A.L. and Fuchs, A. (1970). Direct bioautography on thin-layer chromatograms as a method for detecting fungitoxic substances. *Journal of Chromatography* **51**, 325-327.

Hostettmann, K. and Marston, A. (1987). Plant molluscicide - An update. In *Plant Molluscicides* (ed. K.E. Mott), pp.299-320. Wiley & Sons Ltd., Chichester.

Jakupovic, J., Baruah, R.N., Bohlmann, F., and Msonthi, J.D. (1986). Further acanthospermolides from *Acanthospermum hispidum*. *Planta Medica* **52**, 154-155.

Marston, A., Msonthi, J.D., and Hostettmann, K. (1984). Naphthoquinones of *Diospyros usambarensis*: their molluscicidal and fungicidal activities. *Planta Medica* **50**, 279-280.

Marston, A., Chapuis, J.-C., Sordat, B., Msonthi, J.D., and Hostettmann, K. (1986). Anthracenic derivatives from *Psorospermum febrifugum* and their *in vitro* cytotoxicities to a human colon carcinoma cell line. *Planta Medica* **52**, 207-210.

Marston, A., Potterat, O., and Hostettmann, K. (1988). Isolation of biologically active plant constituents by liquid chromatography. *Journal of Chromatography* **450**, 3-11.

Marston, A., Hamburger, M., Sordat-Diserens, I., Msonthi, J.D., and Hostettmann, K. (1993). Xanthones from *Polygala nyikensis*. *Phytochemistry* **4**, 809-812.

Marston, A., Msonthi, J.D., and Hostettmann, K. (1995). Isolation of antifungal and larvicidal constituents of *Diplolophium buchanani* by centrifugal partition chromatography. *Journal of Natural Products* **58**, 128-130.

Messana, I., Msonthi, J.D., De Vicente, Y., Multari, G., and Galeffi, C. (1989). Mononyasine A and mononyasine B: Two glucosides from *Hypoxis nyasica*. *Phytochemistry* **28**, 2807-2809.

Msonthi, J.D., Toyota, M., Marston, A., and Hostettmann, K. (1990). A phenolic compound from *Hypoxis obtusa*. *Phytochemistry* 29, 3977-3979.

Potterat, O., Stoeckli-Evans, H., Msonthi, J.D., and Hostettmann, K. (1987). Two new antifungal naphthoxirene derivatives and their glucosides from *Sesamum angolense* Welw. *Helvetica Chimica Acta* **70**, 1551-1557.

Potterat, O., Msonthi, J.D., and Hostettmann, K. (1988). Four iridoid glucosides and a phenylpropanoid glycoside from *Sesamum angolense*. *Phytochemistry* **27**, 2677-2679.

Rahalison, L. (1994). *Mise au point et applications d'une méthode de dépistage d'activité antifongique (Candida albicans) dans des extraits végétaux.* Ph.D. Thesis. University of Lausanne, Switzerland.

Rahalison, L., Hamburger, M., Monod, M., Frenk, E., and Hostettmann, K. (1994). Antifungal tests in phytochemical investigations: Comparison of bioautographic methods using phytopathogenic and Human pathogenic fungi. *Planta* Medica **60**, 41-44.

Tomas-Barberan, F.A., Msonthi, J.D., and Hostettmann, K. (1988). Antifungal epicuticular methylated flavonoids from *Helichrysum nitens*. *Phytochemistry* **27**, 753-755.

Toyota, M., Msonthi, J.D., and Hostettmann, K. (1990). A molluscicidal and antifungal triterpenoid saponin from the roots of *Clerodendrum wildii*. *Phytochemistry* **29**, 2849-2851.

Villegas, M., Vargas, D., Msonthi, J.D., Marston, A., and Hostettmann, K. (1988). Isolation of the antifungal compounds falcarindiol and sarisan from *Heteromorpha trifoliata*. *Planta Medica* **54**, 36-37.

Watt, M.J. and Breyer-Brandwijk, M.G. (1962). The Medicinal and Poisonous Plants of Southern and Eastern Africa, 2nd edn. E. & S. Livingstone Ltd., London.

Wolfender, J.-L., Hamburger, M., Msonthi, J.D., and Hostettmann, K. (1991). Xanthones from *Chironia krebsii*. *Phytochemistry* **30**, 3625-3629.

10. The Burseraceae and their resins: ethnobotany and serendipity

P.G. WATERMAN

Strathclyde Institute for Drug Research, and The Phytochemistry Research Laboratories, Department of Pharmaceutical Sciences, University of Strathclyde, Glasgow G1 1XW, Scotland, U.K.

Introduction

In the semi arid bushlands of East Africa and South-West Africa the Burseraceae form a major component of the tree flora. In particular the *Acacia-Commiphora* or 'Nyika' woodlands of northern Kenya, southern Ethiopia and Somalia, a vegetation type dominated by the genera *Acacia* (Leguminosae) and *Commiphora*, and to a lesser extent *Boswellia* (both Burseraceae), represents a major floristic unit, about which we still know relatively little.

I initially became interested in the Burseraceae through my participation in the National Museums of Kenya and Royal Geographical Society 'Ecological Inventory of the Kora National Reserve, Kenya, 1982-1985' (Coe and Collins 1986). Of the numerous species of *Commiphora* and one *Boswellia* species recorded in the Reserve most were conspicuous for the production of resinous exudates the characteristics of which (colour, texture, volatile content, water solubility) varied considerable from species to species. Several of the resins produced by the family have long been and still remain significant items of commerce; notably myrrh, a product of *Commiphora* species, and frankincense, from *Boswellia.* Although Kenya is not a significant producer of either of these products a number of the resins are collected and traded and some of these appeared close to myrrh in form and, perhaps, in chemistry. However, there was, and still is, virtually no quality control on what is collected and our attempts to understand what was being collected and marketed as part of the Kora survey only confused us! The starting point of our interest in the family was, therefore, a desire to investigate what the resins of commerce offered in and around the Kora National Reserve. From that beginning a growing interest in a number of the other resins we encountered inevitably followed.

The 'myrrh-like' resins of Kenya

It is now generally accepted that the origin of myrrh or 'molmol' in Somalia is *Commiphora myrrha* (Nees) Engl. while a second product known as scented myrrh (bissabol *or habak hadi*) is now considered to arise from *C. guidotii* Choiv. and not from *C. erythraea* (Ehrenb.) Engl. which is now considered to be a synonym of *C. holtziana* Engl. (Thulin and Claeson 1991). Various studies have shown myrrh to contain furanosesquiterpenes (see for example Brieskorn and Noble 1982) and *C. guidotti* had also been found to contain a furanoelemane (Craveiro *et al.* 1983).

We had the opportunity to investigate a series of resins attributable *to C. holtziana* (4 samples) and to *C. myrrha* (4 samples, including one from a European commercial source) of which all except the commercial sample were collected in Kenya. The range of compounds isolated (Provan *et al.* 1987) are shown in Figure 10.1.; the furanogermacranes, furanoeudesmanes and furanoelemanes being the most striking products and diagnostic of myrrh type resins. Chemical variability between all the samples investigated was considerable. Three of the four samples from *C. myrrha* showed the presence of all three furanosesquiterpene types while this was true of only one of the samples from *C. holtziana*. The inconclusive nature of the study revealed the complexity of the problem facing any who attempt to use chemical parameters to quality control these materials.

In a further study on the resin of a sample of *C. holtziana* collected from near the township of Garba Tula, Kenya, a further novel furanogermacrane (**1**) was obtained as a result of a bioassay-guided chase based on weak anti-bacterial activity (Cavanagh *et al.* 1993). Indeed all of the furanosesquiterpenes present in these resons appears to exhibit some antibiotic activity.

Very recently Dolara *et al.* (1995) have reported that the furanoelemane (**2**) and the furanoeudesmane (**3**), isolated from myrrh, exhibited analgesic activity which was shown to be caused by their interaction with the brain opioid receptor mechanism. Thus, despite its use for thousands of years myrrh is still being found to have new bioactivities.

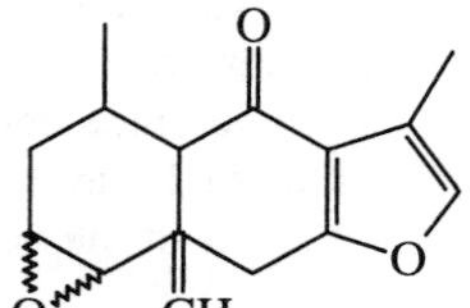

O CH$_2$ O O

(**1**) 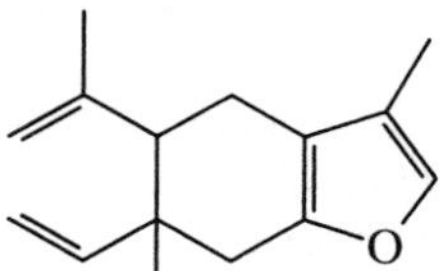 (**2**) 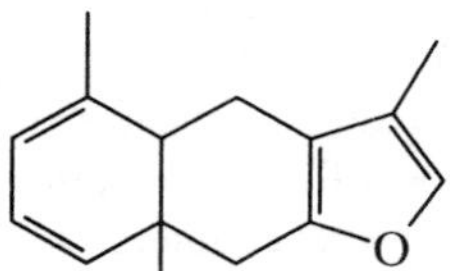 (**3**)

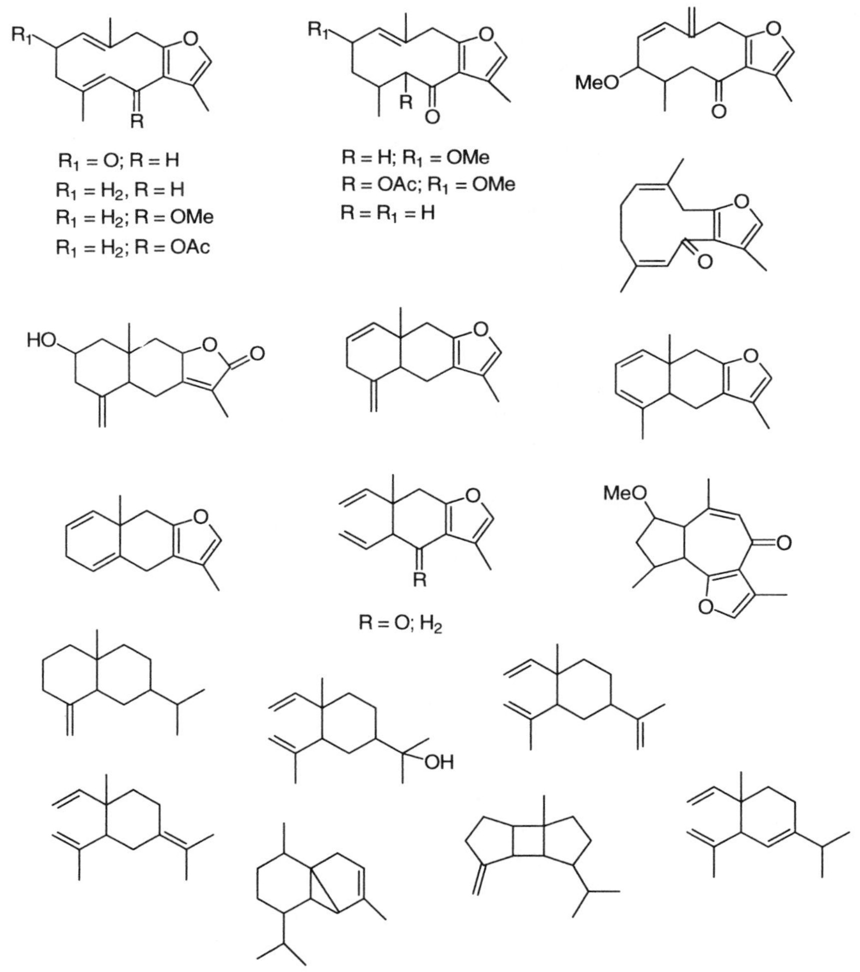

$R_1 = O; R = H$
$R_1 = H_2, R = H$
$R_1 = H_2; R = OMe$
$R_1 = H_2; R = OAc$

$R = H; R_1 = OMe$
$R = OAc; R_1 = OMe$
$R = R_1 = H$

$R = O; H_2$

Fig. 10.1. Typical sesquiterpenes isolated from myrrh-like resins of Kenyan origin.

Highly volatile resin oils from *Commiphora* species

For anyone 'sampling' the *Commiphora* species of northern Kenya one of the most readily recognised is *C. rostrata* Engl. This small tree has an attractive smooth grey bark and at certain times of the year is covered, for a short period, with bright green acid-tasting leaves. The fruit is atypical of the genus. But what really makes this species stand out is the copious amounts of highly liquid resin which lies just under the surface of the bark and can be released by bending twigs causing it to spurt out. This resin can be harvested, rather like rubber, by tapping the major branches or trunk.

The resin gives off an unpleasant odour and rapidly dries to form a 'skin' on the surface around the area where damage has occurred. An analysis of the resin revealed a simple composition made up of a series of methylalkyl ketones with 2-decanone as the major component with lesser but appreciable amounts of 2-undecanone and 2-dodecanone (McDowell *et al.* 1988). McDowell *et al.* have proposed a protective function for this material. Some antifungal activity was demonstrated and it was noted that this species was unusually free from damage by wood-boring insects or browsing mammals. Some local field trials demonstrated that the resin tended to repel ants and termites from the vicinity and there is an appreciable literature (cited by McDowell *et al.* 1988) demonstrating that similar compounds are feeding deterrents to insects.

Currently we know of no other species of *Commiphora* producing a comparable resin. Two species collected in Southern Ethiopia, *C. terebinthina* Vollesen and *C. cyclophylla* Chiov., initially excited interest in that they also yielded highly volatile oils that could be collected in the same manner. However, in both of these species the resin produced proved to be largely terpenoid in nature, the resin oil containing about 50% limonene (Abegaz *et al.* 1989) for both species.

The anti-inflammatory resin of *Commiphora kua*

During studies in Kora National Reserve one of the most conspicuous resin-producing species was a small intensely spiny tree with a grey exfoliating bark. This species, which was provisionally identified as *Commiphora incisa* Chiov on the basis of sterile collections produced, when cut, copious amounts of a yellow viscous resin which rapidly hardened to give large lumps of material. At the time nothing was known about the chemistry of this species and local people did not indicate any uses, other than as an emergency cure for holes in car radiators.

Chemical analysis of the resin first yielded two lignans, polygamain (**4**) and polypicrogamain (**5**) (Provan and Waterman 1985) and then a series of three very unusual triterpene derivatives which were based on the dammarane skeleton but with the C-17 side chain missing. The structure elucidation of mansumbinone (**6**) made extensive use of NMR. The ^{1}H NMR spectrum revealed the typical -CH_2-CH_2-C(O)-$C(Me)_2$ spin system of a 3-oxotriterpene and another system derived from -CH-CH=CH-CH_2-C(Me). The later was interesting in showing long-range W-bond coupling between the methylene and the methyl, a feature often observed between the C-15 protons and C-14 methyl of tetracyclic triterpenes. The position of the double bond was fixed at C-16/C-17 by first oxidising it to the epoxide and then ring-opening to form the corresponding dicarboxylic acid (Scheme 10.1).

This resolved the structure of **6** and the corresponding alcohol, mansumbinol (**7**) and the A-ring opened derivative mansumbinoic acid (**8**) followed.

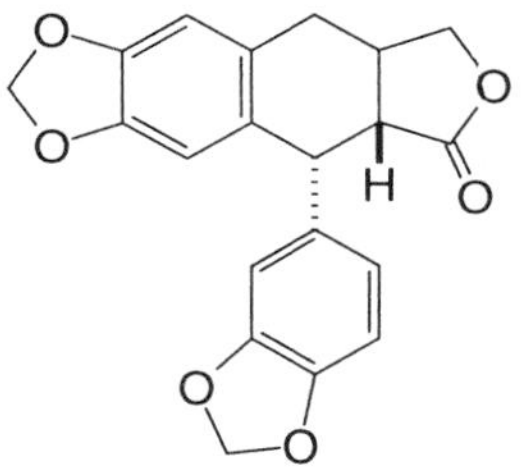

polygamain (**4**)

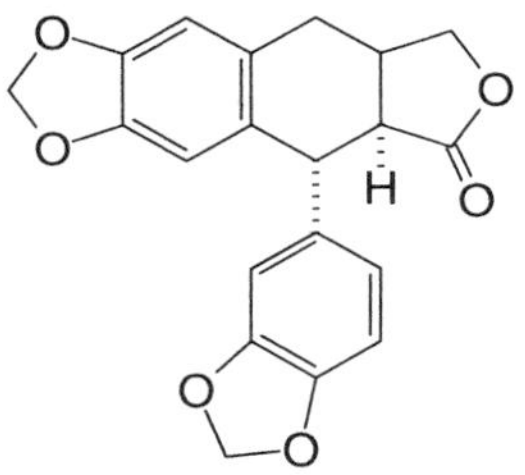

picropolygamain (**5**)

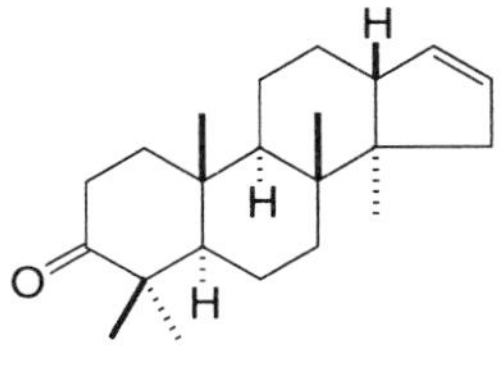

mansumbinone (**6**)

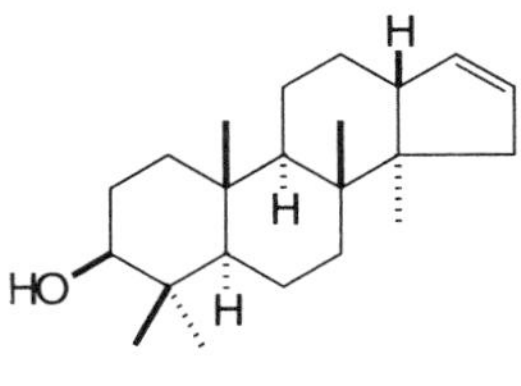

mansumbinol (**7**)

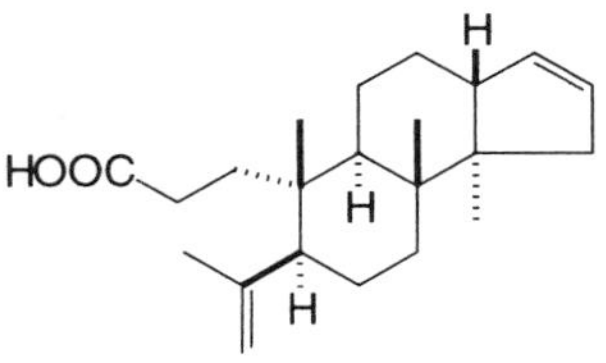

mansumbinoic acid (**8**)

The mansumbinones obviously must derive from a normal dammarane triterpene with the loss of the side-chain. A minor derivative isolated from this resin was 16β,20α-dihydroxydammar-24-en-3-one (**9**) (Provan and Waterman 1985) seemed to be a plausible precursor. This was confirmed by removal of the side chain using sodium ethoxide, which resulted in the formation of mansumbinone in a 25% yield (see below).

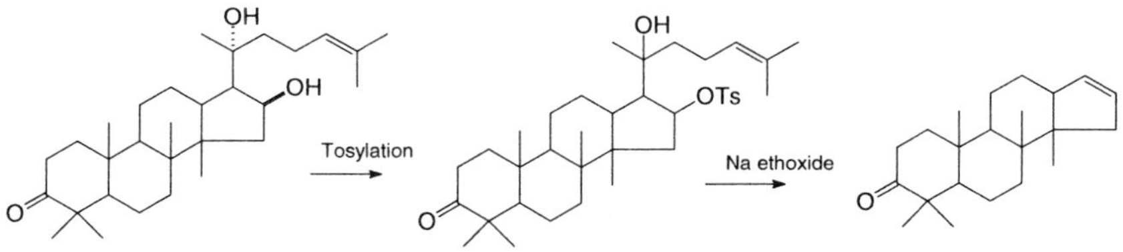

Tosylation

Na ethoxide

(**9**)

(**6**, 25%)

A further collection identified as *C. incisa*, and this time based on fertile material, was obtained in 1987. To our consternation an analysis of the morphologically very similar resin yielded different types of triterpenes, notably 1α-acetoxy-3β-hydroxycycloart-24-ene (**10**) and 1α,2α,3β-trihydroxy-30-nor-lanost-8,24-diene (**11**), both of which were novel (Provan and Waterman 1988) but no trace of the octanordammaranes. By contrast material collected at the same time and attributable to *C. kua* Vollesen was found to yield the mansumbinane octanordammaranes. Subsequent analysis of several collections from both species confirmed that the original identification based on sterile material was incorrect and the true source of the mansumbinone type triterpenes was *C. kua*.

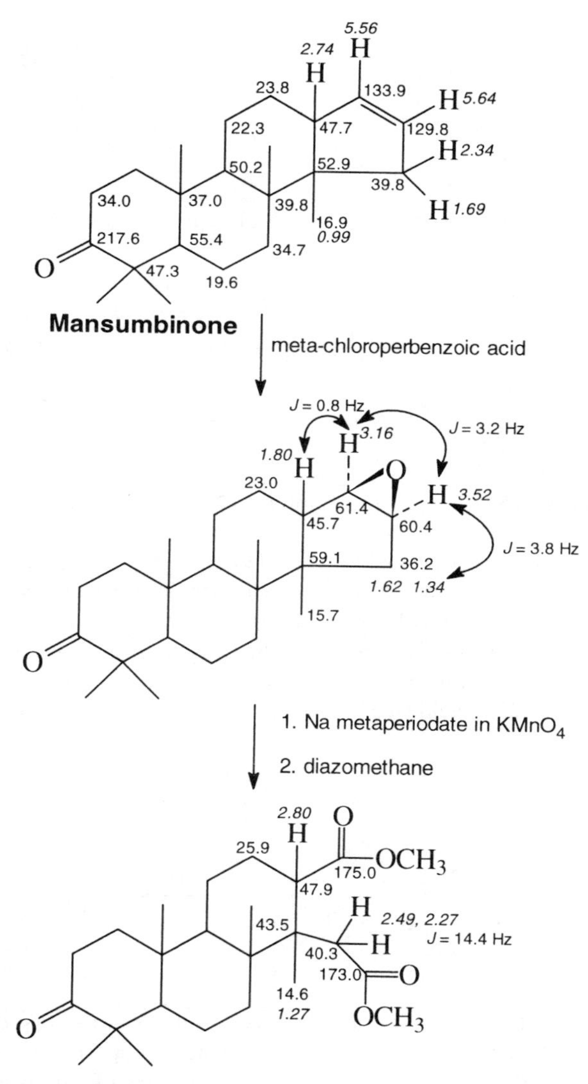

(plain text: ^{13}C-NMR chemical shifts; italic: ^{1}H-NMR chemical shifts)

Scheme 10.1. Chemical modification used to support the proposed structure for mansumbinone giving relevant NMR chemical shift data.

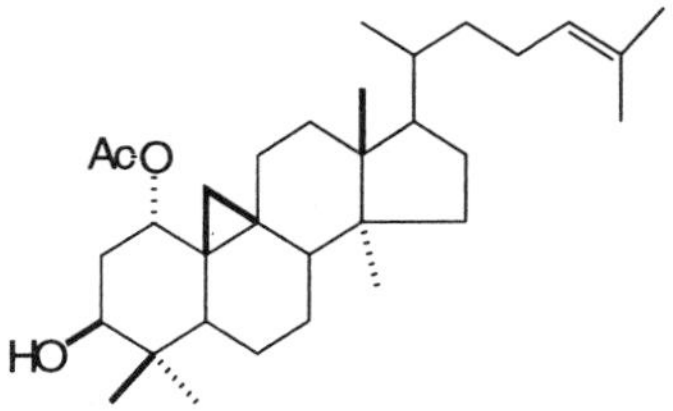

(10)

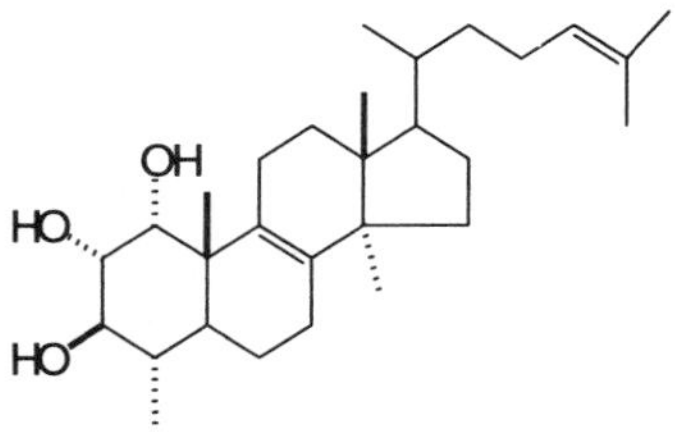

(11)

As well as the three major octanordammaranes isolated from the resin the stem bark of *C. kua* contained several other compounds in smaller amounts, some of which were rather unstable. These were eventually isolated and characterised and proved to be 16-oxygenated octanordammaranes with a double bond between C-13 and C-17 (Provan *et al.* 1992). The presence of oxygen at C-16 was reminiscent of **9**. The most stable of these minor compounds was mansumb-13(17)-en-3,16-dione (**12**) while a series of 16ξ-hydroperoxy compounds, typified by (**13**), proved to be unstable.

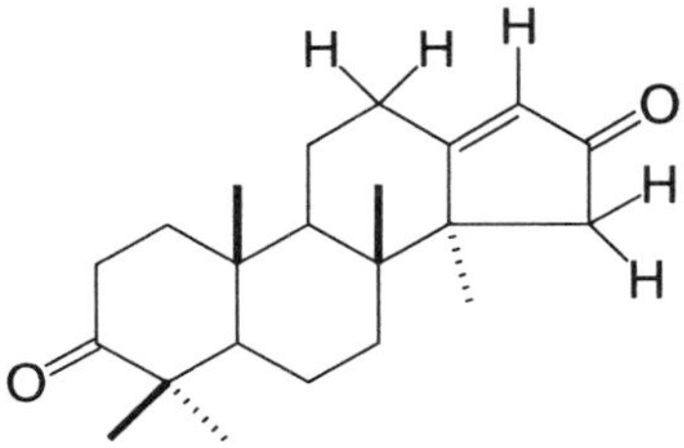

(12)

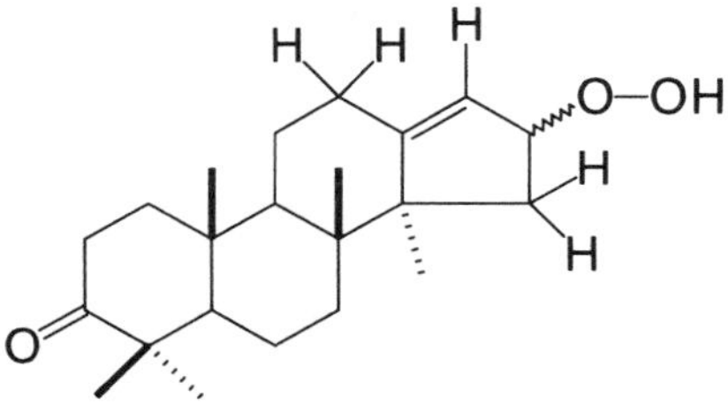

(13)

The anti-inflammatory activity of C. kua resin

The oleo-resin of the Indian species *Commiphora mukul* Engl. (gum guggul or Indian myrrh) is known to exhibit significant anti-arthritic and anti-inflammatory activity and the acidic fraction of an aqueous extract was found to suppress carrageenan-induced acute rat-paw oedema and to inhibit the formation of secondary lesions in adjuvant arhtritis (Oliver-Bever 1986). Side effects from the resin were reported to be negligible in comparison with those due to β-methasone and the activity was eventually traced to steroidal compounds (Gujral *et al.* 1960; Arora *et al.* 1972).

Given the activity of this Indian resin Oliver-Bever (1986) suggested that an examination of the African products for similar activity should be undertaken. Accordingly we undertook the analysis of a number of resins which were to hand, comparing them with a sample of gum guggul (Duwiejua *et al.* 1993). To our surprise the aqueous ethanolic extract of *C. kua* proved to have an activity in the rat paw oedema assay that was as strong as that of gum guggul and as a result be undertook a bioassay-guided separation to determine the active compounds.

The anti-inflammatory activity of the resided lay almost entirely in the two previously reported compounds mansumbinone (**6**) and mansumbinoic acid (**8**). Figure 10.2a. shows the time course of an experiment in which mansumbinone or mansumbinoic acid were administered at a dose level of 2.5 x 10^{-4} mol kg^{-1} and oedema was induced by injection of carrageenan after 2 hours (Duwiejua *et al.* 1993). While mansumbinone did show some reduction in oedema when compared with control animals the effect was far greater with mansumbinoic acid (Fig. 10.2b.) which gave a 48% reduction in swelling.

a)

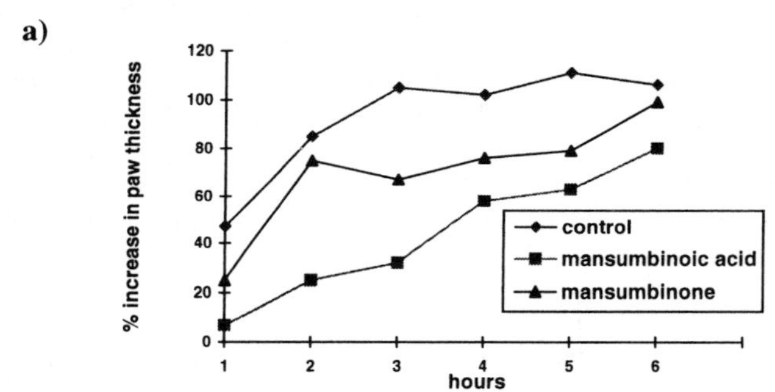

Fig. 10.2a. The effect of 2.5 x 10^{-4} mol kg-1 of mansumbinone and mansumbinoic acid on the carageenan-induced rat paw oedema (time course of increase in paw thickness).

b)

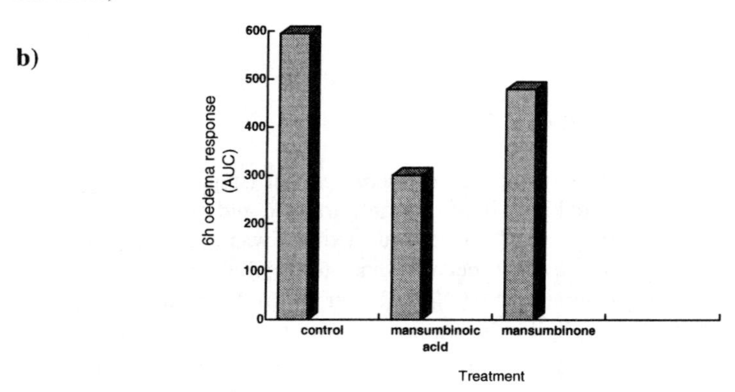

Fig. 10.2b. Total oedema response (in Fig. 2a) expressed as area under curves.

A dose-response curve for mansumbinoic acid suggested an approximate Ed_{max50} of 1.5 x 10^{-4} mol kg^{-1} and the activity was estimated to be within an order of magnitude of that of indomethicin while mansumbinone was more than 20 times less potent. The therapeutic potential of mansumbinoic was evident from its ability to significantly reverse the oedema induced in rats by injection of a dose of 2.5 x 10^{-4} mol kg^{-1} two hours after the oedema had been induced by carageenan (Fig. 10.3.).

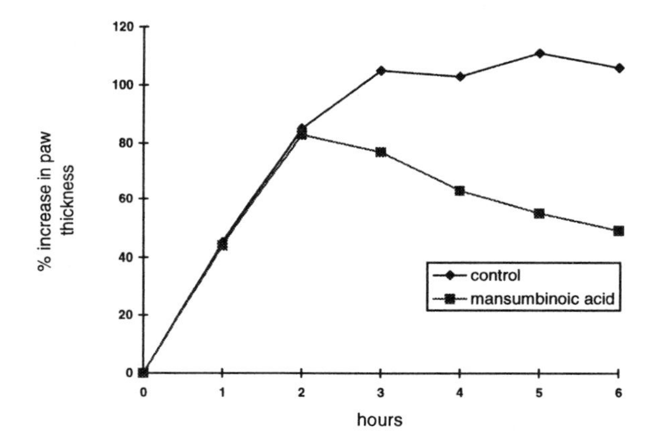

Fig. 10.3. The influence of 2.5 x 10^{-4} mol kg^{-1} of mansumbinoic acid administered p.o. 2 h after carrageenan injection on the time course of established oedema. Results presented as mean n = 6.

The activity of mansumbinoic acid against adjuvant arthritis was also studied. This cell-mediated chronic inflammatory response was induced by the standard method (using Freund's adjuvant) to rats which had been pre-loaded with either indomethicin or mansumbinoic acid at a dose of 1.5 x 10^{-4} mol kg^{-1}. Again the compound showed itself capable of significant reduction in joint swelling, in both the injected and contralateral limbs which indicated that it had potential anti-arthritic value as well as anti-inflammatory activity.

As a result of this investigation it was concluded that the resin of *C. kua*, and in particular the compound mansumbinoic acid (**8**), was an effective anti-inflammatory with a marginally greater activity than guggul resin from *C. mukul.* Other dammarane triterpenes (cabraleadiol and congeners, isofouquierone), which had previously been isolated from *Commiphora dalzielii* Hutch. collected in Ghana (Waterman and Ampofo 1985) showed relatively little activity.

Studies on other African genera of Burseraceae

While *Commiphora* is the most diverse genus of Burseraceae in Africa it is confined to arid or semi-arid areas. Several other genera are found in the rain forests of west Africa and stimulated by our findings with *C. kua* we have looked for opportunities to study these. Our investigations to date have yielded a diverse range of triterpenes but nothing with activity comparable to mansumbinoic acid.

Santiria trimera (Oliv.) Aubrev. collected in Cameroon proved to be a source of lanostane derivatives including the unusual 6β-acetoxy-3,23-dioxo-9β,20β-lanost-7, 24-dien-26-oic acid (**14**), the relative stereochemistry of which was established by an X-ray study (da Silva *et al.* 1990). The resin of *Aucoumea klaineana* Pierre from Gabon likewise yielded triterpenes with a 7/8 centre of unsaturation but with the different stereochemistry at C-9, C-13, C-14 and C-17 of the tirucallane skeleton (Guang-Li *et al.* 1988; 1989); for example 3,23-dioxo-22α-hydroxy-9α,20α-tirucall-7,24-diene (**15**). The tirucallane skeleton is the starting point for the proliferation of limonoid and quassinoid compounds in the Meliaceae, Rutaceae and Simaroubaceae, families to which the Burseraceae is thought to be allied.

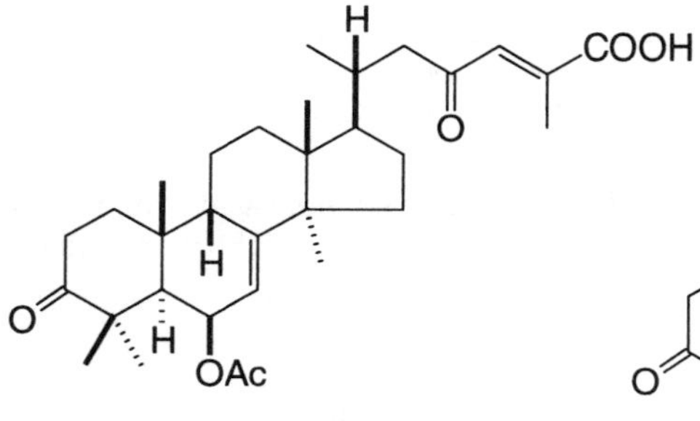

H COOH O H O H OAc

(**14**)

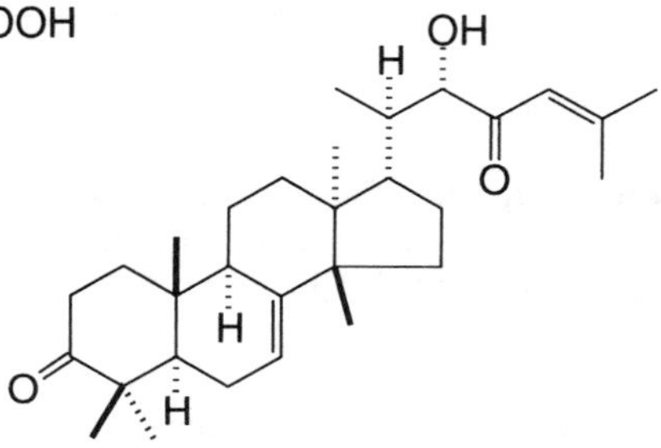

H OH O O H H

(**15**)

The resin from *Dacryodes normandii* Aubrev. et Pellegr. originating from Gabon proved to be chemically more mundane in yielding only ursane and oleanane triterpenes. The two most interesting compounds were 21-oxo-3,4-*seco*-olean-12-en-4(23), 12 dien-3-oic acid and the corresponding ursane (Parsons *et al.* 1991), both of which showed the same A-ring fission that occurs in mansumbinoic acid.

Ethnobotany and serendipity

The studies described here illustrate quite well the problems that confront research groups assaying nature for new leads of biological activity for pharmaceutical development. We are constantly being faced with two seemingly distinct approaches to follow; that of ethnobotany or of what is now fashionably called bioprospecting, which is effectively the academic equivalent of the natural products based random screening used by some major pharmaceutical companies.

The studies we have undertaken on the Burseraceae illustrate how, when in an academic environment where the phytochemist has the luxury of being surrounded by a wide range of pharmacological expertise, both approaches can be going on side by side without conflict. Thus our studies on the myrrh-like resins of Kenya were stimulated by ethnobotanical knowledge and this is what drove us to the isolation of the new sesquiterpene (**1**). On the other hand the finding of the anti-inflammatory activity of mansumbinoic acid (**8**) had no basis in local usage and is an example of serendipity based on access to suitable screening facilities and concomitant pharmacological skills.

Acknowledgements

This work would not have been carried out without the logistical support of The National Museum of Kenya, and in particular the staff of the East African Herbarium, and The Royal Geographical Society, London. Financial support for fieldwork came from The Royal Society, The Carnegie Trust for the Universities of Scotland and the Man and Biosphere Programme of UNESCO. Support for laboratory research came from the Science and Engineering Research Council, The Royal Pharmaceutical Society of Great Britain, The Henry Lester Trust. The Association of Commonwealth Universities and The Research and Development Fund of the University of Strathclyde. The IOCD is thank for support toward the costs of attending this meeting.

References

Abegaz, B., Dagne, E., Bates, C., and Waterman, P.G. (1989). Monoterpene-rich resins from two Ethiopian species of *Commiphora. Flavour Fragrance* Journal **4**, 99-101.

Arora, R.B., Kapoor, V., Gupta, S.K., and Shanma, R.E. (1972). Isolation of a crystalline steroidal compound from *Commiphora mukul* and its anti-inflammatory activity. *Indian Journal for Experimental Biology* **9**, 403-4.

Brieskorn, C.H. and Noble, P. (1982). Inhaltsstoffe des etherischem Öls der Myrrhe. *Planta Medica* **44**, 87-90.

Cavanagh, I.S., Cole, M.D., Gibbons, S., Gray, A.I., Provan, G.J., and Waterman, P.G. (1993). A novel sesquiterpene, 1,2-epoxyfurano-10(15)-germacren-6-one, from the resin of *Commiphora holtziana* Engl. *Flavour Fragrance Journal* **8**, 39-41.

Coe, M. and Collins, N.M. (Eds.) (1986). *An Ecological Inventory of the Kora National Reserve, Kenya*. Royal Geographical Society, London.

Craveiro, A., Corsano, S., Proietti, G., and Strappaghetti, G. (1983). Constituents of essential oil of *Commiphora guidotti. Planta Medica* **48**, 97-98.

Da Silva, M.F.G., Francisco, R.H.P., Gray, A.I., Lechat, J.R., and Waterman, P.G. (1990). Lanost-7-en triterpenes from stem bark of *Santiria trimera. Phytochemistry* **29**, 1629-1632.

Dolara, P., Luceri, C., Ghelardini, C., Monserrat, C., Alolli, S., Luceri, F., Lodovici, M., Menichetti, S., and Romanelli, M.N. (1995). Analgesic effects of myrrh. *Nature* **379**, 29.

11. The chemistry of the Meliaceae of South Africa and Namibia

D.A. MULHOLLAND

Natural Products Research Group, Deptartment of Chemistry, University of Natal, Private Bag X10, Dalbridge, 4014, Durban, South Africa

Introduction

The chemistry of the Meliaceae or Mahogany family has been studied extensively. Meliaceae species are used in traditional medicine worldwide, including South Africa. Extracts from the Meliaceae have yielded a wide range of compounds including coumarins, flavonoids, diterpenoids, tetra- and pentacyclic triterpenoids, alkaloids and limonoids, a class of tetranortriterpenoids in which the triterpenoid sidechain has been modified into a β-substituted furan ring at C-17α. The structure of a simple limonoid is shown below (**1**). Rings A-D may be oxidatively opened to give a wide range of structural types.

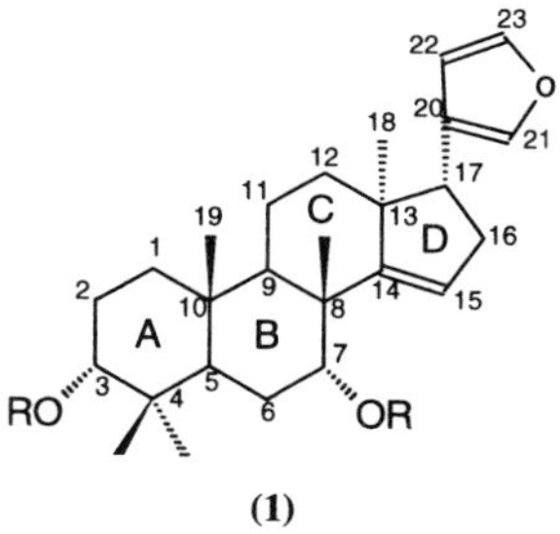

(1)

Limonoids have been shown to exhibit insect antifeedant properties (reviewed by Champagne *et al.* 1992), anti-malarial properties (Rochanakij *et al.* 1985), tumour-preventative properties (Lam and Hasegawa 1987) and anti-cell adhesion properties (Musza *et al.* 1994)

The Meliaceae are represented in South Africa by six genera, *Entandrophragma, Nymania, Turraea, Ekebergia, Trichilia* and *Pseudobersama.* The species, *Entandrophragma spicatum* occurs in Namibia. Figure 11.1. illustrates the grouping of the six South African genera into tribes and subfamilies within the Meliaceae family.

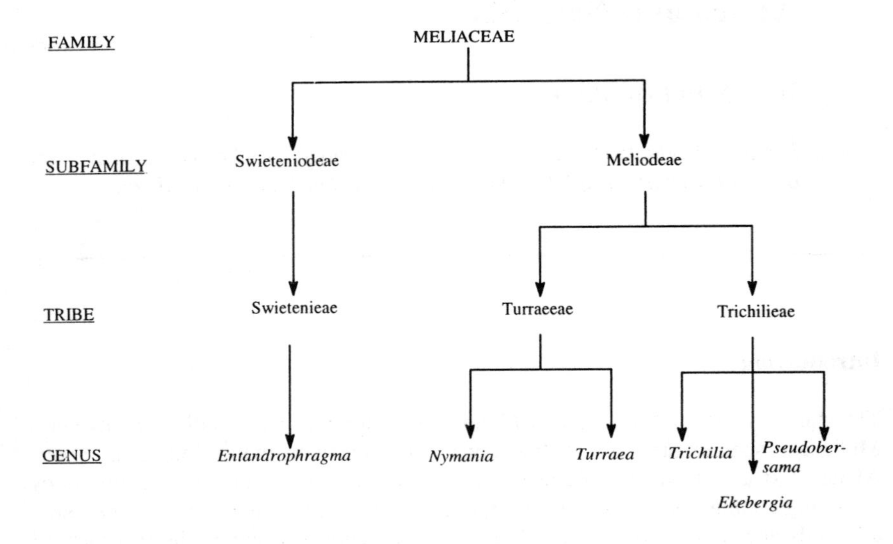

Fig. 11. 1. The grouping of the six South African genera within the Meliaceae family.

The genus *Entandrophragma*

Entandrophragma caudatum Sprague, or the wooden banana, so called because of the appearance of the fruit which, when ripe splits giving the appearance of a stiff, peeled banana, grows in the northern and north-eastern Transvaal, its range extending into Zimbabwe and Mozambique. The bark was found to contain mixed phragmalin esters, which, when subjected to hydrolysis, gave phragmalin (**2**) (Arndt and Baarschers 1972). The wood contained entandrophragmin (**3**), a limonoid that has been isolated previously from West African members of the *Entandrophragma* genus (Akinsanya *et al.* 1960). The seed yielded the protolimonoids melianone (**4**), 3-*epi*melianol (**5**) and four phragmalin esters (**6-9**) (Ansell and Taylor 1988).

Entandrophragma spicatum Sprague, also known as the Ovambo wooden banana, is a rare mahogany growing in Ovamboland and southern Angola. The timber of this species has been shown to contain entandrophragmin (**3**), (**10**), an 11-isobutyryl analogue of bussein and the novel protolimonoid, spicatin (**11**) (Connolly *et al.* 1981).

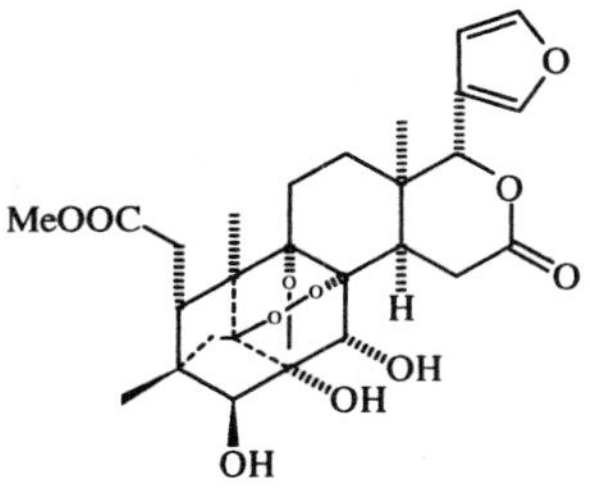

O
MeOOC
O
O O
O
H
OH
OH
OH
O

(2)

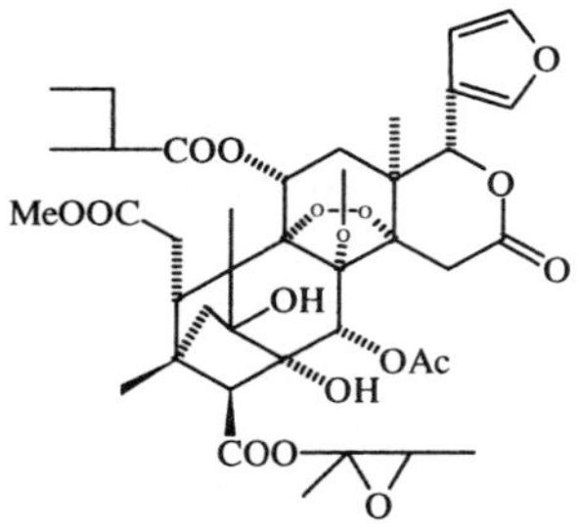

O
COO
MeOOC
O O
O
O
O
OH
OAc
OH
COO
O

(3)

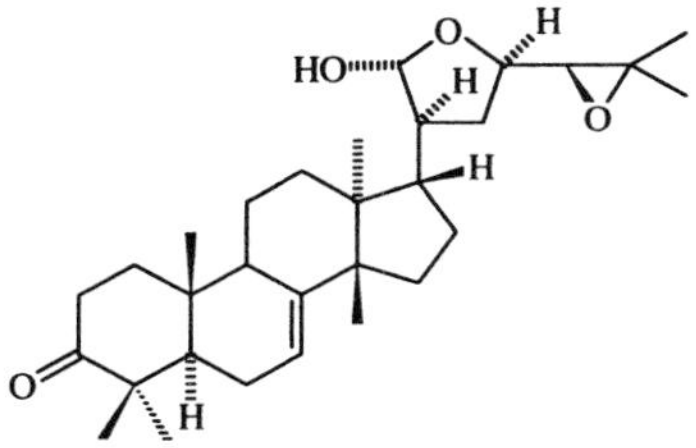

HO
O
H
H
O
H
O
H

(4)

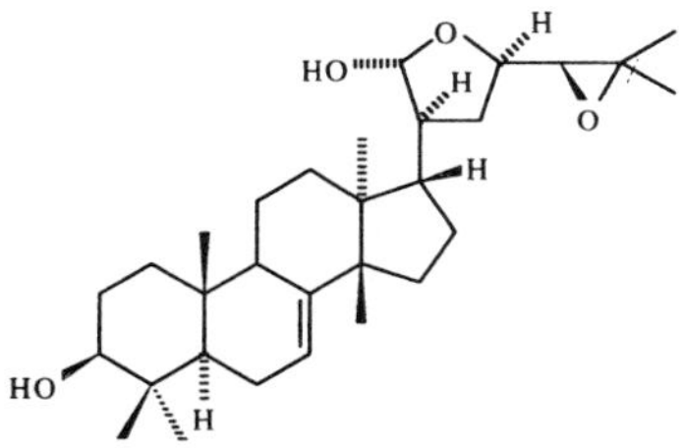

HO
O
H
H
O
H
HO
H

(5)

O
OR_3
MeOOC
O
O O
O
H
OR_2
OH
OR_1
O

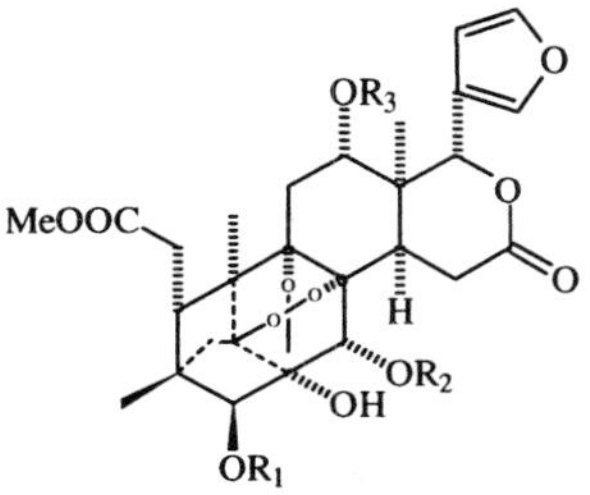

(6) $R_1 = R_2 =$ *iso*PrCO
(7) $R_1 =$ *iso*PrCO, $R_2 =$ EtCO
(8) $R_1 = mC_5H_5NCO$, $R_2 =$ *iso*PrCO, $R_3 =$ MeCO

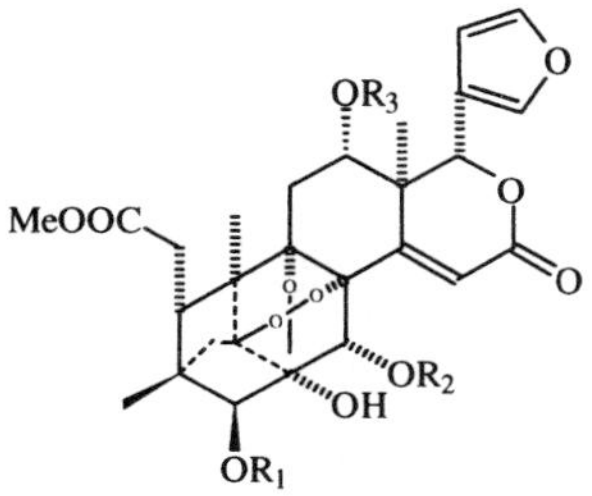

O
OR_3
MeOOC
O
O O
O
OR_2
OH
OR_1
O

(9) $R_1 = mC_5H_5NCO$, $R_2 = R_3 =$ *iso*PrCO

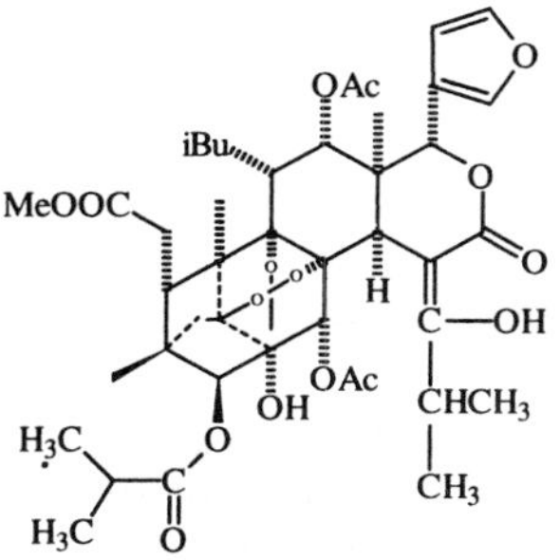

O
OAc
iBu
MeOOC
O
O O
O
H
O
C—OH
OAc
OH
CHCH$_3$
CH$_3$
H_3C
O
C
H_3C
O

(10)

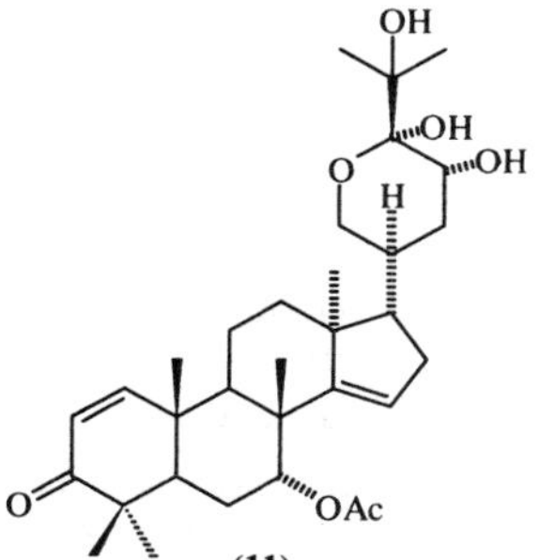

OH
OH
OH
O
H
O
OAc

(11)

The genus *Nymania*

The genus *Nymania* is represented in South Africa by one species, *Nymania capensis* Lindb., which is found in the drier parts of the country. The plant is well known for its colourful inflated pink calyces which surround the ripe fruit, giving the common name of 'Chinese lanterns'. The taxonomic position of this plant has long been disputed, but it is now considered to belong to the Meliaceae family where it is placed in the tribe Turraeeae. The bark and wood of *Nymania capensis* have yielded prieurianin (**12**) and nymania compounds (I-IV), (**13-16**) (MacLachlan and Taylor 1982). We have recently isolated nymania 1 (**13**) from *Turraea obtusifolia* (Fraser *et al.* 1995) and this supports the inclusion of *Nymania* in the Turraeeae.

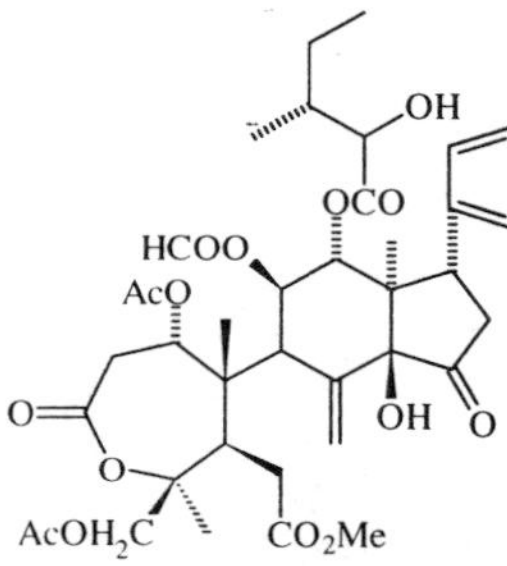

(**12**)

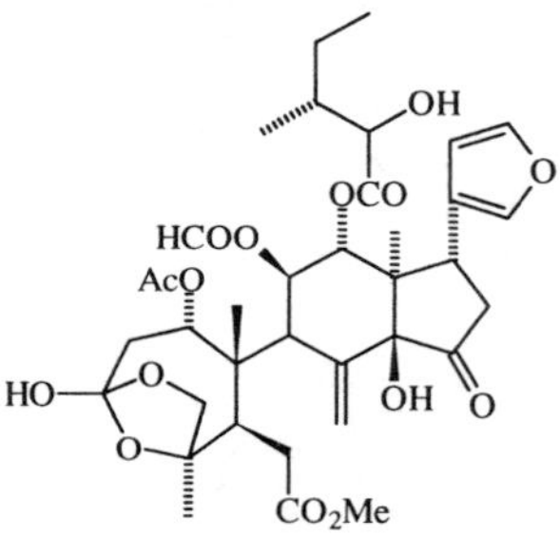

(**13**)

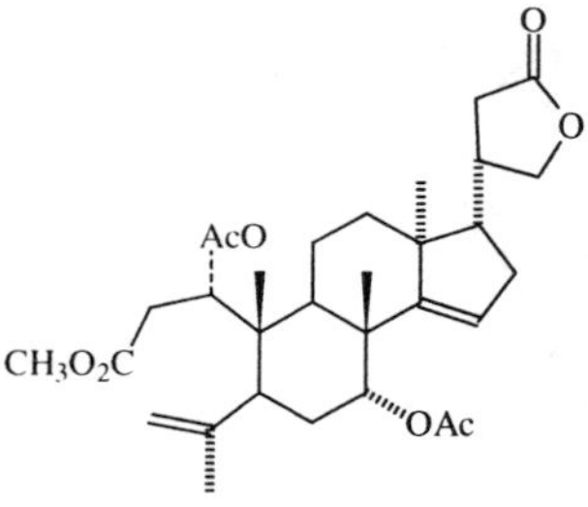

(**14**)

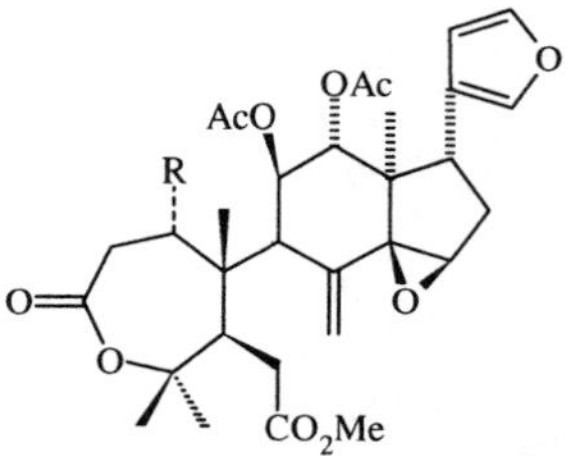

(**15**) R= H
(**16**) R= AcO

The genus *Turraea*

The genus *Turraea* is represented by four species, *T. floribunda, T. nilotica, T. obtusifolia* and *T. pulchella*.

Turraea floribunda Hochst., also known as the Wild Honeysuckle Tree, is distributed throughout East Africa. It is a medium-sized tree with white flowers which give off a powerful scent at night. In African medicine, rheumatism, dropsy and heart disease are treated with a preparation from the root and bark. The bark alone is used to induce a trance prior to performing divining dances. An overdose is said to be poisonous (Coates Palgrave 1977). The bark of this species yielded three limonoids, (**17-19**) (Akinniyi *et al.* 1986). The wood yielded stigmasterol and a further limonoid (**20**) (Akerman, MSc thesis 1990). The seed yielded three novel limonoids which contained a 8,30- 14,15- conjugated diene system. (**21-23**).

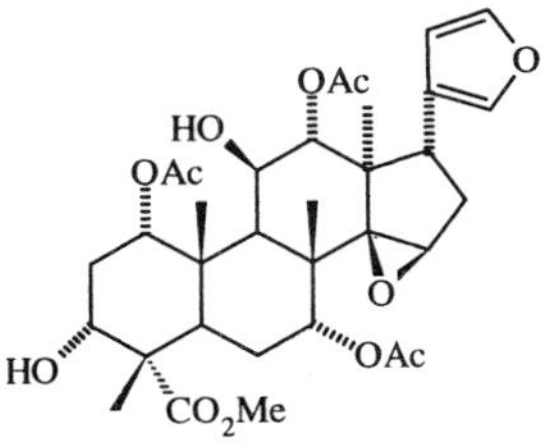

(**17**)

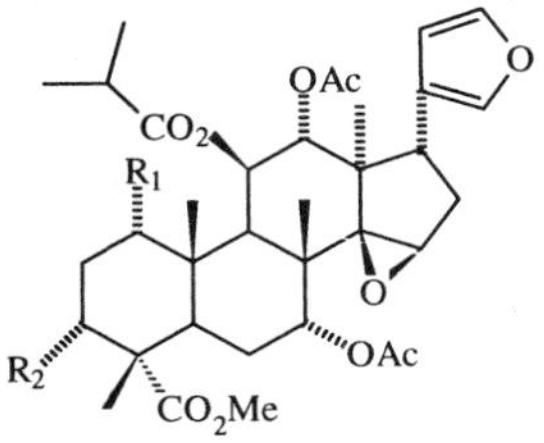

(**18**) R_1= OAc, R_2= OH
(**19**) R_1= OH, R_2= OAc

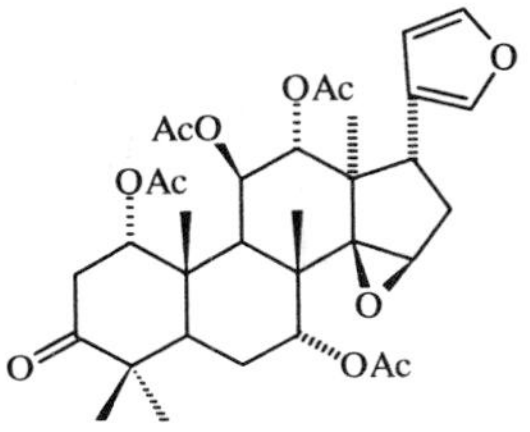

(**20**)

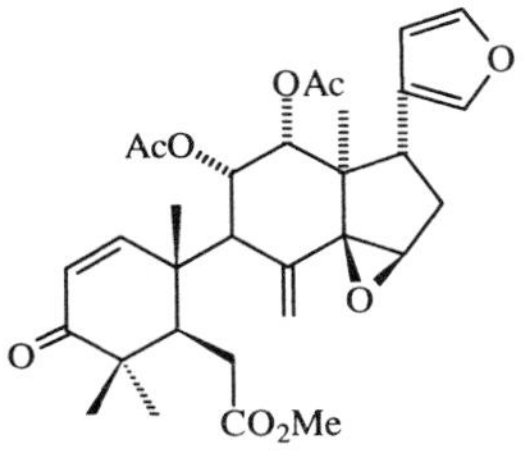

(**21**)

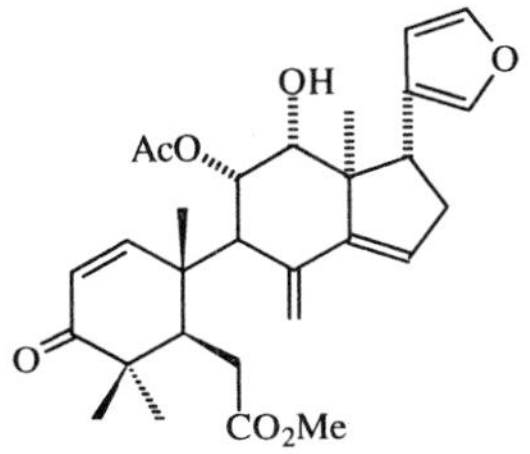

(**22**)

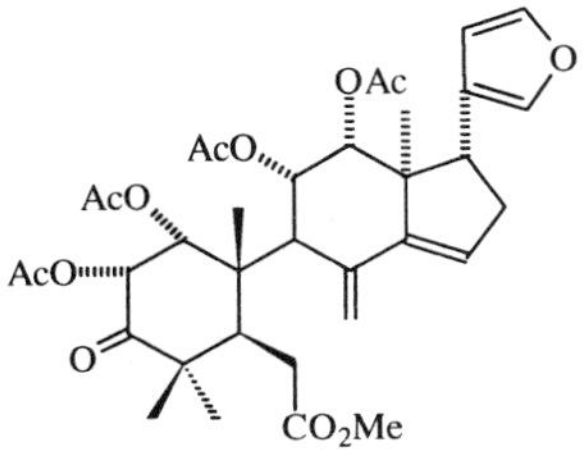

(**23**)

Turraea obtusifolia Hochst., or the Small Honeysuckle Tree, is a small bush. The leaves have yielded the protolimonoids melianone (**4**) and the 3-keto anologue of sapelin F (**24**). The wood has yielded melianone (**4**), melianodiol (**25**), melianotriol (**26**) and 3α-acetoxy-7α-deacetoxyglabretal (**27**) (Akerman 1990). The seed has yielded a complex-mixture of prieurianin-type compounds. This extract is under investigation at present and has yielded prieurianin (**12**), prieurianin acetate and nymania I (**13**) (Fraser *et al.* 1995).

The wood of *Turraea nilotica* Kotschy and Peyr., the bushveld honeysuckle tree, contains niloticin (**28**) and related compounds (Mulholland and Taylor 1988). Niloticin was a proposed intermediate in the biosynthesis of limonoids but had not been found previously.

Turraea pulchella, was thought to be extinct until a small population was found outside Durban. It is a small plant, 2-3 cm high. Extraction of a small amount of leaf and shoot material gave a mixture of protolimonoids containing the 13,14,18-cyclopropane ring structure as in (**27**) isolated from *T. obtusifolia*. (Mulholland, unpublished work).

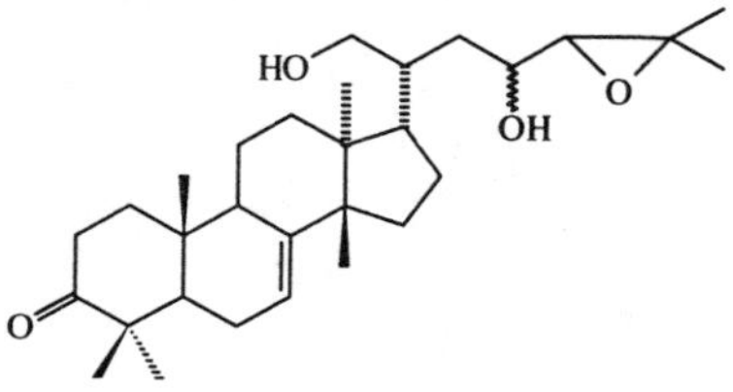

(**24**)

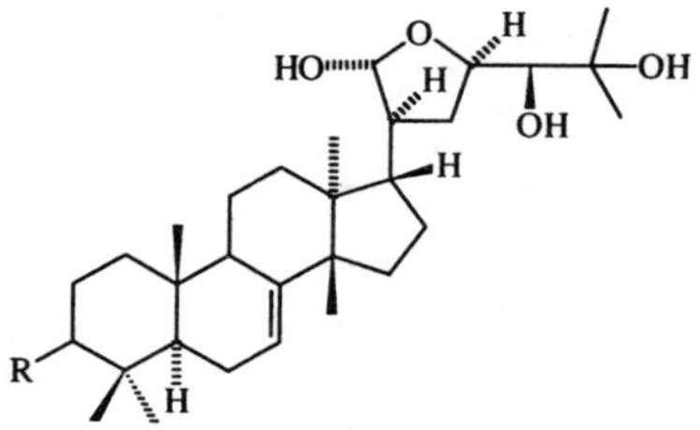

(**25**) R= =O
(**26**) R= β OH

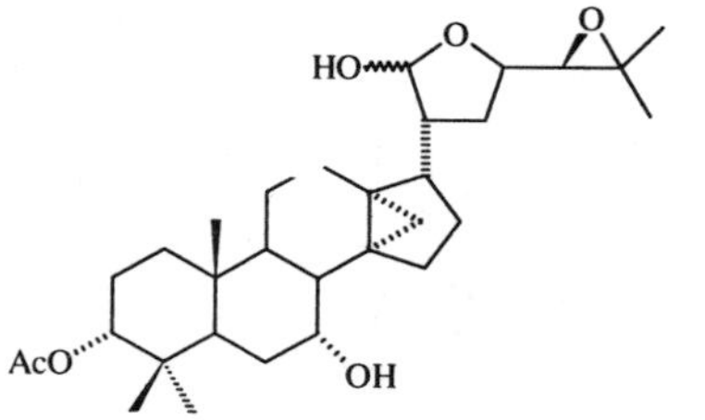

(**27**)

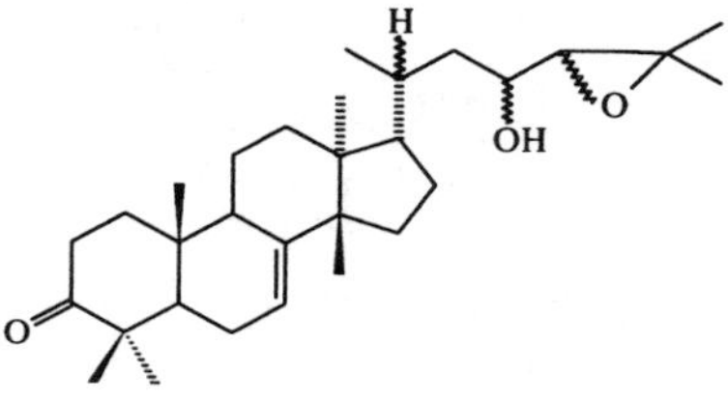

(**28**)

The genus *Ekebergia*

Ekebergia capensis and *Ekebergia pterophylla* have been extensively investigated in our laboratory over the past few years.

Ekebergia capensis Sparrm, or the Cape Ash, is a large tree. The wood of this species makes attractive furniture, being light, soft and straw coloured with an even grain. The bark is used as an emetic and in the treatment of dysentry and also for tanning. An extract of the root is said to relieve headaches and chronic coughs, while the leaves are used to treat instestinal worms. It is planted along the streets in Port Elizabeth (Coates Palgrave 1977).

The leaves of this species have yielded kaempferol 3-O-glucoside (**29**) and lupeol (**30**) (Monkhe 1991) and the bark has yielded methyl-2,4-dihydroxy-3,6-dimethylbenzoate (methyl ester of atraric acid) (**31**), sitosterol, lupeol, oleanolic acid and 3-*epi*oleanolic acid. The wood produces oleanolic acid and 3-*epi*oleanolic acid (Iourine 1995). Extracts from the seed have yielded oleanolic acid and the limonoids ekebergin (**32**) (Taylor 1981) and capensolactones (**33-36**), (Mulholland and Iourine, unpublished work). These limonoids occur as mixed esters and are extremely difficult to separate. Similar compounds were obtained from the West

African species *E. senegalensis*, which is now considered conspecific with *E. capensis*. The methanol extract of the seed was found to contain over 30% of sucrose (Mulholland and Iourine, unpublished work).

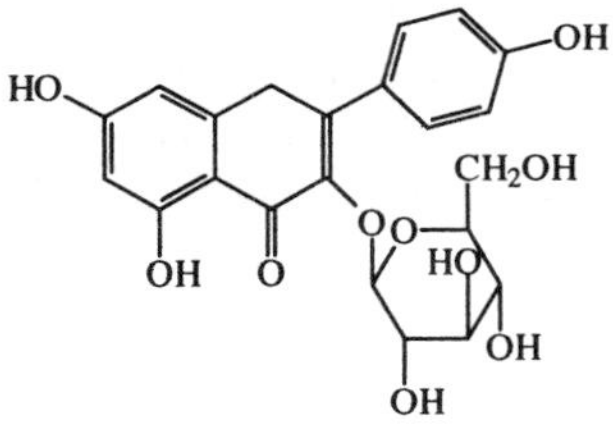

(29)

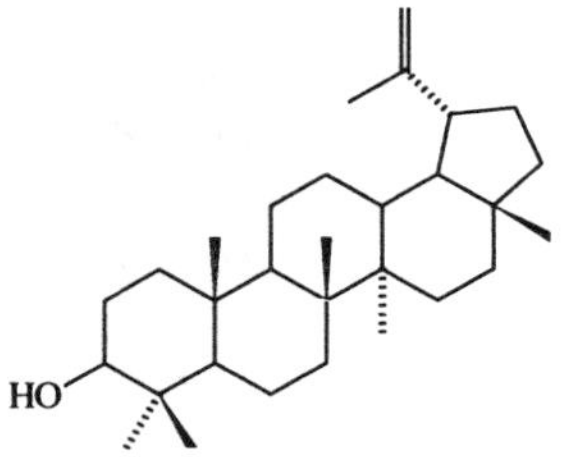

(30)

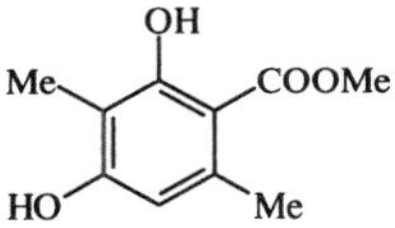

(31)

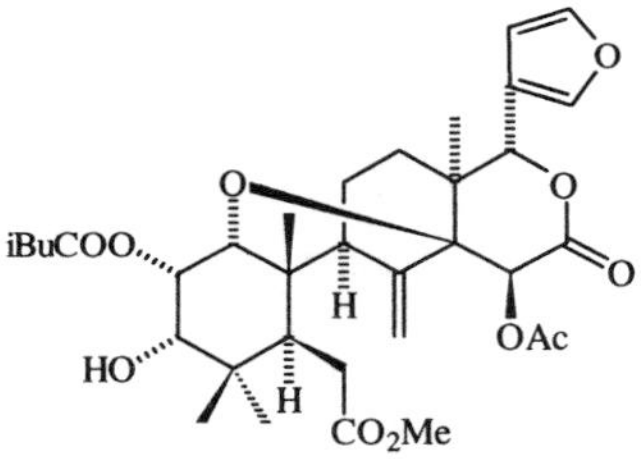

(32)

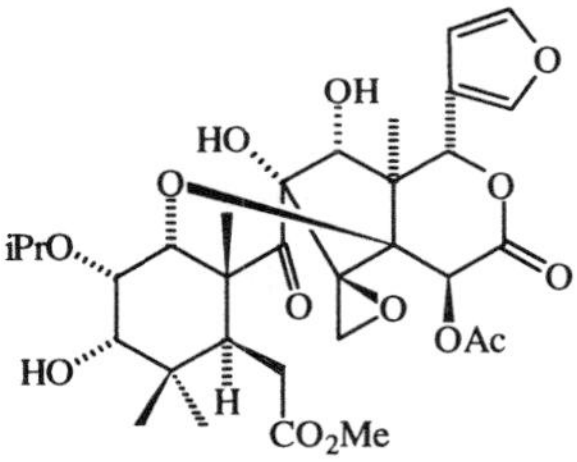

(33)

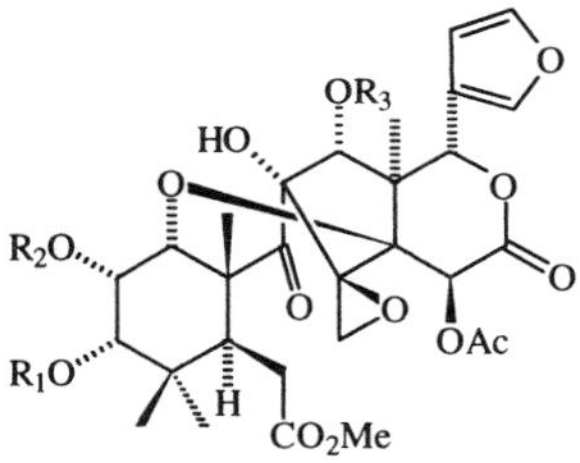

R_1=Nic, R_2=2-MeBu, R_3=Ac

(34)

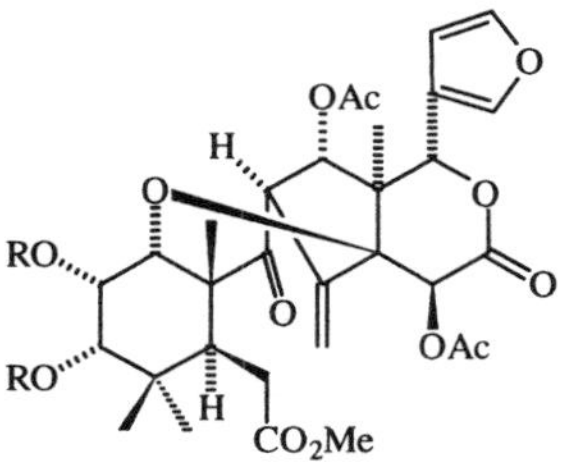

R=Nic, 2-MePr

(35)

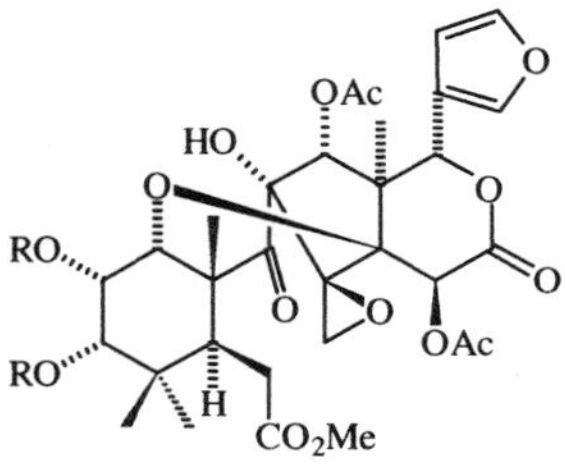

R=Nic, 2-MeBu

(36)

Ekebergia pterophylla (C.D.C.) Hofmeyr, or the Rock Ash, is a bush that grows on outcrops of Table Mountain sandstone. Studies of the seed of this species have yielded prieurianin (**12**), ekebergin (**32**) and the ekebergolactones (**37-42**), (Taylor and Taylor 1984, Kehrli and Taylor 1989). The leaves contain lupeol (**30**). The bark and wood are a rich source of 5-methylcoumarins. The bark produces the methyl ester of atraric acid (**31**), sitosterol, oleanolic acid, β-amyrin, β-amyrone and the coumarins pterophyllin 1 (**43**) and 2 (**44**). The coumarins pterophyllin 3 (**45**), 4 (**46**), 5 (**47**) and 6 (**48**) have been isolated from the bark (Mulholland and Iourine, unpublished work).

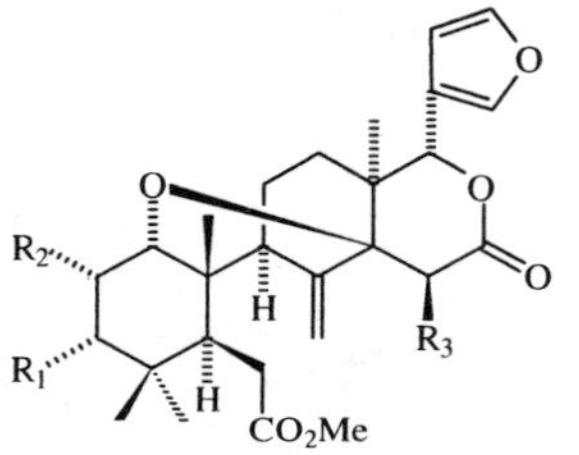

(**37**) R_1=OH, R_2=OAc, R_3=H
(**38**) R_1=OH, R_2=H, R_3=H
(**39**) R_1=OH, R_2=OAc, R_3=OAc
(**42**) R_1=OH, R_2= H, CO_2
R_3=OH

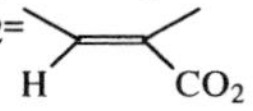

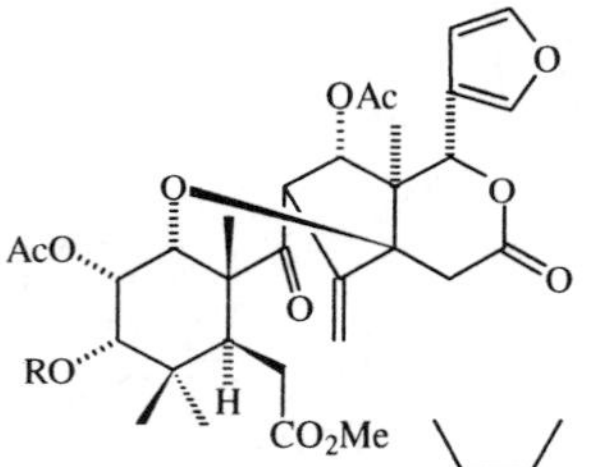

(**40**) R= O=C H

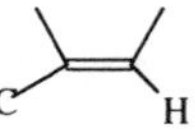

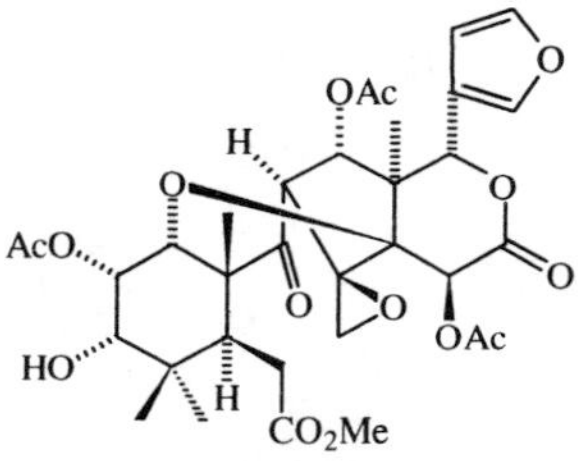

(**41**)

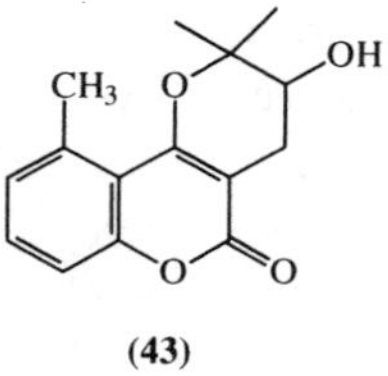

(**43**)

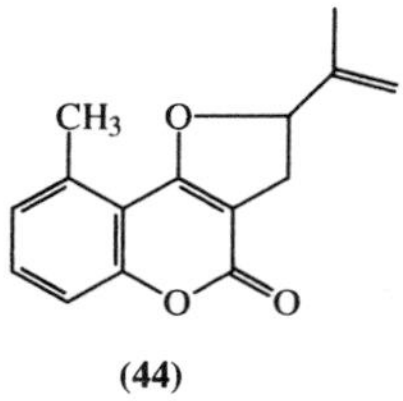

(**44**)

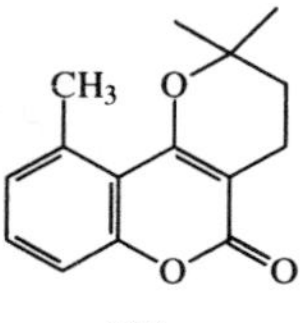

(**45**)

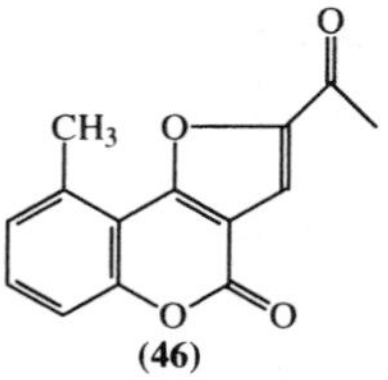

(**46**)

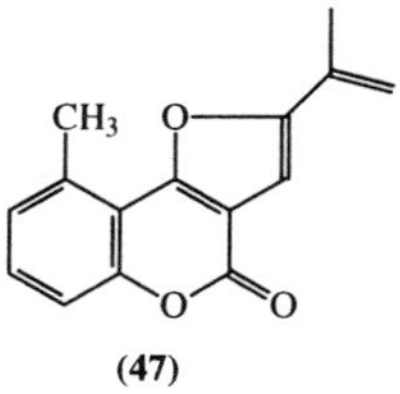

(**47**)

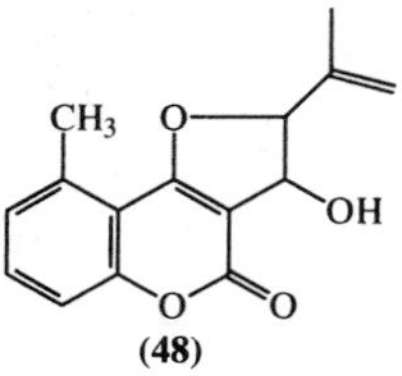

(**48**)

The genus *Trichilia*

Trichilia dregeana Sonder or the Forest Natal Mahogany is an impressive tree used to line streets in Durban. *Trichilia emetica* Vahl, the Natal Mahogany, is similar in appearance and the two species are distinguishable by the absence of a stripe connecting the fruit capsule to its stalk in the former but by having a distinct stripe in the latter. It is thought that both species are used similarly for medicinal purposes. The bark is soaked in warm water and the liquid is used as an emetic. Seeds are ground and boiled in water. The surface of the water is skimmed to separate the oil which is rubbed into cuts made on a fractured limb to hasten healing and to annoint the body generally. The oil is taken internally to treat rheumatism and can be made into good quality soap. The leaves are said to induce sleep when placed in ones bed at night . A hot infusion of the leaves is applied to bruises as a soothing lotion (Coates Palgrave 1977).

The leaves of *Trichilia dregeana* produce 13-*epi*manoyloxide (**49**) (Monkhe, M.Sc thesis, 1991) and the seeds produce complex limonoids of the prieurianin type (**50-55**) (Mulholland and Taylor 1981).

The seed of *Trichilia emetica* have been shown to contain the trichilins A-D (**56-59**) (Nakatani *et al.* 1981) which exhibit anti-feedant activity against the Southern Army Worm.

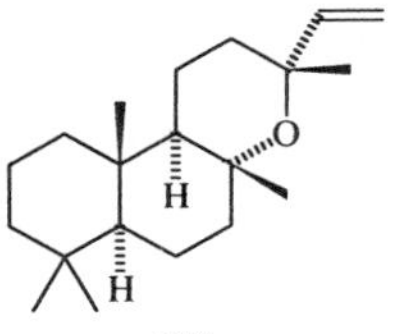

(**49**)

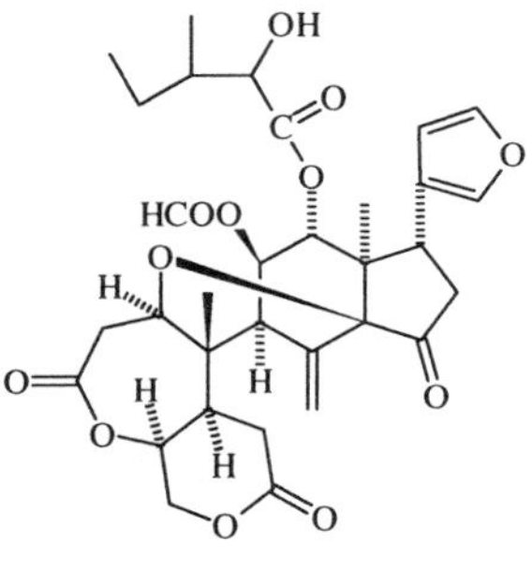

(**50**)

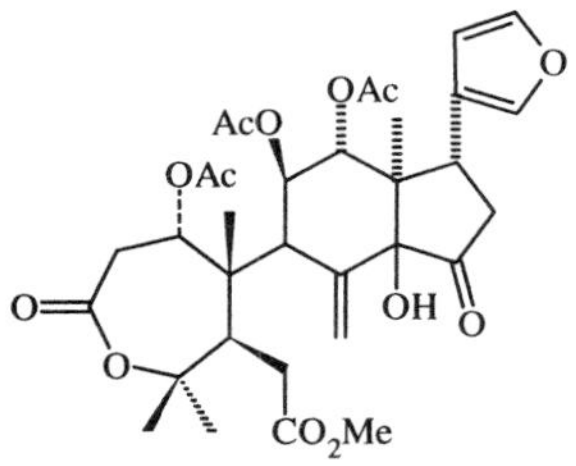

(**51**)

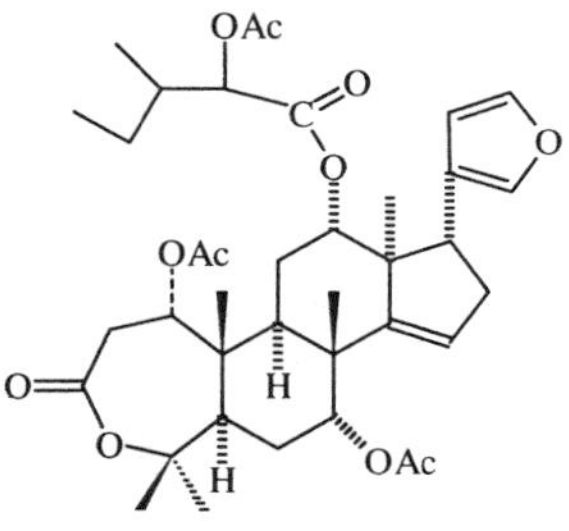

(**52**)

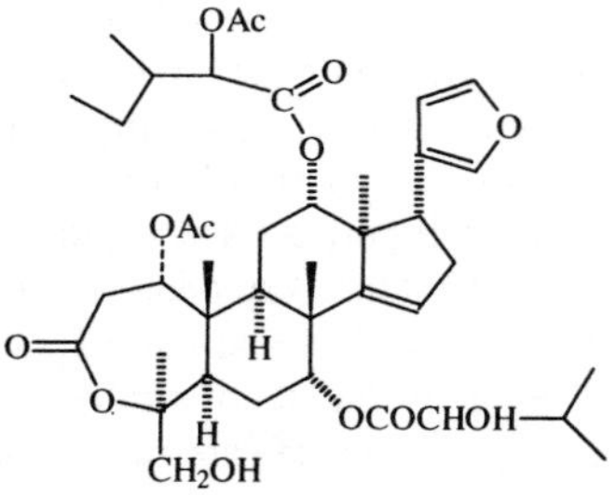

(53)

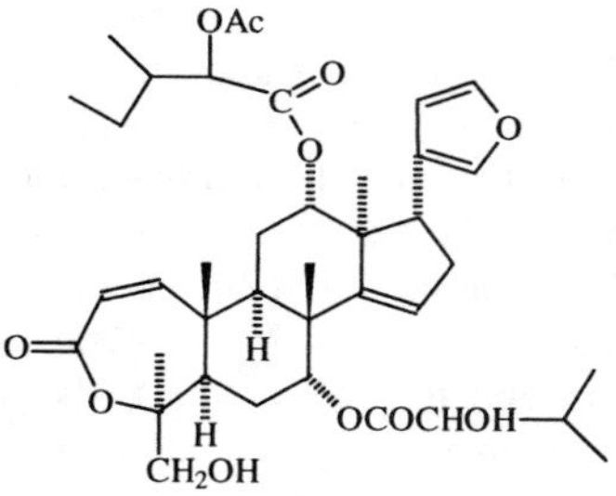

(54)

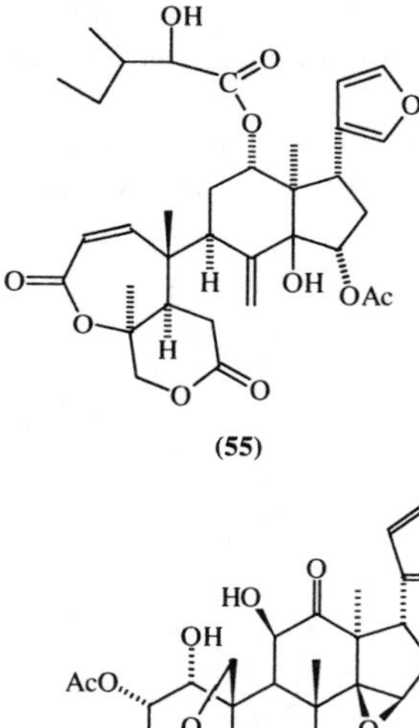

(55)

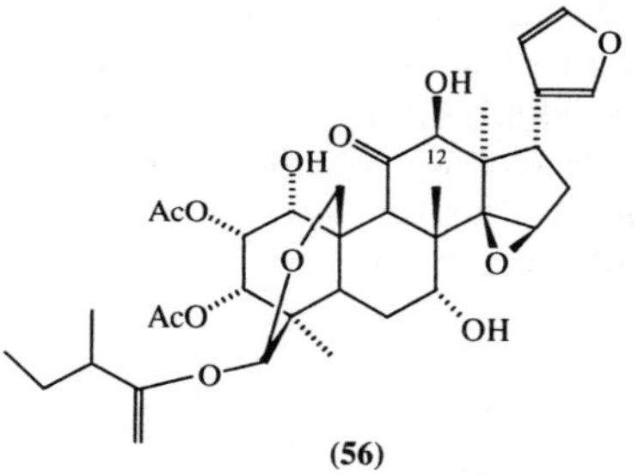

(56)
(57) 12α-OH

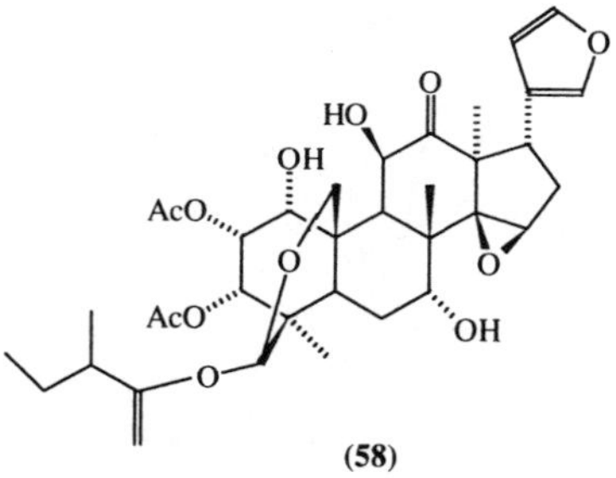

(58)

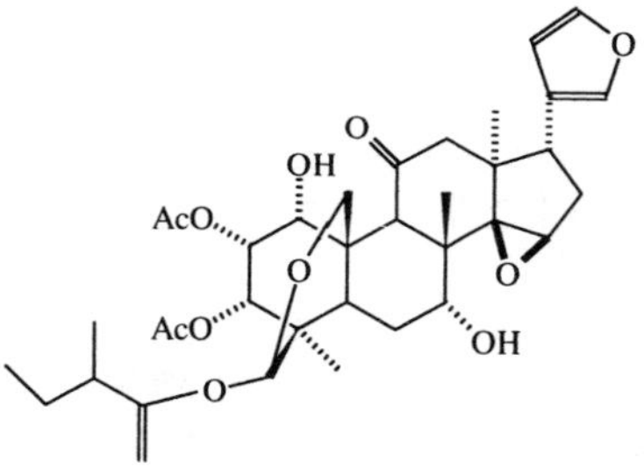

(59)

The genus *Pseudobersama*

Pseudobersama mossambicensis grows in the most northerly tip of Kwazulu/Natal and extends into Mozambique. An investigation of the twigs and leaves of this species has yielded 24-methylene cholest-5-ene-3,7-diol (**60**) and related compounds (**61**) and (**62**) (Gunatilaka *et al.* 1991). This species requires further investigation.

HO OH

(60) R=

(61) R=

(62) R=

HO

OH

Conclusion

The Meliaceae family of South Africa has been extensively investigated and has produce a range of interesting compounds. Most work has, however been done on the less polar extracts of these plants. Meliaceae extracts are used extensively by the Zulu people of Natal, but they make use of aqueous extracts. The aqueous extracts of some of the more commonly used species require investigation.

References

Akerman, L.A. (1990). M.Sc. thesis, University of Natal, Durban, South Africa.

Akinniyi, J.A., Connolly, J.D., Mulholland, D.A., Rycroft, D.S., and Taylor, D.A.H. (1986). Limonoid extractives from *Turraea floribunda*, and *T. obtusifolia*. *Phytochemistry* **25**, 2187-2189.

Akisanya, A., Bevan, C.W.L., Hirst, J., Halsall, T.S., and Taylor, D.A.H. (1960). West African timbers. Part III. Petroleum extractives from the genus *Entandrophragma*. *Journal of the Chemical Society,* 3827.

Ansell, S.M. and Taylor, D.A.H. (1988). Limonoids from the seed of *Entandrophragma caudatum*. *Phytochemistry* **27**, 1218.

Arndt, R.R. and Baarschers, W.H. (1972). The structure of phragmalin, a meliacin with a norbornane part skeleton. *Tetrahedron* **28**, 2333.

Champagne, D.E., Koul, O., Scudder, G.G.E., and Towers, N.G.H. (1992). Biological activity of limonoids from the Rutales. *Phytochemistry* **31**, 377.

Coates Palgrave, K. (1977). Trees of Southern Africa. Struik, Cape Town.

Connolly, J.D., Phillips, W.R., Mulholland, D.A., and Taylor, D.A.H. (1981). Spicatin, a protolimonid from *Entandrophragma spicatum*. *Phytochemistry* **20**, 2596.

Fraser, L.A., Mulholland, D.A., and Taylor D.A.H. (1995) The chemotaxonomic significance of the limonoid, nymania-I, in *Turraea obtusifolia*. *South African Journal of Botany* **61**, 281.

Iourine, S. (1995) Ph.D. thesis, University of Natal, Durban, South Africa.

Kehrli, A.R.H., Taylor, D.A.H., and Nivan, M. (1989). Limonoids from *Ekebergia pterophylla* seed. *Phytochemistry* **29**, 153.

Lam, L.K.T. and Hasegawa S. (1987). *Nutrition and Cancer* **12**, 43.

MacLachlan, L.K. and Taylor D.A.H. (1982). Limonoids from *Nymania capensis*. *Phytochemistry* **21**, 1701.

Monkhe, T.V. (1991) M.Sc. thesis, University of Natal, Durban, South Africa.

Mulholland, D.A. and Taylor, D.A.H. (1980). Limonoids from the seed of the Natal Mahogany, *Trichilia dregeana*. *Phytochemistry* **19**, 2421.

Mulholland, D.A. and Taylor, D.A.H. (1988). Protolimonoids from *Turraea nilotica*. *Phytochemistry* **27**, 1220.

Musza, L.L., Killar, L.M., Speight, P., McElhiney, S., Barrow, C.J., Sillum, A.M., and Cooper, R. (1994). Potent new cell adhesion inhibitory compounds from the roots of *Trichilia rubra*. *Tetrahedron* **50**, 11369.

Nakatani, M., James, J.C., Nakanishi, K. (1981). Isolation and srtuctures of trichilins, antifeedants against the Southern Army Worm. *Journal of the American Chemical Society,* **103**, 1228.

Rochanakij, S., Thebtaranonth, Y., Yenjai, C, and Yuthavong, Y. (1985) *Southeast Asian Journal of Tropical Medicine and Public Health* **16**, 66.

Taylor, D.A.H. (1965). Extractives from East African timbers. Part I. . *Journal of the Chemical Society,* 2495.

Taylor, D.A.H. (1981). Ekebergin, a limonoid extractive from *Ekebergia capensis*. *Phytochemistry* **20**, 2263.

Taylor, A.R.H. and Taylor, D.A.H. (1984). Limonoids from *Ekebergia pterophylla*. *Phytochemistry* **23**, 2676.

12. Development of ethical phytomedicines for Togo, West Africa

M. GBEASSOR[1], H.K. KOUMAGLO[1], D.V.C. AWANG[2], T. DURST[3], S. MACKINNON[3] AND J.T. ARNASON[3]

[1]*Faculté des Sciences, Université du Bénin, Lomé, Togo;* [2]*MediPlant, P.O. Box 8693, Stn. T, Ottawa, Ontario, Canada K1J 3G1;* [3]*Institutes of Chemistry and Biology, University of Ottawa, Ottawa, Ontario Canada K1N 6N5*

Introduction

The use of traditional remedies based on plant, animal and mineral extracts is a common practice in West African countries, especially in the rural areas where approximately 70-80% of the population lives. In Togo for example, ethnobotanical surveys indicate uses for >350 species of plants, 24 species of animals and 5 mineral preparations. (Adajanohoun *et al.* 1996). The remedies are prepared by healers, according to their ancestral knowledge. The healers are well known in their villages and the population has a great regard for most of them. In Togo, they often have their own clinics and are well organized into a national association.

It is very common to see in West Africa, people using both modern and traditional drugs together. A survey of community attitudes in Ghana indicated that many patients believe that "a complete treatment consists of both" and that "modern drugs give fast relief but herbal remedies act slowly and provide the final cure" (Le Grand and Wondergem 1990). The great diversity of plant resources in West Africa is a great natural resource that can provide health care products locally at a fraction of the cost of modern pharmaceuticals. Unfortunately most traditional drugs have not been scientifically tested and physicians with western training are hesitant to recommend them even though they recognize their potential. This is because biological efficacy and toxicological tests may not have been performed with the traditional drugs or the mode of action and concentration of active principles is not known.

Development of ethical phytomedicines

We are now attempting to bridge the gap between traditional and modern medicine by verification of selected medicinal plants as "ethical phytomedicines" in order to provide low cost medicines for the primary health care system in Togo. The different stages of development of an ethical phytomedicine are outlined in Table 12.1.

Table 12.1. Stages in development of ethical phytomedicines

1. Ethnobotanical surveys
2. Priority of medicinal plants according to the list of pathologies
3. Review of the literature
4. Botanical studies: collection, certification , vouchering and culture of the plant material, proper storage and stabilization to retain activity
5. Mode of action studies, isolation of active principles and toxicology
6. Galenic processing
7. Clinical trials, drug authorization, marketing and pharmacovigilance

Ethnobotanical surveys

Primary surveys of medicinal plants usage have been completed in Togo (Adajanhoun *et al.* 1996) and some neighboring West African countries (Iwu 1995). However, detailed knowledge of formulations and administration of effective remedies are lacking and require further ethnobotanical work. In West Africa, ethnobotanists must fully gain the confidence of healers who are reluctant to reveal details of species composition, preparation and administration of remedies.

Although it is a commonly held view that ethnobotany is often the best approach to drug discovery, there is little in the way of statistical data on the success of this approach. To test the ethnobotanical approach, we recently completed a quantitative analysis of 17 traditional malaria remedies and 12 control plants that showed that malarial remedies were more significantly active in bioassays than controls (Leaman *et al.* 1995). Local consensus among healers regarding malaria remedies, measured by an importance value index, was also highly predictive of those malaria remedies, with the highest activity in bioassays against *Plasmodium falciparum.*

Priority medicinal plants

The Ministry of Public Health in Togo has established a priority list of pathologies for which drugs are needed. These priorities include malaria, diabetes, diarrhea and hypertension. Plants that have been selected for investigation are traditionally used for these pathologies. A list of some of the more commonly used plants, their pathologies and phytochemical markers are given in Table 12.2. Species of more interest from a drug discovery point of view are expected to be found in primary and secondary forests and savannah areas, rather than the well studied pan-tropical weed flora. Unfortunately, those natural areas that are the best sources of phytochemical resources are also areas most threatened by human activity. We are therefore collaborating with the IUCN to develop a rescue strategy for biodiversity resources.

Table 12.2. Commonly used plants with potential for development

Medicinal Plant	Use	Phytochemical marker
Psidium guajava (guava, goyavier)	diarrhoea	6-gingerol
Zingiber officinale (ginger, gingembre)	nausea	quercetin glycosides
Carica papaya (papaya, papayer, adibati)	antihelmintic	benzyl isothiocyanate
Euphorbia hirta (spurge, notsigbe)	antiamoebic	ellagic acid
Momordia charantia (Bitter melon, aduka)	diabetes	mormordicins, charantin
Azadirachta indica (neem, nim, margose)	antimalarial	gedunin
Picrilima nitida	antimalarial, antileischmanial, antitrypanosomal	akuammine
Alstonia boonei	antipyretic, antimalarial	echitamine
Lippia multiflora (thé de Gambie)	gingivitis	monoterpene profile
Piper guineense (Guinea pepper, poivrier)	hypotensive	piperine

Review of the literature

Using the Natural Product Alert (NAPRALERT) database we have located and reviewed the primary literature with respect to priority plants (Awang and Arnason 1996). Two species, namely ginger and guava, have been the subject of sufficient investigation to be considered currently as ethical phytomedicines, because of the availability of clinical and toxicological data in addition to efficacy and phytochemical data. Ginger has been reviewed elsewhere and guava is considered in detail in a following section of this article. Other species in Table 12.1. may be advanced to ethical phytomedicines (or rejected) more rapidly than others not included here because of the considerable phytochemical and pharmacological information available. For example, the current state of our investigations of neem are also considered in this paper.

Botanical studies

Identification of chemotypes and influence of environmental conditions on actives can be made more predictable by the application of chemical ecology principles to botanical studies. For example, Koumaglo's group has found that *Lippia multiflora* expresses at least three chemotypes with distinctly different profiles of volatiles present in the essential oil. Another critical point is post harvest treatment, since many secondary metabolites are unstable if not handled properly.

Mode of action, phytochemical and toxicological studies

The first concern here is not the isolation of active principles but to confirm or reject in appropriate bioassays the indications of the remedy in traditional medicine. When the total extracts confirm these indications, the next steps are the activity guided isolation of the active principles and the study of the toxicology of the extracts. A priority for toxicology testing laboratory is to control environmental conditions to meet international standards.

Galenic processing

The phytomedicine may be presented as an infusion, syrup, lotion etc., as required by therapeutic indications, but many plant materials may be delivered as a powder or extract in gel capsules or as a tablet Standardization of dose at this stage is essential and depends on the identification of a suitable phytochemical marker (Marles *et al.* 1992), preferably, the active principle, as described in Table 12.2. The structures of several of these markers are illustrated Figure 1. We have recently developed or adapted straightforward HPLC procedures for the determination of 6-gingerol in ginger, quercetin in guava and gedunin in neem, which are operational in the Laboratoire des Extraits Végétaux et Arômes Naturels, Université du Bénin. Other analyses are readily accessible if pure standards can be obtained. Another quality control issue addressed in the Laboratoire de Microbiologie, Université du Bénin, is monitoring plant material for microbial contamination.

Clinical trials, approval, marketing and pharmacovigilance

Clinical trials are the critical step towards approval of an ethical phytomedicine and along with the biological and phytochemical studies are the basis for approval by the appropriate body in the Ministry of Health. After release, a workshop for pharmacists and medical doctors is planned during which the results of studies will be presented along with the new drug. Following release of the product, a pharmacovigilance committee will monitor any problems associated with the drug.

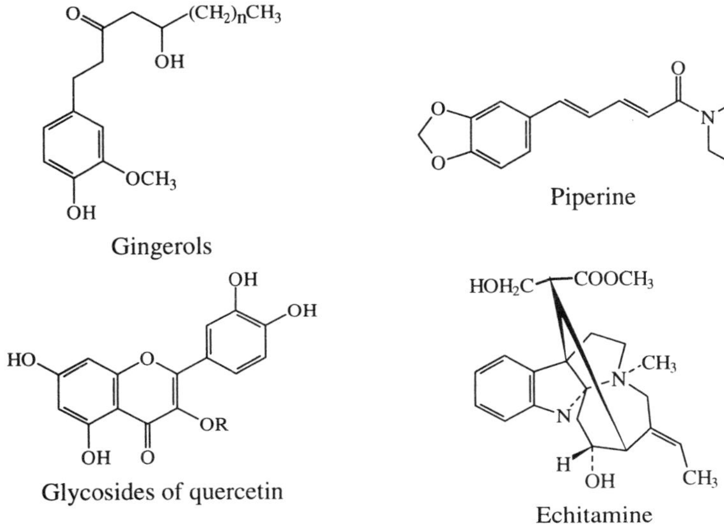

Gingerols

Piperine

Glycosides of quercetin

Echitamine

Fig. 12. 1. Phytochemical markers for standardization of phytomedicines.

Guava leaf as an example of an ethical phytomedicine

Guava (*Psidium guajava*, Myrtaceae) is a low cost prophylactic for acute diarrhoea (AD) that has been used world-wide in traditional medicine. AD is a leading cause of mortality in children under two in developing countries. In Togo, guava leaves are used traditionally to prepare a decoction that is administered orally for AD. Following toxicological studies and clinical use in Mexico, Instituto Mexicano de Securidad Social (IMSS), has given approval to this herbal remedy as a replacement therapy for more costly pharmaceuticals. Guava leaf extracts have antimicrobial activity to a variety of pathogens associated with AD (Iwu 1995). In addition, Lutterodt (1989) demonstrated that alcoholic Soxhlet extracts of guava leaf inhibited both the spontaneous and electrically coaxially stimulated contraction of isolated guinea pig ileum. These results suggested that the antidiarrhoeal activity of the extract is in part due to an inhibition of peristalsis. The inhibition is similar to that observed with morphine, one of the most effective clinically used drugs for treatment of AD.

In our phytochemical studies, a bioassay guided isolation of active fractions was undertaken and the active fractions analysed. Fourteen fractions were separated on a poylvinylpyrrolidine column using a water-methanol gradient. The greatest

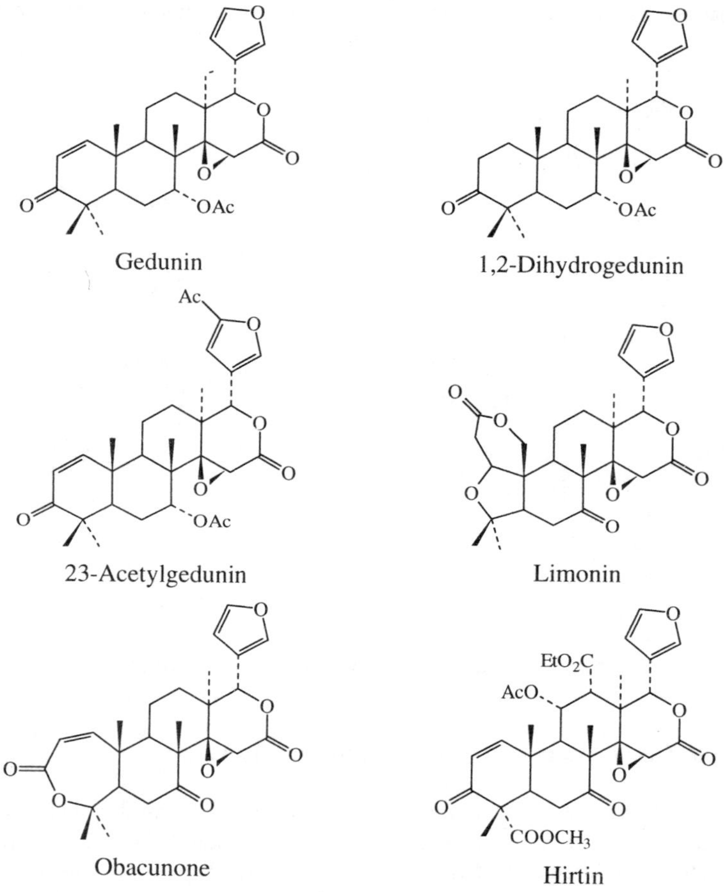

Gedunin

1,2-Dihydrogedunin

23-Acetylgedunin

Limonin

Obacunone

Hirtin

Fig. 12. 2. Representative limonoids and derivatives tested.

inhibition of guinea pig ileum was found with two fractions which contained only flavonol glycosides. The compounds were identified as quercetin-3-O-α-L-arabinoside (guaijavarin); quercetin-3-O-β-D-glucoside (isoquercitrin); quercetin-3-O-β-D-galactoside; quercetin-3-O-β-L-rhamnoside (quercitrin) and quercetin-3-O-gentiobioside. The quercetin glycosides were found to be hydrolysed to the aglycone in the gut and the aglycone was the most active compound in the bioassay. Therefore we have recently developed a method for the standardization of guava leaf based on total hydrolysed quercitin concentration by HPLC which will be used as a basis for controlling dosage as an ethical phytomedicine in Togo.

Development of an antimalarial phytomedicine from neem

Azadirachta indica A. Juss. (Meliaceae) is widespread in Africa and is commonly known as neem in English, nim or margose in French and kiniti in Ewe. In the traditional pharmacopoeia it is used as therapy for malaria and fevers, as well as diabetes, hypertension and other conditions. Previous studies have suggested the *in vivo* antimalarial activity of some neem extracts and neem derived principles (Khalid *et al.* 1986, 1989).

Our investigations of neem extracts from Togo, in collaboration with Pezzuto and Angerhoffer at the University of Illinois, Chicago, showed that they did indeed have good antimalarial activity, with IC_{50}'s below the threshold for promising extracts (20 μg/gm) for both chloroquine sensitive (D6) and chloroquine resistant (W2) clones of *Plasmodium falciparum* (MacKinnon *et al.* *Journal of Natural Products*, submitted). In fact, unrefined neem extracts from Togo were the most active of 60 extracts of 22 species of Meliaceae which we evaluated for antimalarial activity against *Plasmodium falciparum* (MacKinnon *et al.* 1996). The neem extract contained approximately 0.1% of the recognized active principle, gedunin as measured by HPLC. A related species of Meliaceae, *Cedrela odorata*, available as a plantation tree in West Africa, also contains gedunin in ethanol or toluene extracts and showed activity in the screen (Table 12.3.).

Table 12.3. Antimalarial potential of gedunin-containing extracts as determined with *Plasmodium falciparum* clones in culture

Species or standard drug	extract	IC_{50} [μg/mL] Clone D6	IC_{50} [μg/mL] Clone W2
Azadirachta indica	leaf in ethanol	2.50	2.48
Cedrela odorata	wood in ethanol	3.88	3.26
Cedrela odorata	wood in toluene	9.29	2.77
chloroquine		3.4×10^{-3}	0.101
quinine		9.5×10^{-3}	38.2×10^{-3}
mefloquine		2.7×10^{-3}	0.9×10^{-3}
artemisinin		4.5×10^{3}	2.2×10^{-3}

Evaluation of the antimalarial potential of pure gedunin (Figure 2) demonstrated that this limonoid has considerably higher activity than extracts, with an IC_{50} of 39 and 20 ng/mL against the D6 and W2 clones, respectively. Gedunin was more effective than five related limonoids tested (data for limonin, hirtin and obacunone are shown in Table 12.4.). With the resistant clone, gedunin was more effective than chloroquine or quinine but less active than artemisinin and mefloquine. The IC_{50} for cytotoxicity of gedunin to KB cells was 2300 ng/mL suggesting a selectivity of > 100 for the resistant parasite compared to the human cell line. Although the limonoid hirtin has 20% as much activity as gedunin, the selectivity for the parasite compared to KB cells was only 5.

In an attempt to improve on the activity of gedunin, we prepared a series of 10 derivatives, two of which are illustrated in Figure 2 and Table 12.4. Investigation of the antimalarial activity of these gedunin derivatives did not yield a compound with a higher activity than gedunin but was useful in defining the structural features contributing to activity of the molecule (MacKinnon *et al. Journal of Natural Products*, submitted).

Table 12.4: Antimalarial IC_{50} values of limonoids and standard antimalarials

Compounds	*Plasmodium falciparum* Clone D6 IC_{50} [ng/mL]	*Plasmodium falciparum* Clone W2 IC_{50} [ng/mL]
Gedunin	39 (100)[#]	20 (100)[#]
1,2-dihydrogedunin	>10000 (<0.39)	>840 (2.38)
23-acetylgedunin	832 (4.69)	156 (12.8)
Obacunone	>10000 (<0.39)	>10000 (<0.0005)
Limonin	>10000 (<0.39)	>10000 (<0.0005)
Hirtin	173 (22.6)	96 (20.8)
Chloroquine	1.3 (3000)	29.5 (67.8)
Quinine	14.8 (264)	34.9 (57.3)
Mefloquine	7.5 (521)	1.4 (1429)
Artemisinin	1.8 (2170)	0.5 (4000)

[#] Toxicity relative to the potency of gedunin (=100%)

Because of the traditional use and widespread acceptance of neem extracts, we believe it may be feasible to develop neem leaf preparations which are standardized based on gedunin content measured by HPLC. Further *in vivo* testing and toxicology and clinical trials will be required before such a medicine can be considered.

Intellectual property

The question of intellectual property rights is not fully resolved in Togo as in many West African countries. Medicinal plants are generally regarded as a community resource, but individual healers jealously guard their individual recipes. It is clear, however that benefits derived from phytomedicines should be shared with the traditional healers.

Acknowledgement

This project is funded by the Canadian International Development Agency's PRIMTAF program, managed by COGESULT. Research funds in Canada were granted by the Natural Sciences and Engineering Research Council of Canada.

References

Adjanohoun, E.J., Ahyi, M.R.A., Ake Assi, L., Akpagana, L., Chibon, P., Al-Hadji, A., Eyme, J., Garba, AM., Gassita, J.N., Gbeassor, M., Goudote, E., Guinko, S., Hodouto, K.K., Hougnon, P., Keita, A., Keoula, Y., Kluga-Ocloo, W.P., Lo, I., Siamevi, K.M., and Taffame, K.K. (1986). *Contribution aux Etudes Ethnobotaniques et floristiques du Togo*, Agence de la Coopération Culturelle et Technique, Paris.

Awang, D.V.C. and Arnason, J.T. (1996). Plantes prioritaires pour le développement des phytomédicaments, Proc. Symposium International sur les phytomédicaments éthiques, Lomé, Université du Bénin, Lomé.

Iwu, M. M. (1995). *Handbook of African Medicinal Plants*. CRC Press, Boca Raton.

Khalid, S.A., Farouk, A., Geary, T.G., and Jensen, J.B. (1986). Potential antimalarial candidates from African plants. *Journal of Ethnopharmacology* **15**, 201-209.

Khalid, S.A., Duddeck, H., and Gonzalez-Sierra, M. (1989). Isolation and characterization of an antimalarial agent from the neem tree. *Journal of Natural Products* **52**, 922-927.

Leaman, D.J., Arnason, J.T., Razali, Y., Sangat-Roemantyo, H., Soedjito, H., Angerhofer, C., and Pezzuto, J.M. (1995). Malaria remedies of the Apo Kayan, E. Kalimantan, Indonesian Borneo. *Journal of Ethnopharmacology* **49**, 1-16.

Le Grand, A. and Wondergem, P. (1990). *Herbal Medicine and Health Promotion*. Royal Tropical Institute, Amsterdam.

Lozoya, X., Meckes, M., Abou-Zaid, M., Tortoriello, J., Nozzolillo, C., and Arnason, J.T. (1994). Quercetin glycosides in extracts of *Psidium guajava* leaves and determination of a spamolytic principle, *Archivos Medico de. Mexico* **25**, 11-15.

Lutterodt, G.D. and Maleque, A. (1989). Effect on mice locomotor activity of a narcotic-like principle from guava leaves. *Journal of Ethnopharmacology* **24**, 219-231.

Marles, R. , Kaminski, J., Arnason, Pazos, L., Heptinstall, S., Fischer, N., Crompton, C.W., Kindack, D., and Awang, D.V.C. (1992). A bioassay for inhibition of seretonin release: screening for migraine prophylactics. *Journal of Natural Products* **55**, 1044-1056.

13. Research, development and production of plant-based medicines in Africa. Example of Rwanda.

L. VAN PUYVELDE

Department of Organic Chemistry, Faculty of Agricultural and Applied Biological Sciences, University of Gent, Coupure links 653, B-9000 Gent, Belgium

---

Introduction

Rwanda is one of the smallest and poorest countries in Central Africa. It possesses a wealth of traditional medicine that is still very much alive.

This traditional medicine can be viewed as a rich resource and as having real potential in resolving the country's various health problems. But, sadly, in the countries where this form of medicine flourishes, it cannot find the socio-technical support needed to exploit it and get the most out of it. The World Health Organization has recognized the part that this medicine must play in its health program for the year 2000.

In Rwanda, as in other developing countries, around 80% of the population has recourse to this medicine which is packed with a great mass of knowledge that is still secret and that needs to be saved quickly if it is not to disappear altogether. This is made all the more urgent by the fact that the arsenal of so-called modern medicines can no longer cope with the many different health problems and that these medicines are less and less available. It is very fortunate that peoples or groups of people who have no other recourse to medicine than the traditional variety have not abandoned it. It makes it possible to come upon new combined resources that guarantee endogenous development while opening up new capacities for further progress and co-development.

If, for peoples of the least developed countries like Rwanda, medicinal plants and traditional medicine retain such an important place, not only for their own benefit but also for that of the whole human race, are there not grounds for preserving this heritage or, better still, exploiting and developing it as a response to bad development in the South and in the North? Indeed is there not the obligation to do so? That at least was the idea that motivated certain researchers at the National University of Rwanda (N.U.R.) in their efforts to (re)direct their research activities and give them greater social usefulness with sustainable development as the goal.

The Center of Research on Pharmacopoeia and Traditional Medicine in Rwanda: CURPHAMETRA

It was with the aim of first preserving the health of the Rwandese people and rationally exploiting its natural resources that research work in this domain was undertaken at the end of 1971 at the Faculty of Medicine of the National University of Rwanda, in Butare, by a small network of young researchers. As a result of the growing interest shown by researchers in other faculties, an interfaculty and multidisciplinary research group was set up in 1977.

The development of this research work and the promising results obtained prompted the National University of Rwanda in 1980 to set up within its precincts the University Center of Research into Pharmacopoeia and Traditional Medicine (Curphametra) as an autonomous and multidisciplinary research center. The principle goals assigned to Curphametra were at the start :

- to study and exploit traditional medicine;
- to undertake phytochemical and pharmacological research into Rwandese medicinal plants;
- to produce medicines from medicinal plants and other commercially viable products from local raw materials;
- to help in producing a Rwandese pharmacopoeia;
- to take part in providing university training in the domain of natural products.

Such objectives were difficult to get accepted in a university milieu given the prevailing mentality. A university center of research simply does not go in for production. As there was no pharmaceutical industry in Rwanda, the researchers decided on their own initiative that, in order to give great practical value to the results of their research and to help with the development of a country lacking in medicines, they would create a pilot production department, even if that meant taking up arms against academia.

Apart from the involvement of the healer of which we will say more later, the center has been able to extend its many intersectorial activities which include interdisciplinary research that goes beyond the strict bounds of chemistry and biomedicine, the development of new drugs, production and marketing.

At present, Curphametra forms part of the Institute of Scientific Research and Technology. This institute was created in 1989 by the Rwandese government with the aim of stimulating and favoring programs of research necessary for the development of the country.

Nevertheless, the center remains localized within the university campus so as to remain closely linked to education and research (training advanced students, associating academics from different disciplines with the work of the center, etc.).

The production of medicines from local raw material, *i.e.* traditional prescriptions and medicinal plants, offers several significant advantages:

- it validates knowledge that has been handed down from previous generations;
- it reduces imports and so cuts the external debt;
- it creates jobs at several levels;
- it makes worthwhile medicines available to the population at a reasonable price;
- it opens up the possibility of exports;
- it protects and preserves the natural environment.

In order to achieve these objectives, Curphametra has as its disposal a dispensary of traditional medicine, an interdisciplinary research program and a unit for the production of plant derived drugs (see Fig. 13.1.).

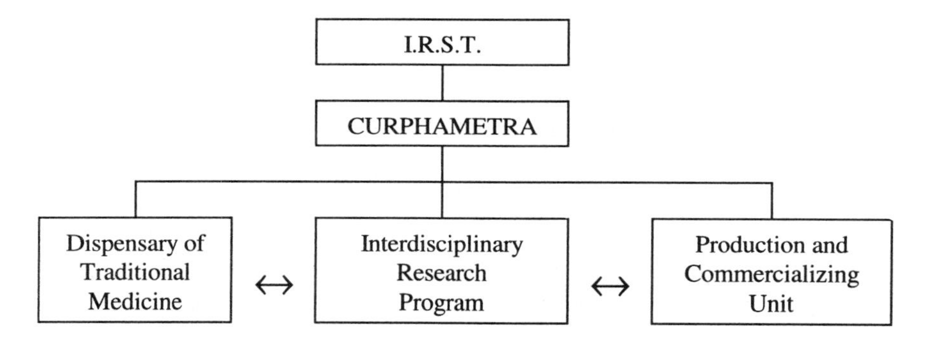

Fig. 13.1. Chart illustrating Curphametra's structure and interaction.

In order to attain as far as possible the objective considered of prime importance by the researchers (the development and production of drugs derived from local raw materials), a six-stage work plan was developed :

1) The promotion of traditional Rwandese medicine.
2) The use of officinal medicinal plants for the production.
3) The acclimatization of foreign medicinal plants
4) The production of solvents and excipients.
5) The development of commercially viable plants.
6) Research and development of new medicines.

Promotion of traditional Rwandese medicine

To help in the promotion of traditional Rwandese medicine we started *systematic ethnobotanical investigations*, we created a *Dispensary of Traditional Medicine*, and we began *checking on the efficacy of traditional remedies*.

Ethnobotanical Investigations - A growing interest in these investigations led the researchers of Curphametra to systematize them in order to have an inexhaustible source of information. This was progressively conceived in respect of certain geographical areas (Van Puyvelde *et al.* 1975, 1977a and 1982). Other ethnobotanical investigations were carried out in the prefecture of Kibuyé (1978, all communes), Butare (1981, all communes), Gisenyi (1987, 10 communes), and Kibungo (1988, 6 communes).

This ethnobotanical inventory consists of information on medicinal flora collected from the healers themselves by a multidisciplinary team composed often of botanists, pharmacists, chemists and physicians.

In the course of these investigations, the healer explained what part of the plant was used, in what form he administered his medication and how he prepared it; likewise he gave the indications and contra-indications. Specimens were collected to make up the collections of the herbarium.

These investigations led to the scientific identification of several plant species known to our healers and which they employ in making up their prescriptions. This inventory serves as a tool for phytochemical and pharmacological research into plants appearing to have a particular value.

Dispensary of Traditional Medicine - In order to benefit from the information in the possession of the holders of traditional know-how, Curphametra formed a Dispensary (or Community Clinic) of Traditional Medicine where healers and doctors could carry out consultations with patients. This dispensary began with six healers who formed part of the medical team. The role of the medical team was restricted solely to directing and following the progress of the patients and monitoring the therapy. It also gave advice on certain matters of general concern : cleanliness of the working surroundings and on the making up of the medicines, the size of the doses and the period of conservation.

Attempts to make diagnostic comparisons with what the healers think to be appropriate are in progress. This collaboration, which has been strengthened by the desire to have things organized, has allowed our healers to learn more about other health activities. The confidence of the consultant doctors has also increased. The healers themselves have agreed to reveal their secrets.

Research into traditional medicine must lead to an improvement in the way medicines are made up and their posology, while studying their toxicity. It will also have to contribute to the self-organization of the healers which will facilitate training in the area of diagnosis.

Efficacy of traditional remedies - The results obtained (see below : Research and development of new medicines) by Curphametra from studying several medicinal plants constitutes additional proof of the value of the plants exploited by traditional Rwandese therapy. A new drug, an antimycotic, has even been developed as a result of information obtained from the healers. In the near future we should be able to propose a series of traditional remedies for integration into the system of basic health care.

The link, or better the network, of researchers and local healers, which grew out of personal contacts and developed in a climate of confidence, was the first necessity in the rational study of Rwandese medicinal plants. Without access to traditional data, it is difficult to obtain tangible results very quickly, even though it is possible to study plants selected at random and attain results. It has, however, been found that, by studying the prescriptions traditionally used, the number of medicinal plants found to be active is much higher (Boily and Van Puyvelde 1986). And what is more, it is the more complex approach allowed by the exchange and reciprocal training between doctors and local healers that is a determining factor in the rapid success of the work.

Use of officinal medicinal plants for the production and acclimatization of foreign medicinal plants

We preferred to begin producing medicines from plants already known and described in pharmacopoeias for the simple reason that it took less time and did not require long and costly research.

These plants are "Officinal Medicinal Plants" listed in the different pharmacopoeias : European, Belgian, French, English, etc. In this case, all one has to do is to follow the methods of dosage, analysis, extraction, etc., already described in these pharmacopoeias.

Moreover, if the medicines proposed are well known ones to be found in official pharmacopoeias, the doctor who has been trained in the modern system and who often knows nothing of the origin of the medicines he prescribes, has no reason to refuse these drugs. Subsequently, when the center has been able to gain the doctor's confidence through its medicines which are therefore regarded as of good quality, it is easier to introduce new medicines derived from medicinal plants.

We made an inventory of the officinal medicinal plants that already grow in Rwanda and the following plants are grown or picked :

- *Datura stramonium* : the tincture is prepared from the leaves (antispasmodic),
- *Eucalyptus globulus* : the tincture and essential oil are produced from the leaves (pulmonary disinfectant),
- *Capsicum frutescens* : the liquid extract is prepared from the fruit (counter-irritant),
- *Plantago lanceolata* : the leaves are used for preparing the liquid extract (antitussive).

With a view to increasing the number of medicinal plants found in pharmacy, we introduced several well-known plants. At present, three plants are being grown and used for production purposes :

- tincture and liquid extract from the flower petals of *Calendula officinalis* (anti-inflammatory, healing substance),
- tincture and liquid extract from the green part of *Thymus vulgaris* (antitussive),
- essential oil from the leaves of *Mentha saccharinensis*.

For the production of these different extracts, an extraction and distillation unit was set up. For this type of production, we use dryers, crushing machines, percolators, extractors, concentrators and distillers.

For the production of medicines, we have pharmaco-technical equipment at our disposal for making linctuses, solutions, tablets, ointments and powders. There is also a well-equipped laboratory for purposes of analysis and checking and we also have 100 hectares of land for growing medicinal plants.

Curphametra produces several medicines using various plant extracts :

- a mouth disinfectant
- two cough syrup and a cough solution
- an antispasmodic syrup
- an anti-inflammatory healing ointment
- an antirheumatic solution
- four herbal teas

All these medicines are sold wholesale to the hospitals, health centers and community clinics or dispensaries and retail to the private pharmacies. A second series of medicines also derived from well-known plants will be developed and marketed in the near future.

After these different stages, the medicine is packed (for wholesale or retail delivery) and marketed. Even if the quality of the drug is the primary objective, it is absolutely essential, despite the low income of local consumers, that the packaging be suitable and attractive for sale in private pharmacies so that it can compete with imported medicines. Also explanatory leaflets for the different medicines have been produced for the benefit of doctors, pharmacists and paramedical staff. What is more, one of the pharmacists from Curphametra makes regular visits to the private pharmacies, doctors and hospitals in order to follow the progress of the medicines and do some advertising to promote the drugs.

It should be noted that several new medicines have been developed and manufactured (see below : *Research and development of new medicines*).

Table 13.1. shows the evolution of our small-scale production. This evolution in production and marketing has enabled us to program and start production on a large scale. We have begun to promote intensively our products and we foresee that from 1992-1993 we shall be able to start large-scale production so as to satisfy the Rwandan market.

Table 13.1. Annual evolution of the production of the various pharmaceutical forms

Products	Quantities		
	86/87	87/88	88/89
- Syrups	481 kg	2690 kg	5763 kg
- Solutions	34 kg	114,5 kg	788 kg
- Ointments	-	50 kg	315 kg
- Alcohol	1362 L	3540 L	9260 L

As can be seen, there are three pharmaceutical forms being produced : syrups, solutions and ointments. Other pharmaceutical forms will be developed and manufactured, namely tablets, powders, suppositories and capsules.

Self-financing for the production unit will be reached around 1994. What is more, by not just increasing production capacity but also by broadening the range of medicines and developing new drugs, we aim to become financially independent both in respect to production and in development research by around 1997.

Production of solvents and excipients

The development and the production of medicines do not only involve the use of active principles but also require solvents and excipients which account for a large part of the production cost.

Solvents - In the production of various extracts from medicinal plants (*e.g.* tinctures, liquid and soft extracts) a mixture of water and ethanol is used. These two solvents are prepared in the production unit. - A simple distiller serves to produce distilled water. Ethyl alcohol is obtained from the molasses at the sugar refinery (Mungarulire *et al.* 1980). After the diluted molasses has fermented, it is distilled a first time (ethyl alcohol 60-80°), then the alcohol is passed through a carbon active reactor to remove the impurities and finally the alcohol is rectified to 95°. For the moment, the production capacity is 12,000 L of 95° ethyl alcohol; this capacity will be extended to reach a minimal capacity of 45,000 L a year. This alcohol does not serve only as a solvent for extraction purposes but is also commercialized as a disinfectant. For the needs of the laboratory we distill, on a small scale, raw petroleum to obtain petroleum ether.

Excipients - Excipients often play a large part in the production of medicines. The few possibilities that can be explored in the near future are starch, which can be produced from cassava; it serves among other things in the manufacture of tablets. Beeswax, which is produced to some extent everywhere in the country, can be used in the preparation of certain ointments.

It is absolutely essential to start research projects for developing solvents and excipients using local raw material to reduce importation costs in the manufacture of medicines.

Exploitation of economically viable plants

Here it is a question of plants which provide essential oils, tannins, dyes, alkaloids, etc., and that have a certain economic value.

- Several plants have been analyzed for their tannin content with a view to finding exploitable plants. These tannins could replace the plant tannins imported by the local tanneries (Dubé *et al.* 1979).
- For the food industry and the textile industry, plants containing dyes must be found. In Rwanda several sorts of *Indigofera* exist, plants that produce indigo, a well-known dye for fabrics. A new biological dye, which colors the material yellow, has been developed from the fungus *Pisolithus arhizus* (Van Puyvelde *et al.* 1983).
- Alkaloids: a method for extracting industrially berberine from the roots of *Thalictrum rynchocarpum* has been developed (Dubé and Van Puyvelde 1979). Berberine is an anti-amoebic drug much used in Asia.
- Essential oils : one of Curphametra's research programs is concerned with the exploitation of essential oils for local use and for exportation. The essences have a variety of applications : in pharmacy, hygiene, food, cosmetics, etc.

At present, Curphametra produces the oils of *Eucalyptus globulus*, *E. smithii* and *Mentha saccharinensis*, used in the preparation of several medicines. In the production of an antimosquito cream and candle, the following plants are used : *Pelargonium graveolens, Eucalyptus citriodora* and *Cymbopogon citratus*. The substantial production of essential oils for the Rwandese cosmetics industry is also envisaged.

Research and development of new medicines

At Curphametra, the studies carried out on medicinal plants are for the most part based on the information provided by the local healers in the course of the *ethnobotanical investigations.* These plants are the subject of a *botanical study* in order to determine the exact scientific name of the plant. Subsequently, the extracts of these plants, prepared with the help of various solvents, are submitted to a series of *biological and pharmacological tests* to discover one or other useful biological activity.

At present, Curphametra is able to carry out several biological and pharmacological activities : antimicrobial (Boily and Van Puyvelde 1986; Van Puyvelde *et al.* 1983), antipyretic (Hakizamungu 1981), hepato-protective (Van

Puyvelde *et al.* 1989), antispasmodic (Van Puyvelde *et al.* 1987a), antimitotic (Van Puyvelde *et al.* 1988), insecticide-pesticide (Weaver *et al.* 1992), acaracidal (Van Puyvelde *et al.* 1985), antiprotozoan (Hakizamungu *et al.* 1992), cardiotonic (Uwilingiyimana *et al.* 1990), anticonvulsive (Mukarugambwa *et al.* 1990), antidiarrhoeal (Maikere-Fanyo *et al.* 1989), antiscabies (Heyndrickx *et al.* 1992)

After a study of the literature, *phytochemical bio-guided studies* are carried out on those plants that display useful biological and pharmacological activity in order to obtain the active principles possessing the previously identified activity.

After the chemical and pharmacologico-toxicological studies of the medicinal plants, *clinical trials* started. If the results are positive, the necessary steps can then be taken to obtain *legal recognition* of the new medicines.

Three new products were developed :

- an antiscabies solution and ointment from the tubers of *Neorautanenia mitis*
- a pesticide from the leaves of *Tetradenia riparia*
- an antimycotic ointment from the roots of *Pentas longiflora*

A drug to treat scabies - The Rwandese farmer employs the powder of the tubers of *Neorautanenia mitis* (Leguminosae), mixed with butter, to treat his calves for scabies.

In a biological screening of several medicinal plants we found that the tuber of *N. Mitis* possessed acaricidal and antiscabies activity (Heyndrickx *et al.* 1992). The bio-guided phytochemical study led to the isolation of several isoflavones of which 12a-hydroxyrotenone was responsible for the biological activity (Van Puyvelde *et al.* 1987b). Here it should be noted that in the past, in western medicine, rotenone was used to treat scabies. An antiscabies solution and an antiscabies and disinfecting ointment were prepared from an alcohol extract (total extract) from the tuber of *N. mitis*.

After preliminary trials on rabbits infected with scabies, a clinical study on humans was carried out. More than 500 patients suffering from scabies were treated with the solution or the ointment. A comparative study of these drugs with a known antiscabies drug and a double blind study formed part of these clinical studies (Van Puyvelde *et al.* 1990). All the persons treated were cured (Figs. 13.2. and 13.3.). We found no side effects, either with the solution or with the ointment.

The solution and the antiscabies ointment were commercialized by Curphametra.

Antiscabies solution:	Tincture of Neorautanenia	
Antiscabies and disinfectant ointment:	Liquid extract of Neorautanenia	5.0 g
	Vioform	1.0 g
	Lanoline	10.0 g
	Vaseline *s.q. ad.*	100.0 g

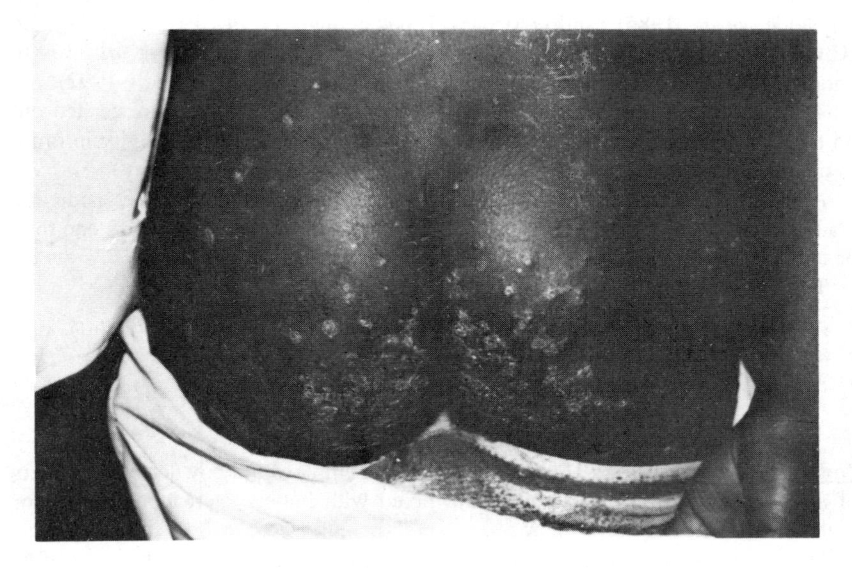

Fig. 13.2. Patient with infected scabies before treatment

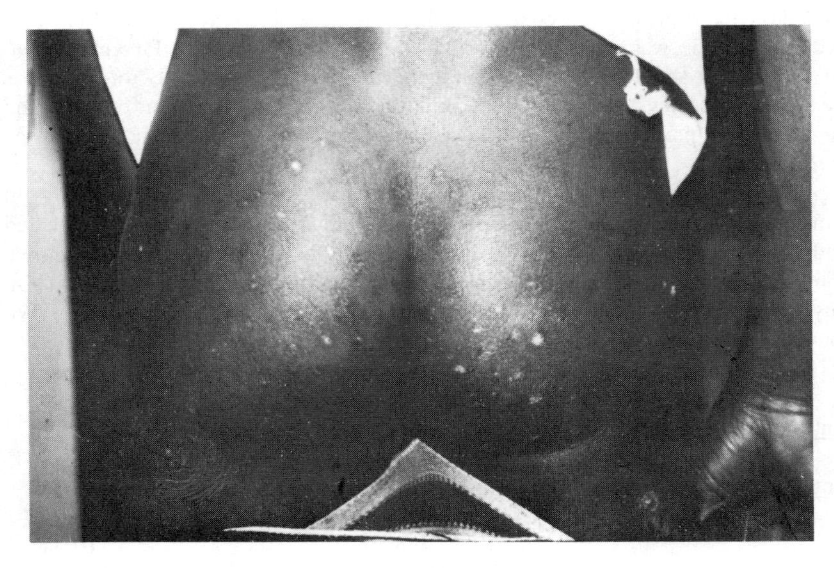

Fig. 13.3. Same patient, 1 week after treatment with *Neorautanenia mitis* preparation .

A pesticide - For Rwanda, the development of pesticides from local drugs, *i.e.* medicinal plants, is acquiring capital importance, given the quantity of pesticides being imported to support the priority program for self-sufficiency in food and, in contrast, the poor purchasing power of the population. In 1985, the importation of pesticides exceeded 1,400 tones. Moreover, these pesticides are reserved almost only for development projects.

The first new pesticide was developed from the leaf of *Tetradenia riparia* (Lamiaceae). This plant is well-known in traditional medicine. It is grown around houses for medical reasons, but it is also used for preserving beans in traditional silos (Van Puyvelde *et al.* 1975). Biological screening showed that the leaves of *Tetradenia riparia* possessed several useful biological and pharmacological activities (Boily and Van Puyvelde 1986; Van Puyvelde *et al.* 1988; Hakizamungu *et al.* 1992). The bio-guided phytochemical study led to several new molecules (Van Puyvelde *et al.* 1979 1981 and 1987c; De Kimpe *et al.* 1982) of which the 8(14),15-sandaracopimaradiene-7a,18-diol is the most active principle. This new diterpenediol displayed several activities : an antimicrobial activity (Van Puyvelde *et al.* 1986), an antispasmodic activity (Van Puyvelde *et al.* 1987), an antitrichomonas activity (Hakizamungu *et al.* 1988) and inhibited the growth of wheat rootlets (Van Puyvelde *et al.* 1988). The antimicrobial activity (Hakizamungu *et al.* 1988) has been used against *Pseudomonas solanacearum*, an important potato pest in Rwanda.

A formulation with an alcohol extract from the leaf of *T. riparia* has been perfected. *In vitro* and *in vivo* trials have been carried out with the formulation. After trials in the greenhouse and in the fields (*in vivo*), we found that our product completely prevented the development of the bacteria and resulted in an increase in the potato yield of roughly 35% compared with the sets that were not inoculated and not treated and of roughly 120% with those sets that were inoculated and not treated (Butare 1987).

The formulated extract, which can be used in a spray or in a simple solution, is ready to be marketed.

Pesticide solution (concentrated)	Dry extract of Tetradenia	200 g
	Tensiofix B	50 g
	Ethanol 80°	950 ml
	(Dilution in water 200 x)	

An antimycotic - One of the local healers from our Dispensary of Traditional Medicine used the powder from the root of *Pentas longiflora* (Rubiaceae) to treat pityriasis versicolor, a dermatological disease of fungal origin. In a study involving several plants used as antifungals in traditional medicines, a high level of activity was found in the roots of *P. longiflora*. This led to the isolation of the active principle, a naphthoquinone. An ointment manufactured from the total extract of the roots of this plant was tested clinically on 80 persons suffering from pityriasis versicolor. In this case too, all the persons treated with our new antimycotic ointment were healed, without any undesirable effects being observed (Van Puyvelde *et al.* 1994).

<u>Antimycotic ointment</u>	Dry extract of Pentas	5.0 g
	Lanoline	10.0 g
	Vaseline *s.q. ad.*	100.0 g

Final observations

It goes without saying that for this research to be brought to a satisfactory conclusion there was a great deal of scientific collaboration both within the country and with the outside world.

Several academics from the Faculty of Medicine, Science and Agronomy at the National University of Rwanda and researchers from other institutions like the I.S.A.R. (*Institut des Sciences Agronomiques du Rwanda*) carry out research projects within an in collaboration with Curphametra. The latter also takes students from several faculties who do their degree thesis on one or several Rwandese medicinal plants.

There exists, moreover, intense collaboration with the Faculty of Economic and Social Sciences and Management to help the Center develop the commercial section of the production unit.

Scientific collaboration between the Laboratory of Organic Chemistry at the Faculty of Agronomy (University of Gent - Belgium) and Curphametra exists in order to determine the structure of the molecules (active principles) which have been isolated in Rwanda from medicinal plants. Several biological trials currently being conducted at Curphametra, such as the search for antiprotozoan agents, have been developed at the Institute of Tropical Medicine in Antwerp (Belgium) where several Rwandese researchers took a six-month training course. The Institute of Pharmacognosy and Phytochemistry, of the University of Lausanne participated in developing Curphametra.

All this collaboration, inside and outside the country, has led to joint publications. It has to be noted that the Pilot Plant for the production was financed by the Rwandese Government, the Belgian Administration for Development Cooperation, the United Nations Industrial Development Organization and the World Health Organization.

References

Boily, Y. and Van Puyvelde, L. (1986). Screening of medicinal plants of Rwanda (Central Africa) for antimicrobial activity. *Journal of Ethnopharmacology* **16**, 1-13.

Butare, J.B. (1987). Incidence des traitements aux extraits végétaux sur l'épidémiologie de la bactériose vasculaire de la pomme de terre (due au *Pseudomonas solanacearum* E.I. Smith) au Rwanda. *Mémoire de fin d'études*. UNR, Butare (Rwanda).

De Kimpe, N., Schamp, N., Van Puyvelde, L., Dubé, S., Chagnon, M., Borremans, F., Anteunis, M.J.O., Declercq, J.P., Germain, G., and Van Meerssche, M. (1982). Isolation and structural identification of 8(14),15-sandaracopimaradiene-7a,18 diol from *Iboza riparia*. *Journal of Organic Chemistry* **47**, 3628-3630.

Hakizamungu, E. (1981). Recherche d'agents antipyrétiques et/ou antiinflammatoires dans la flore rwandaise. *Mémoire de fin d'études*. UNR, Butare (Rwanda).

Hakizamungu, E., Van Puyvelde, L., Wéry, M., De Kimpe, N., and Schamp, N. (1988). Active principles of *Tetradenia riparia* III. Antitrichomonas activity of 8(14),15-sandaracopimaradiene-7a,18 diol. *Phytotherapy Research* **2**, 207-208.

Hakizamungu, E., Van Puyvelde, L., and Wéry, M. (1992). Screening of rwandese medicinal for antitrichomonas activity. *Journal of Ethnopharmacology* **36**, 143-146.

Heyndrickx, G., Brioen, P., and Van Puyvelde, L. (1992). Study of Rwandese medicinal plants used in the treatment of scabies. *Journal of Ethnopharmacology* **35**, 259-262.

Maikere-Fanyo, R., Van Puyvelde, L., Mutwewingabo, A., and Habiyaremye, F.X. (1989). Study of Rwandese medicinal plants used in the treatment of diarrhoea. I. *Journal of Ethnopharmacology* **26**, 101-109.

Mukarugambwa, S., Van Puyvelde, L., Sibobugingo, B., and Dumont, P. (1990). Activité anticonvulsivante des plantes médicinales rwandaises utilisées dans le traitement de l'épilepsie. *10ème Anniversaire du CURPHAMETRA. Colloque International sur la Recherche et la Production de Médicaments à base de Plantes Médicinales, Kigali (Rwanda)*, 26 février-3 mars 1990.

Mungarulire, J., Dubé, S., and Van Puyvelde, L. (1980). Essais pour la production d'éthanol au Rwanda. *Bulletin Agricole du Rwanda* **3**, 200-203.

Uwilingiyimana, A., Brioen, P., and Van Puyvelde, L. (1990). Study of the cardiotonic effect of rwandese medicinal plants by monitoring the "*in vitro*" spontaneous cardiac frequency. *10ème Anniversaire du CURPHAMETRA. Colloque International sur la Recherche et la Production de Médicaments à base de Plantes Médicinales*, Kigali (Rwanda), 26 février-3 mars 1990.

Van Puyvelde, L., Pagézy, H., and Kayonga, A. (1975). Plantes médicinales et toxiques du Rwanda (I). *Afrique Médicale* **14**, 925-930.

Van Puyvelde, L., Mukarugambwa, S., Rwangabo, P.C., Ngaboyisonga, M., and Runyinya Barabwiliza (1977a). Plantes médicinales et toxiques du Rwanda (II). *Afrique Médicale* **16**, 531-534.

Van Puyvelde, L., Ngaboyisonga, M., Rwangabo, P.C., Mukarugambwa, S., Kayonga, A., and Runyinya Barabwiliza (1977b). *Enquêtes ethnobotaniques sur la médecine traditionnelle rwandaise, tome 1: Préfecture de Kibuyé*. UNR, Butare (Rwanda).

Van Puyvelde, L., Dubé, S., Uwimana, E., Uwera, C., Domisse, Esmans, E.L., Van Schoor, O., and Vlietinck, A.J. (1979). New α-pyrones from *Iboza riparia*. *Phytochemistry* **18**, 1215-1218.

Van Puyvelde, L., De Kimpe, N., Dubé, S., Chagnon, M., Boily, Y., Borremans, F., Schamp, N., and Anteunis, M.J.O. (1981). 1',2'-Dideacetylboronolide, an α-pyrone from *Iboza riparia*. *Phytochemistry* **20**, 2753-2755.

Van Puyvelde, L., Rwangabo, P.C., Runyinya Barabwiliza, Ayobangira, F.X., and Mungarulire, J. (1982). Plantes médicinales et toxiques du Rwanda (III). *Afrique Médicale* **21**, 401-404.

Van Puyvelde, L., Geiser, I., Rwangabo, P.C., and Sebikali, B. (1983). Rwandese herbal remedies used against gonnorrhea. *Journal of Ethnopharmacology* **8**, 279-286.

Van Puyvelde, L., Geysen, D., Ayobangira, F.X., Hakizamungu, E., Nshimyimana, A., and Kalisa, A. (1985). Screening of medicinal plants of Rwanda for acaricidal activity. *Journal of Ethnopharmacology* **13**, 209-215.

Van Puyvelde, L., Nyirankuliza, S., Panebianco, R., Boily, Y., Geizer, I., Sebikali, R., De Kimpe, N., and Schamp, N. (1986). Active principles of *Tetradenia riparia*. I. Antimicrobial activity of 8(14),15-sandaracopimaradiene-7a,18 diol. *Journal of Ethnopharmacology* **17**, 269-275.

Van Puyvelde, L., Lefebvre, R., De Kimpe, N., Mugabo, P., Schamp, N. (1987a). Active principles of *Tetradenia riparia*-II. Antispasmodic activity of 8(14),15-sandaracopimaradiene-7a,18 diol. *Planta Medica* **2**, 156-158.

Van Puyvelde, L., De Kimpe, N., Mudaheranwa, J.P., Gasiga, A., Schamp, N., Declercq, J.P., and Van Meerssche, M. (1987b). Isolation and structural elucidation of potentially insecticidal and acaricidal isoflavone-type compounds from *Neorautanenia mitis*. *Journal of Natural Products* **50**, 349-356.

Van Puyvelde, L., De Kimpe, N., Borremans, F., Zhang, W., and Schamp, N. (1987c). 8(14),15-sandaracopimaradiene-2a,18 diol, a minor constituent of the Rwandese medicinal plant *Tetradenia riparia*. *Phytochemistry* **26**, 493-495.

Van Puyvelde, L., De Kimpe, N., Ayobangina, F.X., Costa, J., Nshimyumukiza, P., Boily, Y., Hakizamungu, E., and Schamp, N. (1988). Wheat rootlet growth inhibition test of Rwandese medicinal plants. Active principles of *Tetradenia riparia* and *Diplolophium africanum*. *Journal of Ethnopharmacology* **24**, 233-246.

Van Puyvelde, L., Kayonga, A., Brioen, P., Costa, J., Ndimubakunzi, A., De Kimpe, N., and Schamp, N. (1989). The hepato-protective principle of Hypoestes triflora leaves. *Journal of Ethnopharmacology* **26**, 121-127.

Van Puyvelde, L., Heyndrickx, G., Brioen, P., Hakizayezu, D., and De Kimpe, N. (1990). Development of anti-scabies drug from the roots of *Neorautanenia mitis*. *International Joint Symposium of Biology and Chelistry of Active Natural Substances,* Bonn, Germany, July 17-22, p. 157.

Van Puyvelde, L., Hakizayezu, D., Brioen, P., De Kimpe, N., De Vroey, C., Bogaerts, J., and Hakizamungu, E. (1994). Development of an antimycotic ointment from the roots of *Pentas longiflora*. *International Congress on Natural Products Research*, Halifax, Nova Scotia, Canada, July 31 to August 4, p. 119.

Weaver, D.K., Dunkel, F.V., Cusker, J.L., and Van Puyvelde, L. (1992). Oviposition patterns in two species of bruchids (Coleoptera : Bruchidae) as influenced by dried lea-ves of *Tetradenia riparia* a perennial mint (Lamiales · Lamiaceae) that suppress population size. *Physiology and Chemical Ecology* **21,** 1121-1129.

14. Production of traditional medicine: preparations accepted as medicines in Mali

D. DIALLO[1], B.S. PAULSEN[2] AND B. HVEMM[3]

[1]*Département de Médecine Traditionnelle, B.P. 1746 Bamako, Mali;* [2]*Institute of Pharmacy, Section of Pharmacognosy, University of Oslo, Oslo, Norway and* [3]*Center for Development and the Environment, University of Oslo, Oslo, Norway*

Introduction

Mali is a West African country without a coastline. Its sanitary situation is characterized by the predominance of parasitic illnesses, infections and nutritional diseases together with an insufficiency of qualified health workers, medicines and equipment.

Due to this situation, 80% of the population use traditional medicine. In the Bandiagara District a study has shown that only 19.4% of current diseases have been cured by conventional medicine. Generally traditional medicine has been through 4 stages :

- the only means of therapy;
- the clandestine practice;
- the tolerance period;
- the creation of a research institute to promote traditional medicine.

In Mali, the government created, in 1968, an Institute named " *Institute of Phytotherapy"* which, in 1973, became the *National Institute of Pharmacopoeia and Traditional Medicine*. Now, this institute is called *Department of Traditional Medicine* (DMT). It is a part of the *National Institute of Research in Public Health*. The main objectives of this Department are to promote traditional medicine, the organization of the traditional system of health care and the production of medicines from local resources.

Methodology

In order to attain the objectives the methodology employed by the DMT includes: bibliographic researches, ethnobotanic surveys, phytochemical and pharmacological studies, formulation of improved traditional medicines, clinical tests (in collaboration with " conventional doctors") and evaluation of the results by a scientific and technical committee.

Results

Since its creation, the Department of Traditional Medicine has been working mainly to bring traditional healers and herborists into associations. In addition, it started the study of the medicinal plants of Mali, through the creation of herbaria and ethnobotanic surveys. Phytochemical, pharmacological and clinical tests were also performed with some of these plants. The Institute has been working on more than 12 different traditional medicines claimed to be effective in the treatment of different diseases, such as malaria, dysentery, diarrhea, jaundice (icterus), ulcers, gastritis, constipation, asthma, cough, diabetes, hypertension, stomatitis and skin diseases. Since 1990, four Improved Traditional Medicines "Médicaments Traditionnels Améliorés" (M.T.A.) have been accepted as medicine in Mali. They are sold in pharmacies like conventional medicines. Those medicines are: Hepatisane®, Laxa cassia®, Dysenteral® and Balembo®.

This paper shows traditional uses and results of phytochemical, pharmacological and clinical studies which have been done on these four different medicines.

Hepatisane®

Hepatisane® is a finished product, in tea-bags containing 10 g of powdered dry leaves of *Combretum micranthum* G. Don. (Combretaceae). This medicine is used against digestive disturbances associated with liver diseases and jaundice (constipation, fat intolerance, sickness, loss of appetite).

The traditional indications of *C. micranthum* in Mali are fever, constipation, indigestion, stomachache, weakness, hypertension, anorexia, headache and stomatitis.

The phytochemical studies done in our laboratory showed that the leaves of *C. micranthum* contain coumarins, carotenoids, flavonoids, catechic and gallic tannins, sterols, terpenes and alkaloids. Other studies have shown the presence of the alkaloids hydroxy-stachydrine, betain and choline (Paris and Moyse-Michon 1956; Bassene *et al.* 1986), of the heterosides of flavones vitexin and orientin (Jentsch *et al.* 1962; Bassene *et al.* 1987), of inositol, sorbitol and mannitol (Bassene *et al.* 1985), and of organic acids: gallic, tartric, malic, citric and oxalic acids (Kerharo and Adams 1974).

The cholagogue effect of the leaves has been studied in our laboratory. In addition the plant has also been classified as tissular and hepatorenal diuretic (Paris and Moyse-Michon 1956; Balansard and Delphaut 1946).

Decoction of Hepatisane® (2 bags/day) has been administrated to voluntary persons suffering from jaundice during many years, at the clinic of the Department of Traditional Medicine. These persons were followed by seric control of bilirubine and transaminases which became normal within 2 and 3 weeks after beginning of treatment. It has to be noticed that Hepatisane® should not be used in cases of obstruction of the choledoc canal or when serious hepatic cellular insufficiency is associated.

Laxa cassia®

Laxa cassia® is a finished product presented in boxes of 4 tea-bags each containing 5 g of the powdered dry leaves of *Cassia italica* Mill. (Caesalpinaceae). This medicine is used against constipation.

The traditional indication of *C. italica* in Mali is constipation through a concept of "cleaning of the stomach". In some cases, intoxication have been described after self-overdosage of this traditionally used plant.

The phytochemical studies done in our laboratory showed that the leaves contained: mucilages, anthracenic derivatives, sterol, carotenoids and coumarins. Sennosides have been identified as the active principle of this drug. In pharmacological studies, cholagogue, laxative and/or purgative effects were established.

Laxa cassia® is used in decoctions of one bag per day. Excess of automedication represents some risk, and a long term use can provoke "laxative disease" characterized by diarrhea with abdominal pain and disorder of electrolytic equilibrium (hypokaliemia, hyponatremia, etc.). For this reason, Laxa cassia® should not be used by pregnant and breast feeding women.

Balembo®

Balembo® is the local name of the plant which is used for its preparation. It is a finished product, presented in syrup children and adults forms, in bottle of 100 ml. is used against cough and prepared with the fruits of *Crossopteryx febrifuga* Benth. (Rubiaceae).

Crossopteryx febrifuga is traditionally used in Mali in the treatment of cough, pneumonia, chest pain, fever, edema, diarrhea and sickness.

The phytochemical studies done at the Department of Traditional Medicine showed that the fruits contain tannins, carotenoids, coumarins, mucilages, sterols, triterpenes, leucoanthocyans and saponins. Of these, two glycosides of oleanolic acid, the crossoptines A and B were easily extracted from the roots of *C. febrifuga* by polar solvents. They exhibit light toxicity but possess antiinflammatory, analgesic and mucolytic proprieties (Pousset 1992). Their structures have been elucidated (Gariboldi *et al.* 1990) and a patent is deposited in Italy.

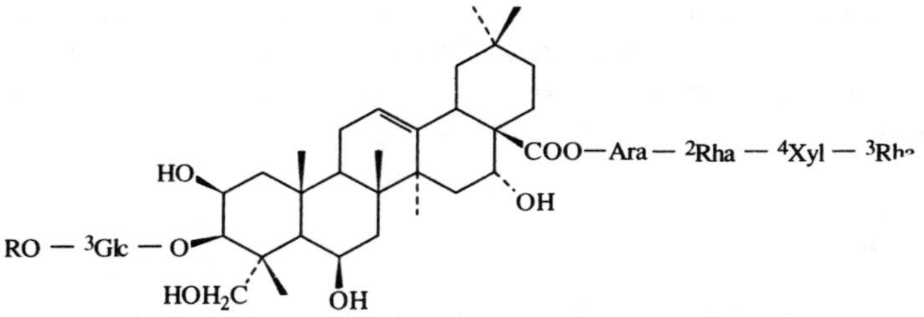

HO

COO—Ara— [2]Rha — [4]Xyl — [3]Rha

OH

RO — [3]Glc — O

HOH_2C OH

R = H Crossoptine A
R = Apiose Crossoptine B

The pharmacological studies done in Mali on Balembo® syrup, showed analgesic, antiinflammatory and expectorant effects. From a toxicological point of view, no evidence has been found that the aqueous extract of *C. febrifuga* was toxic for rabbits after *per os* administration. However, in intraperitoneal injection, it provokes an acceleration of breathing movements, loss of motor coordination and of balance in mice. The animals finally die as a result of respiratory stop. According to Tomas-Barberan and Hostettmann (1988), the dichloromethane extract of trunk bark contain triterpenic derivatives such as betulic acid, that have been shown to be active on colon carcinoma cell-lines.

Clinical studies done in Mali at D.M.T and "*Dispensaire anti-tuberculeux (D.A.T)*" have established the sedative effect of Balembo® syrup. For this reason, this syrup should not be used by children under six months of age. Its normal dosage consists of 1 teaspoon 3 times a day.

Dysenteral®

Amoebic dysentery and diarrhea are dominant child diseases in Mali. Although they often use oral rehydration salts, mothers continue to use medicinal plants like *Euphorbia hirta* L. (Euphorbiaceae). Dysenteral® is a tisane presented in tea bags containing 10 g of the powdered dry entire plant of *E. hirta* and its only indication is dysentery.

The traditional indications of *Euphorbia hirta* in Mali are diarrhea, dysentery and curiously asthma.

The phytochemical studies done at the DMT showed that the plant contains tannins, mucilages, leucoanthocyans, flavonoids, sterols, triterpenes and coumarins. According to Kerharo, previous studies have shown sugars, volatile substances, fatty acids, essential oils and alkaloids (Kerharo and Adams 1974). However the presence of phorbol-ester derivatives in some Euphorbiaceae recognized to be irritant and cocarcinogenetic are detrimental its use. Phenyl acetate-13-deoxy-12 hydroxy-4-phorbol has been isolated from the latex of *E. hirta* (Ayensu 1979; Sofowora 1984). However, it seems that the co-carcigenic effect disappears with drying as is the case with *Euphorbia lathyris* (Bissel *et al.* 1981).

We have mainly run clinical tests because Ndir and Pousset have shown that extracts of *E. hirta* inhibited the proliferation of amoebae (Ndir and Pousset 1982). In the regional center of traditional medicine of Bandiagara in Mali, clinical comparison between metronidazole and Dysenteral® have been performed. After two days administration of Dysenteral® no vegetative forms of dysenteric amoebae were found. An important clinical study undertaken by Ridet and Chartol on 53 patients has shown the efficiency of an extract from *Euphorbia hirta* for stopping an epidemic amoebic dysentery (Ridet and Chartol 1964). In Dakar, Dalil has healed ten cases of amoebic dysentery with 3 times ten grams of lyophilized extract of the plant (Dalil 1984). Dysenteral® is presented in packet of 9 tea bags, and is used as a decoction (1 bag 3 times a day).

Production of M.T.A. (improved traditional medicines)

In a program of research and commercialization our department has a service of M.T.A. production. The objectives of this program are to measure how the population accepts M.T.A.; the identification of the possibility to make a large production and the incitation of the public-pharmacists to sell M.T.A. and of medical doctors to prescribe such medicines . Table 14.1. shows the production of M.T.A. from 1992 to the first half of 1995.

Table 14.1. Production of M.T.A. 1992 - 1995

M.T.A.	Presentation	Quantities/Period 1992	1993	1994	1995	Price per unit (CFA*) DMT	Public
Balembo®	child.	6'000	12'000	19'000	1'0000	415	630
	adult.	4'000	8'000	11'000	7'000	490	690
Hepatisane®	Packet	6'494	129'870	194'805	129'870	588	825
Laxa Cassia®	Packet	2'250	11'000	15'000	8'000	144	200
Dysenteral®	Packet	444	600	100	200	324	450

*100 CFA = 1 FF = *ca* 3.5 US$

We find an increase in the production of improved traditional medicines in the period 1992 to 1995. The only exception here is Dysenteral® which is not very often sold by pharmacists. This exemption can be explained by the common use of *Euphorbia hirta* which can be obtained easily by people in traditional markets. We are presently doing a in-depth toxicological research to evaluate the cocarcinogenetic effect of this plant.

Conclusion

Medicinal plants are an important element of our cultural inheritage, and their use is still expanding. The reasons for their increasing interest are the devaluation of the local currency (CFA), which has made conventional medicines more expensive, the improve infrastructure of the herborists in Bamako markets and of course the production of improved traditional medicines (M.T.A.).

Hepatisane®, Laxa cassia®, Dysenteral® and Balembo® have been retained as essential drugs in Mali. Since 1994, a law has been taken by the Malian government about the practice of private traditional medicine. M.T.A can now be produced in private industries. Each M.T.A has a technical dossier which can be obtained at the D.M.T.

New M.T.A have been proposed for sale, *i.e.* Malarial N°5® against malaria, Gastrosedal® to cure gastritis and gastroduodenal ulcers; Psorospermine® against some skin diseases. We have to continue research to identify active substances and produce pharmaceutical forms, improving quality control and securing sources of supply of plants.

Acknowledgments

Thanks are due to University of Oslo Program Sudan - Sahel - Ethiopia (SSE) for its financial support.

References

Ayensu, E.S. (1978). *Medicinal plants of West Africa.* Michigan, 330p.

Balansard, J., Delphaut. (1946). Sur le principe alcaloïdique du Kinkeliba (*Combretum micranthum* Don.). *Médecine Tropicale* **2**, 139.

Bassene, E., Laurance,A., Olschwang, D., and Pousset, J.L. (1985). Plantes médicinales africaines: Dosage de la vitexine par chromatographie liquide haute performance dans un extrait brut de *Combretum micranthum* G. Don. *Journal of Chromatography* **346**, 428-430.

Bassene, E. Olschwang, D., and Pousset, J.L. (1986). Plantes médicinales africaines Les alcaloïdes de *Combretum micranthum* G. Don (Kinkeliba). *Annales Pharmaceutiques françaises* **44**, 191-196.

Bassene, E., Olschwang, D., and Pousset, J.L. (1987). Plantes médicinales africaines: Flavonoïdes de *Combretum micranthum* G. Don (Kinkeliba). *Plantes médicinales et Phytothérapie* **21**, 173-176.

Bissell, J.M., Nemethy, E. K., Riddle,L., and Calvin, M. (1981). Testing for tumor promoters in *Euphorbia lathyris*, analysis of possible health hazards. *Bulletin Environnement Contamination Toxicologie* **27**, 894-902.

Dalil, M. (1984) Essai thérapeutique d'un décocté lyophilisé de *Euphorbia hirti* L. (mbal) dans le traitement ambulatoire de l'amibiase intestinale. PhD Thesis, Pharmacy, University of Dakar, Senegal.

Tomas-Barberan, F.A. and Hostettman, K. (1988). A Cytotoxic triterpenoid and flavonoids from *Crossopteryx febrifuga*. *Planta Medica* 166-267.

Gariboli, P., Verotta, L., and Gabetta, B. (1990). Saponins from *Crossopterix febrifuga*. *Phytochemistry* **29**, 2629-2635

Jentsch, K., Spiegl, P., and Fuchs, L. (1961). Untersuchungen über die Inhaltsstoffe der Blatter von *Combretum micranthum*. *Planta Medica* **9**, 1-8.

Kerharo, J. and Adam, J.G. (1974). La pharmacopée sénégalaise traditionnelle, plantes médicinales et toxiques. Eds. Vigot Frères, Paris. 1011 p.

Ndir, O. and Pousset, J.L. (1982) Contribution à l'étude pharmacologique et chimique de *Euphorbia hirta*. *Medecine d'Afrique Noire* **29**, 503

Paris, R. and Moyse-Michon, H. (1956). Caractérisation de la choline chez quelques plantes médicinales. *Annales Pharmaceutiques Françaises* **14**, 464-459.

Pousset, J.,L. (1992). Les plantes médicinales africaines: possibilités de développement. Edition Marketing, éditeur de préparation aux grandes écoles de médecine, Tome 2, pp. 57-58 Agence de cooperation culturelle et technique, Paris.

Ridet, J. and Chartol, A., (1964). Les propriétés antidysentériques de *Euphorbia hirta* .*Médecine tropicale* **24**, 119-143.

Sofowora, A. (1984), Medicinal plants and Traditional Medicine in Africa. John Wiley and Sons Ltd, New York.

15. Bioactive metabolites from African medicinal plants

C. SPATAFORA AND C. TRINGALI

Dipartimento di Scienze Chimiche, Università di Catania, Viale A. Doria 6, I-95125 Catania, Italy

---

Introduction

A peculiarity of African culture is that the near totality of ingredients used for the formulation of medical remedies come from plant sources; when subjected to chemical analysis, these plants frequently afford biologically active substances (Iwu 1994). As part of our search for bioactive natural products, we considered as candidates for chemical analysis some African medicinal plants which are used in Guinea, as in other regions, for the so-called 'cure salée' of cattle. In this prophylaxis, bark, leaves or other parts of the plant are ground with salt and administered to the animals. A number of the plants commonly used for the 'cure salée' are known as folk medicines and have been subjected to chemical analysis; nevertheless, no systematic study aimed at the identification of the active principles involved in this treatment has been carried out to date. Thus, we have collected samples of plants used for the 'cure salée' in the region of Boké (Republic of Guinea) with the aim of subjecting them to a systematic chemical study.

In the isolation of bioactive compounds from natural sources, a variety of criteria can be used to select the more interesting constituents from the complex crude extracts normally obtained by treatment of ground material with solvents. One method is to isolate the main components of the extract, in the assumption that they may be responsible for the activity of the plant. An alternative is to use a biological assay as a guide for fractionation, thus obtaining compounds which are certainly biologically active, at least according to the bioassay used. Tests strictly related to the medical utilization of the plant are generally difficult to perform: indeed, medicinal plants are frequently used for a wide range of different afflictions; moreover, bioassays based on the manipulation of human pathogenic microorganisms require specialized equipment and staff trained in microbiological techniques. To overcome this difficulty, a number of bench bioassays which can be carried out in a chemical laboratory have been developed among them, the brine shrimp test (BST) is convenient for a number of reasons. In this simple test larvae of the small crustacean *Artemia salina* are used as an indicator of toxicity.

This activity can be considered, in a broad sense, as an indicator of bioactivity (Meyer *et al.* 1982). Lethality to *A. salina* larvae has been shown to be related to *in vitro* and *in vivo* antitumor activity; in particular, a correlation with cytotoxic activity against KB and P-388 tumoral cells has been observed (Meyer *et al.* 1982; Ferrigni *et al.* 1984; Anderson *et al.* 1991). In addition, toxicity to *A. salina* larvae has been related to insecticidal activity (Michael *et al.* 1956, Rupprecht *et al.* 1986; Li *et al.* 1990). Brine shrimp test has been used for bioactivity-guided fractionation of plant extracts, and is recommended by National Cancer Institute (Bethesda, Maryland) as an in-house test for the search of promising antitumor compounds (Cragg, G.M., personal communications 1992 and 1995).

For the above cited reasons, we subjected the extracts of the plants collected in Guinea to a preliminary screening based on BST. Bark, leaves or pods of the plants were ground and the powder extracted with ethyl acetate (EtOAc), the residue was subsequently extracted with ethanol (EtOH). A list of the plants subjected to this procedure and the relevant LD_{50} values (µg/ml) are presented in Table 15.1.

Table 15.1. Activity against *A. salina* (BST)[a] of African plant extracts

Plant species	material	LD_{50} (EtOAc)[b]	LD_{50} (EtOH)[b]
Anthocleista djallonensis	bark	>1000	>1000
Bombax constatum	bark	>1000	>1000
Crossopterix febrifuga	bark	>1000	72.8
Dialium guineense	leaves	>1000	140.1
Fagara macrophylla	bark	3.3	15.7
Ficus gnaphalocarpa	bark	>1000	315.1
Lophira lanceolata	bark	>1000	30.7
Markhamia tomentosa	bark	231.1	>1000
Nauclea latifolia	bark	>1000	985.3
Parkia biglobosa	bark	>1000	315.2
Pericopsis laxiflora	bark	190.3	570.1
Piliostigma thonningii	pods	>1000	55.8

[a] Brine shrimp test: mortality of larvae of *Artemia salina*, determined after 24 h exposure. See Meyer (1982) and Tringali (1995) for details of the method.
[b] EtOAc = ethyl acetate extract; EtOH = ethanol extract. Values are measured in µg/ml.

In applications of this test to crude extracts from plant material, an LD_{50} value lower than 1000 µg/ml is considered an indication of activity (Meyer *et al.* 1982), and makes the extract worthy of further analysis. The majority of the plants listed in Table 15.1. showed a notable activity in the ethanolic extract, some of them (*Crossopterix febrifuga, Fagara macrophylla, Lophyra lanceolata, Piliostigma thonningii*) with LD_{50} lower than 100 µg/ml. *Markhamia tomentosa* is active only in the ethyl acetate extract. Interestingly, the two species *Fagara macrophylla*

Engl (Rutaceae; syn. *Zanthoxylum macrophyllum* Miq.) and *Pericopsis laxiflora* Benth (Leguminosae; syn. *Afrormosia laxiflora* Harms) display lethality to *A. salina* larvae in both extracts; in particular, the ethyl acetate extract of *F. macrophylla* is by far the most active one, showing an LD_{50} = 3.3 μg/ml. Confirmation of the biological activity of the EtOAc extract of *F. macrophylla* came from cytotoxicity tests on tumoral cell cultures carried out at the University of Nantes (France): this extract revealed activity against lung tumor cells (IC_{50} = 33.8 μg/ml); further cytotoxicity tests have been carried out at the NCI and a broad spectrum cytotoxicity has been observed (IC_{50} not available). Finally, the EtOAc extract of *F. macrophylla* has also been tested at the NCI against the HIV virus, showing a 'moderate activity' (EC_{50} = 23.4 μg/ml, IC_{50} = 57.1 μg/ml).

In addition to these data, mention should be made of the ethnobotanical importance of plants belonging to the genus *Fagara* (or to the allied *Zanthoxylum*), which are used in folk medicine for the cure of different afflictions (Adesina 1987). *F. macrophylla* is also used as a fish poison and arrow poison (Watt and Breyer-Brandwijk 1962). Analogously, *P. laxiflora* is employed for the treatment of various infirmities, among them fever, pain and snake-bites (Watt and Breyer-Brandwijk 1962; Kerharo 1974).

Both *F. macrophylla* and *P. laxiflora* have been previously subjected to chemical studies. *F. macrophylla* (considered also as a synonym of *Zanthoxylum gillettii* Waterm.) afforded the furoquinoline alkaloid skimmianine (Fish and Waterman 1971a), various benzo[*c*]phenatridine alkaloids (Torto and Mensah 1970; Fish and Waterman 1971b) and neutral compounds like *N*-isobutylamides (Kubo *et al.* 1984; Adesina and Reisch 1988). The only report on bark constituents of *P. laxiflora* concerns the alkaloid *N*-methylcytisine (Fitzgerald *et al.* 1976); flavonoids and further neutral metabolites were obtained from the heartwood (Fitzgerald *et al.* 1976) and from the leaves (Sultana and Ilyas 1987).

None of the above cited studies have been accomplished on the basis of bioassay-guided fractionation and they do not allow the sure identification of the active principles responsible for the BST results; thus, we decided to carry out the chemical analysis of the bark extracts of the plants *F. macrophylla* and *P. laxiflora*, using the BST as a guide for fractionation.

Results and Discussion

Fagara macrophylla

The ground bark of *F. macrophylla* was extracted with EtOAc and subsequently with EtOH. The less active (LD_{50} = 15.7 μg/ml) ethanolic extract was partitioned between water and ethyl acetate; the BST of these fractions showed that the activity against *A. salina* was retained by the organic phase (LD_{50} = 9.4 μg/ml; aqueous phase, LD_{50} = 246.4 μg/ml). The EtOAc extracts were joined on the basis of their similarity (TLC) and the whole organic extract was subjected to an activity-guided flash-chromatography on acetyl-polyamide. Fractions showing a similar TLC profile were pooled (A-F) and subjected to the BST. Only subfractions showing LD_{50} lower than 200 μg/ml (B-E) were promoted to further isolation work.

The elucidation of the structures of the metabolites isolated from the active fractions B-E has been based essentially on spectral analysis. The established structures were used for an on-line literature check, which allowed the identification of previously known compounds. When necessary, two-dimensional NMR methods were applied to achieve the complete assignments of ^{1}H and/or ^{13}C NMR spectra of the purified compounds.

The two major constituents obtained from fraction B (LD_{50} = 3.6 μg/ml) were identified as the lignan (-)-sesamin (**1**) and the *N*-isobutylamide γ-sanshoöl (**2**).

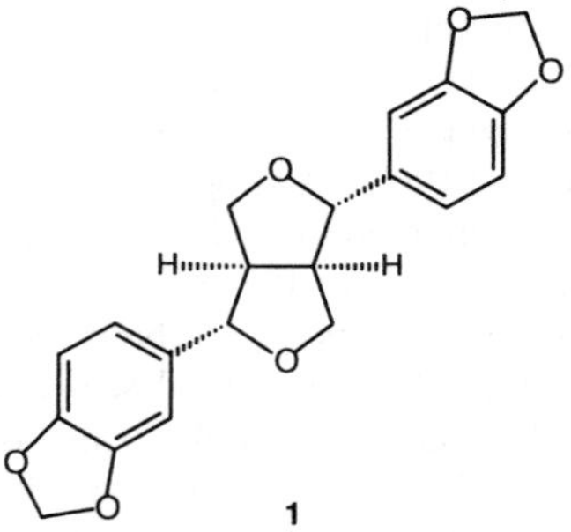

1

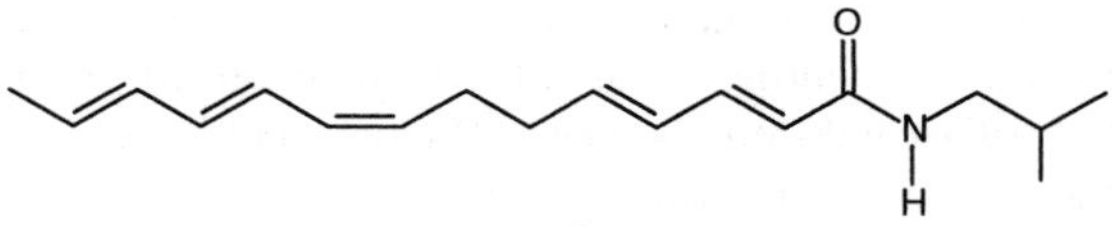

2

The ^{1}H NMR spectrum of γ-sanshoöl (**2**) was previously only partly assigned (Yasuda *et al.* 1981); the complete, revised assignment of this spectrum was an intriguing task, because of the severe overlapping of the low-field signals, and was achieved by the use of advanced homonuclear and heteronuclear 2D-NMR techniques (COSY, TOCSY, NOESY, HMQC and HMBC), performed at 500 MHz. The results of the NOESY experiment, coupled with careful $^{3}J_{H,H}$ measurements, allowed to confirm the stereochemistry of the double bonds, previously assigned on the basis of ^{13}C chemical shift analysis.

Fractions C (LD_{50} = 3.4 μg/ml) and D (LD_{50} = 21 μg/ml), afforded as main components the known acridone alkaloids 1-hydroxy-3-methoxy-*N*-methyl-acridone (**3**) and arborinine (**4**); analogously, xanthoxoline (**5**) was identified in fraction E (LD_{50} = 54.0 μg/ml). From the same fraction, the previously unreported 1-hydroxy-3-methoxy-acridone (**6**) was isolated.

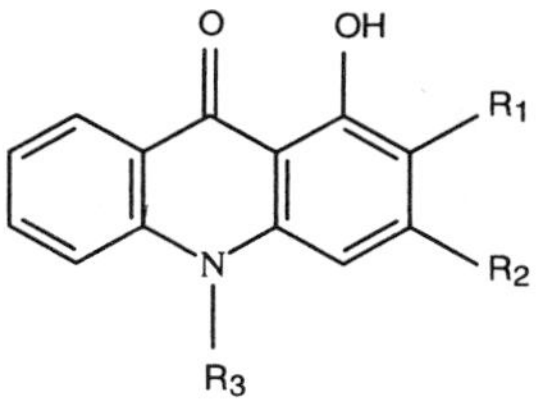

3: R_1 = H; R_2 = OMe; R_3 = Me

4: R_1 = OMe; R_2 = OMe; R_3 = Me

5: R_1 = OMe; R_2 = OMe; R_3 = H

6: R_1 = H; R_2 = OMe; R_3 = H

The structure of **6** was established on the basis of spectral analysis. The molecular formula $C_{14}H_{11}NO_3$ was determined by MS and NMR data. Further physical measurements (IR, UV, ^{1}H and ^{13}C NMR) strongly suggested a structural similarity with the co-occurring acridones **3** - **5**. In particular, the analysis of both ^{1}H and ^{13}C NMR spectra of **6**, showed the presence of only one methoxy group, and the lack of the NMe function; conversely, an NH resonance was discernible in the ^{1}H NMR spectrum at δ 12.10. In the same spectrum, a typical low-field OH resonance was observed at δ 14.22, attributable to an hydroxyl function involved in an intramolecular hydrogen bond with the carbonyl group, and consequently located in C-1. The methoxy group must be located at C-3, because location at C-2 or C-4 would require *ortho*-coupled signals, while typical *meta*-coupled signals appear at δ 6.14 and δ 6.38 (1 H each, *d*, J = 2.0 Hz). These assignments were confirmed by ^{13}C NMR chemical shift calculations.

Complete ^{1}H and ^{13}C NMR assignments were achieved for compound **6**; analogously, the NMR data available from the literature for metabolites **3** - **5** (Fish and Waterman 1971; Bergenthal *et al.* 1979; Reisch *et al.* 1985) were revised and/or completed. The NMR assignments for compounds **2** - **6** will be reported in an extended paper.

Pericopsis laxiflora

In the present study, only the EtOAc extract of the ground bark of *P. laxiflora* was subjected to chemical analysis, in view of its higher BST activity (LD_{50} = 190.3 μg/ml) with respect to the EtOH extract (LD_{50} = 570.1 μg/ml). Fractionation on acetyl-polyamide, followed by TLC analysis and paralleled by BST, afforded an active subfraction (LD_{50} = 130.3 μg/ml), subjected to further chromatography to give as pure compounds the α-methyldeoxybenzoins *R*-(–)-angolensin (**7**) and *R*-(+)-2-*O*-methyl-angolensin (**8**), and the pterocarpan (–)-maackiain (**9**).

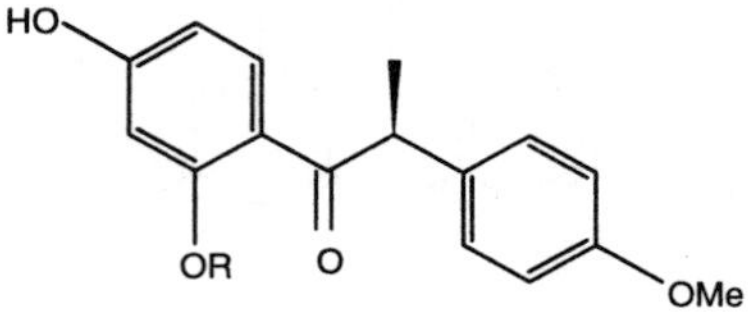

7 R = OH
8 R = OMe

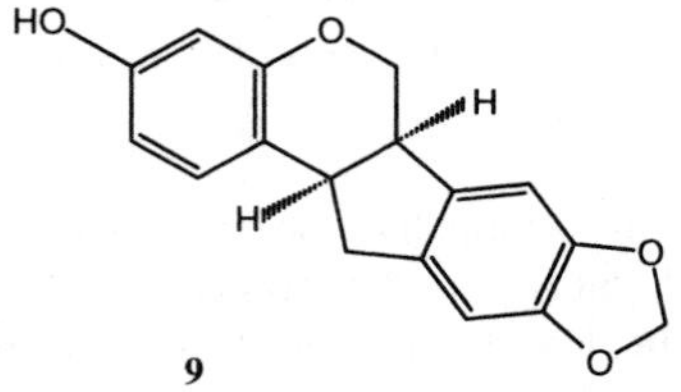

9

The unambiguous identification of compounds **7-9** was based on the concerted use of advanced 2D NMR methods and the preparation of the acetate of **9**. The previously reported ^{1}H NMR data of synthetic (±)-angolensin and (±)-2-*O*-methylangolensin (Jain and Paliwal 1988) were revised and their complete ^{1}H and ^{13}C NMR assignments were determined (Tringali 1995). The identity of **9** with the earlier isolated (–)-demethylpterocarpin was also proved through physico-chemical analysis of its derivative, 3-acetoxy-demethylpterocarpin.

Biological activity of compounds 1-9 and conclusive remarks.

As expected, compounds **1-9** are active against larvae of *A. salina*; their LD_{50} values are reported in Table 15.2.

Table 15.2. Activity against *A. salina* (BST) for compounds **1-9**

Compound	**1**	**2**	**3**	**4**	**5**	**6**	**7**	**8**	**9**
LD_{50} [a]	2.4	7.1	11.3	3.4	12.4	35.1	28.1	36.0	89.2

[a] Values are measured in μg/ml.

Referring to *A. salina* larvae, compounds **1** and **4** are strongly active; compounds **6** - **9** display a mild activity. Nevertheless, in considering these data it should be remembered that there is no direct quantitative relationship between the BST results and related activities like cytotoxicity against tumoral cells (Solis *et* *al.* 1993). With reference to the metabolites obtained from the bark of *F. macrophylla*, mention should be made of the fact that only compound **1** displays an LD_{50} value lower than that determined for the crude EtOAc extract: this suggests that compounds **1** - **6** could act sinergistically.

Interestingly, sesamin (**1**) is known as an insecticidal synergist (MacRae and Towers 1984) and inhibits the growth of *Bombix mori* (Kamikado *et al.* 1975); γ-sanshoöl (**2**) is strictly related to molluscicidal and insecticidal *N*-isobutylamides (Kubo *et al.* 1984); metabolites **3** - **6** belong to the class of acridone alkaloids, an important group of bioactive compounds (Michael 1995, Takemura *et al.* 1995), including antitumor substances (Su and Watanabe 1993); arborinine (**4**) is known for its antispasmogenic activity (Minker *et al.* 1979). No biological data have been previously reported for angolensin **7** and its derivative **8**; the same is true for compound **9**; however, this latter compound is related to bioactive pterocarpans, displaying interesting biological properties, *e.g.* antimicrobial (Kamat *et al* 1981; Mitscher *et al* 1988; Taniguchi and Kubo 1993), antifungal (Perrin *et al* 1972), anti-HIV (Engler *et al* 1993) and cytotoxic (Dagne *et al* 1993) activity.

The above results confirm the reliability of the BST as a guide for fractionation of plant extracts in the search for bioactive compounds. Taking into account the activity against *A. salina* and the above cited biological data, compounds **1** - **6** are currently under investigation for their possible action against insects, HIV virus and tumoral cell cultures.

On the basis of the above discussed data, one may reasonably presume that metabolites **1** - **9** could act as defense substances for *F. macrophylla* and *P. laxiflora* and are probably useful components of the traditional preparations obtained from the bark of these plants.

Acknowledgments

The authors are grateful to Dr. G.M. Cragg and Dr. A.B. Mauger and his staff at the National Cancer Institute, NCI (Bethesda, Maryland), for the human cancer cell line *in vitro* test panel and for the *in vitro* anti-HIV test. Thanks are also due to Prof. J.F. Verbist and Dr. C. Roussakis, SMAB, Université de Nantes (France) for the cytotoxicity test against lung tumoral cells. Acknowledgments are also due to Dr. M. Girelli, Zooconsult Srl (Roma, Italy) for the gift of the plant material, and to Dr. E. Timmermans and Mr. K. Camara (Boké, Republic of Guinea) for the collection and preparation of this material. The authors also acknowledge Prof. N. Longhitano and Dr. E. Cirino, Dipartimento di Botanica, Università degli Studi (Catania, Italy) for the identification of the plants. The work described here was financially supported by the Consiglio Nazionale delle Ricerche (Roma, Italy) and by the Ministero della Pubblica Istruzione, Università e Ricerca Scientifica e Tecnologica (Roma, Italy).

References

Adesina, S.K. (1987). Further new constituents of *Zanthoxylum leprieurii. Fitoterapia* **58**, 123-126.

Anderson, J.E., Goetz, C.M., McLaughlin, J.L., and Suffness M. (1991). A blind comparison of simple bench-top bioassays and human tumor cell cytotoxicities as antitumor prescreens. *Phytochemical Analysis* **2**, 107-111.

Bergenthal, D., Mester, I,. Rozsa, Z., and Reisch, J. (1979). ^{13}C-NMR-Spektren einiger Acridon-Alkaloide. *Phytochemistry* **18**, 161-163.

Dagne, E., Gunatilaka, A.A.L., Kingston, D.G.I., Alemu, M., Hofmann G., and Johnson, R.K. (1993). Two bioactive pterocarpans from *Erythrina burana. Journal of Natural Products* **56**, 1831-1834.

Engler, T.A., Lynch, Jr., K.O., Reddy, J.P., and Gregory, G.S. (1993). Synthetic pterocarpans with anti-HIV activity. *Bioorganic Medicinal Chemistry Letters* **3**, 1229-1232.

Ferrigni, N.R., McLaughlin, J.L., Powell, R.G., and Smith, C.R. (1984). Use of potato disc and brine shrimp bioassays to detect activity and isolate piceatannol as the antileukemic principle from the seeds of *Euphorbia lagascae. Journal of Natural Products* **47**, 347-352.

Fish, F. and Waterman, P.G. (1971a). A note on the chloroform-soluble alkaloids of *Fagara macrophylla. Journal of Pharmacy and Pharmacology* **23**, 67.

Fish, F. and Waterman, P.G. (1971b). Chloroform-soluble alkaloids of *Fagara leprieurii. Phytochemistry* **10**, 3322-3324.

Fitzgerald, M.A., Gunning, P.J.M., and Donnelly, D.M.X. (1976). Phytochemical examination of *Pericopsis* species. *Journal of the Chemical Society, Perkin Transactions I*, 186-191.

Iwu, M.M. (1994). African medicinal plants in the search for new drugs based on ethnobotanical leads. In *Ethnobotany and the search for new drugs*, Ciba Foundation Symposium 185, pp. 116-126. John Wiley & Sons, New York.

Jain, A.C. and Paliwal P. (1988). A facile synthesis of α-methyldesoxybenzoins including racemates of natural angolensin, 2-O-methylangolensin and 4-O-methylangolensin. *Indian Journal of Chemistry* **27B**, 985-988.

Kamat, V.S., Chuo, F.Y., Kubo, I., and Nakanishi, K. (1981). Antimicrobial agents from an East African medicinal plant *Erythrina abyssinica. Heterocycles* **15**, 1163-1170.

Kamikado, T., Chang, C.F., Murakoshi, S., Sakurai, A., and Tamura, S. (1975). Isolation and structure elucidation of growth inhibitors on silkworm larvae from *Magnolia kobus* DC. *Agricultural and Biological Chemistry* **39**, 833-836.

Kerharo, J. (1974). In La Pharmacopée Sénégalaise Traditionelle (Plantes Médicinales et Toxiques), pp 441-443. Editions Vigot Frères, Paris.

Kubo, I., Matsumoto, T., Klocke, J.A., and Kamikawa, T. (1984). Molluscicidal and insecticidal activities of isobutylamides isolated from *Fagara macrophylla. Experientia* **40**, 340-341.

Li, X.H., Hui, H.; Rupprecht, J.K., Liu, M., Wood, K.V., Smith, D.L., Chang, C.J., and McLaughlin, J.L. (1990). Bullatacin, bullatacinone, and squamone, a new bioactive acetogenin, from the bark of *Annona squamosa. Journal of Natural Products* **53**, 81-86.

MacRae, W.D and Towers, G.H.N. (1984). Biological activities of lignans. *Phytochemistry* **23**, 1207-1220.

Meyer, B.N., Ferrigni, N.R., Putnam, J.E., Jacobsen, L.B., Nichols, D.E., and McLaughlin, J.L. (1982). Brine shrimp: a convenient general bioassay for active plant constituents. *Planta Medica* **45**, 31.

Michael, A.S., Thompson, C.G., and Abramovitz, M. (1956). *Artemia salina* as a test organism for bioassay. *Science* **123**, 464.

Michael, J.P. (1995). Quinoline, quinazoline and acridone alkaloids. *Natural Products Report* **12**, 77-89.

Minker, E., Bartha, C., Rozsa, Z., and Szendrei, K. (1979). Antispasmogenic effect of Rutamarin and arborinine on isolated smooth muscle organs. *Planta Medica* **37**, 156-160.

Perrin, D.R., Whittle, C.P., and Batterham, T.J. (1972). The structure of phaseollidin. *Tetrahedron Letters* **17**, 1673-1676.

Reisch, J., Adesina, S.K., and Bergenthal, D. (1985). Constituents of *Zanthoxylum leprieurii* fruit pericarps. *Pharmazie* **40**, 811-812.

Rupprecht, J.K., Chang, C., Cassady, J.M., McLaughlin, J.L., Mikolajczack, K.L., and Weisleder, D. (1986). Asimicin, a new cytotoxic and pesticidal acetogenin from the pawpaw, *Asimina triloba* (Annonaceae). *Heterocycles* **24**, 1197-1201.

Su, T.L. and Watanabe, K.A. (1993). Anticancer Acridone Alkaloids, in *Studies in Natural Products Chemistry* **13**, Atta-ur-Rahman Ed., Elsevier London, 347-382.

Sultana, S. and Ilyas, M. (1987). A new isoflavone from leaves of *Afrormosia laxiflora* Harms. *Indian Journal of Chemistry* **26B**, 799.

Taniguchi, M. and Kubo, I,. (1993). Ethnobotanical drug discovery based on medicine men's trial in the African savanna: Screening of East African plants for antimicrobial activity II *Journal of Natural Products* **56**, 1539-1546.

Torto, F.G. and Mensah, I.A. (1970). Alkaloids of *Fagara macrophylla. Phytochemistry* **9**, 911-914.

Tringali, C. (1995). Identification of bioactive metabolites from the bark of *Pericopsis (Afrormosia) laxiflora. Phytochemical Analysis* **6**, 289-291.

Watt, J.M, and Breyer-Brandwijk, M.G. (1962) In *The Medicinal and Poisonus Plants of Southern and Eastern Africa*, p. 730. E. & S. Livingstone Ltd., Edinburgh.

Yasuda, I., Takeya, K., and Itokawa, H. (1981). Two new pungent principles isolated from the pericarps of *Zanthoxylum ailanthoides. Chemical and Pharmaceutical Bulletin* **29**, 1791-1793.

16. Polyphenolic constituents of *Brackenridgea zanguebarica* (Ochnaceae) and their biological activities

A. MARSTON[1], J.D. MSONTHI[2], K. HOSTETTMANN[1]

[1]*Institut de Pharmacognosie et Phytochimie, BEP, Université de Lausanne, CH-1015 Lausanne, Switzerland and* [2]*Chemistry Department, University of Swaziland, Private Bag 4, Kwaluseni, Swaziland*

Introduction

Brackenridgea zanguebarica Oliv. (Ochnaceae) is a shrub or small tree which is distributed over southern and central Africa. It is used by the Nyamwezi in East Africa as an antidote for snake bite and in the treatment of wounds (Watt and Breyer-Brandwijk 1962). Haerdi (1964) reports that infusions of the roots are also used in traditional medicine for the treatment of malaria and leprosy, while bark extracts speed the healing of sores.

A particularity of the tree is that the stem bark has a bright yellow tissue layer immediately underneath the grey outer coat. The heartwood is reddish in colour. Since the yellow bark colour is often indicative of interesting polyphenolic constituents, an investigation of *B. zanguebarica* was undertaken.

The original batch of plant material was collected in Tanzania, near Ifakara in the Kilombero District. Initially, the tree was identified as *Cordia goetzei* (Boraginaceae). However, collection of an authentic sample of *B. zanguebarica* from Malawi suggested that the material from Tanzania had been incorrectly named. For this reason, a detailed analysis of the the two batches was undertaken, together with the isolation of 5 polyphenolic compounds (**1** - **5**) from the Tanzanian plant material.

HPLC analysis of the plant material

Samples of the yellow inner stem bark of both Tanzanian and Malawian batches of *B. zanguebarica* were first extracted with dichloromethane and then with methanol. HPLC analysis of the two methanol extracts which contained the polyphenolic constituents was performed, giving virtually identical traces (Fig. 16.1). In conjunction with TLC analysis, it was concluded that one and the same plant was being studied.

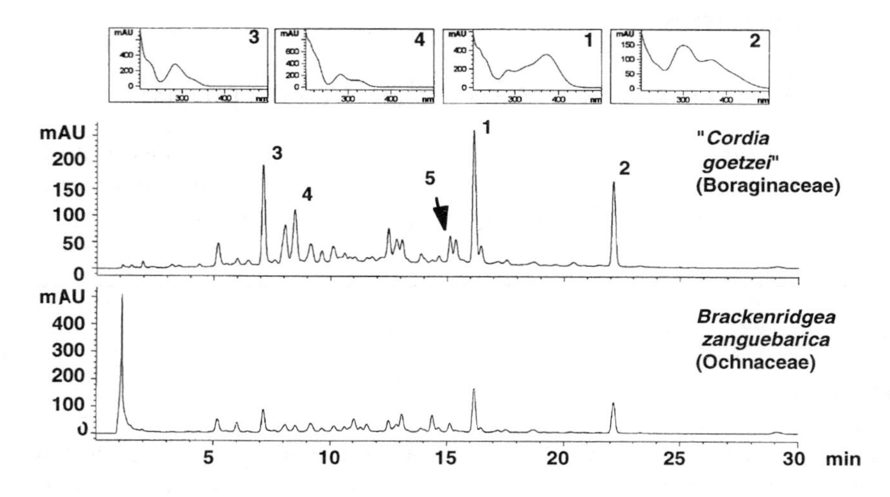

Fig. 16.1. HPLC analysis of Tanzanian (named "*Cordia goetzei*"; upper trace) and Malawian (lower trace) batches of *Brackenridgea zanguebarica* yellow stem bark methanol extracts.

HPLC conditions: NovaPak C_{18} (150 x 3.9 mm); MeOH-H_2O 40:60 to 70:30 in 20 min, then MeOH-H_2O 70:30 for 10 min; flow-rate 1.5 ml/min.

Isolation of polyphenols 1 - 5 from *B. zanguebarica* stem bark

The methanol extract of the stem bark was found to be active against the plant-pathogenic fungus *Cladosporium cucumerinum* in a TLC bioautographic assay (Homans and Fuchs, 1970). Bioactivity-guided fractionation led to the isolation of four polyphenols (**1-4**) (Marston *et al.* 1988). Two of these (**1, 2**) were previously isolated by Drewes *et al.* (1987). The all-liquid method of countercurrent chromatography proved to be a very useful means of obtaining these bioactive components and initial experiments were performed with droplet countercurrent chromatography (DCCC). Subsequent work, however, used the more rapid technique of centrifugal partition chromatography (CPC) (Marston and Hostettmann 1994). With a Sanki LLN cartridge instrument (total capacity 250 ml in 12 cartridges), the pure polyphenols were all obtained in a maximum of two separation steps, starting with 2g of crude extract (Fig. 16.2.).

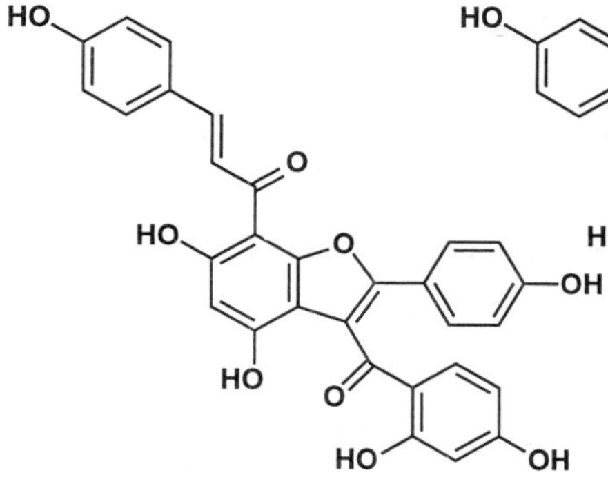

1

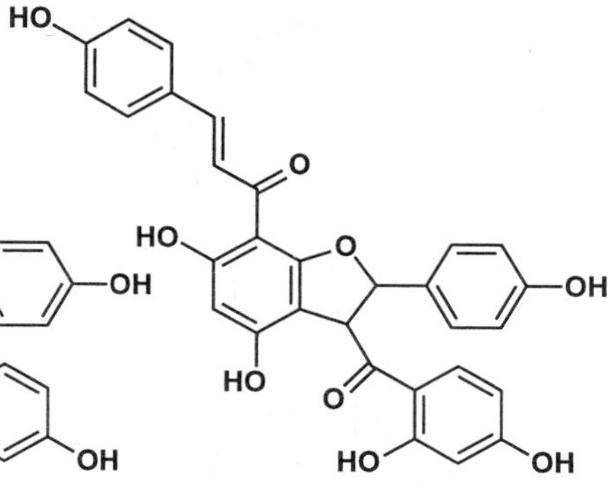

2

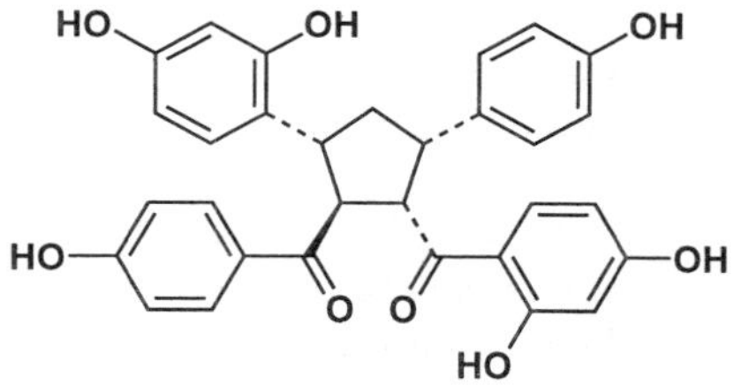

3

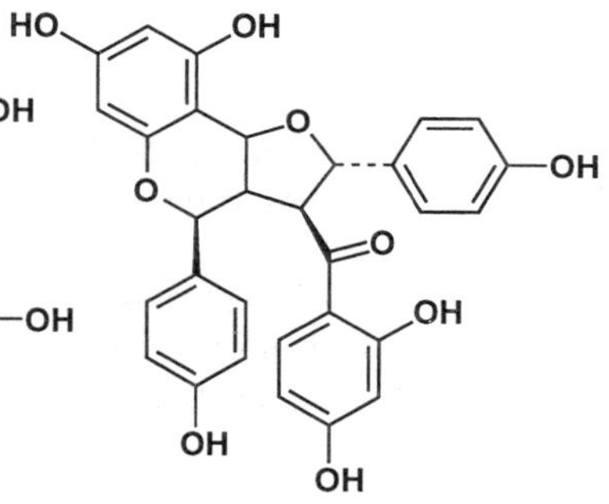

4

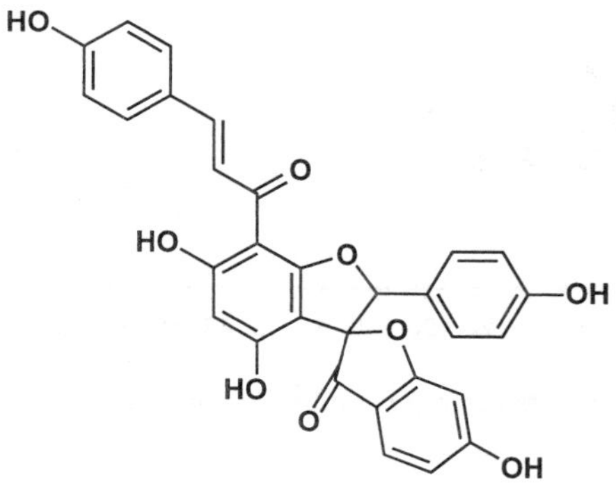

5

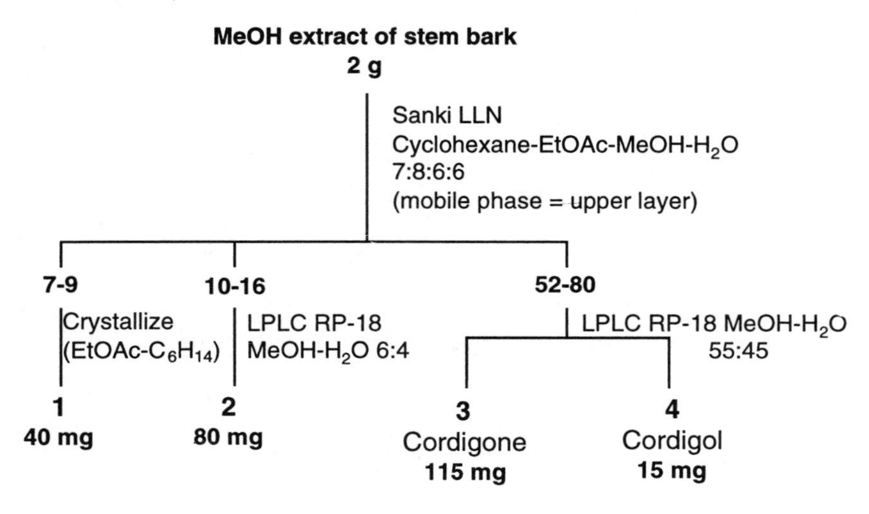

Fig. 16.2. Isolation of polyphenols **1** - **4** from *B. zanguebarica* stem bark by centrifugal partition chromatography on a cartridge instrument.

Progress of the CPC separation was monitored both by a UV detector and by TLC. Since the polyphenols possessed strong chromophores, the UV detector was saturated and more efficient analysis of the different fractions was possible by TLC (Fig. 16.3.).

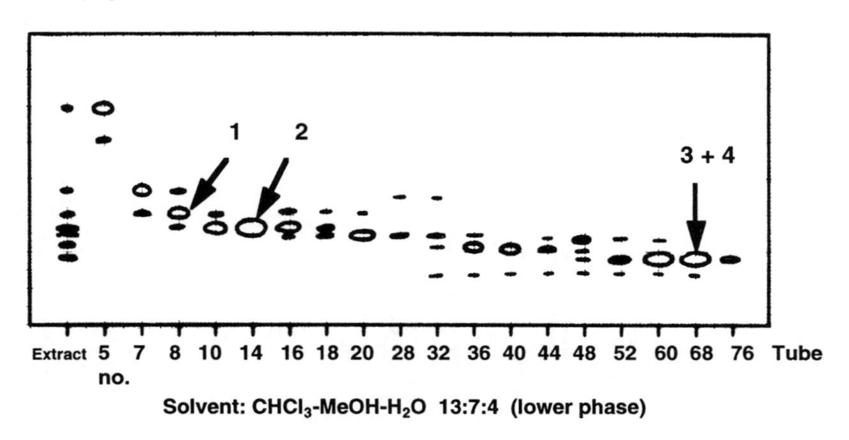

Fig. 16.3. TLC monitoring of fractions from the CPC of *B. zanguebarica* stem bark methanol extract.

There was a very efficient preliminary separation of the antifungal compounds by CPC. The TLC chromatogram shows that the extract was composed of many closely-related compounds. Some of these are very difficult to resolve by more traditional liquid-solid chromatographic techniques. In the case of **3** and **4**, however, CPC was not sufficient to give a separation and recourse to low-pressure liquid chromatography (LPLC) was necessary.

The benzofuran **1** was obtained by crystallization of fractions 7 - 9 of the CPC separation, while dihydrobenzofuran **2** required a final LC step on RP-18.

Using a different CPC instrument, the Pharma-Tech CCC-1000 rotating coil chromatograph, a scale-up of the separation was possible. By this method, it was possible to separate 20 g of methanol extract in one charge. The corresponding TLC analysis of the different fractions obtained is shown in Fig. 16.4. The polyphenols **1** - **4** could be purified by the same procedure as shown above.

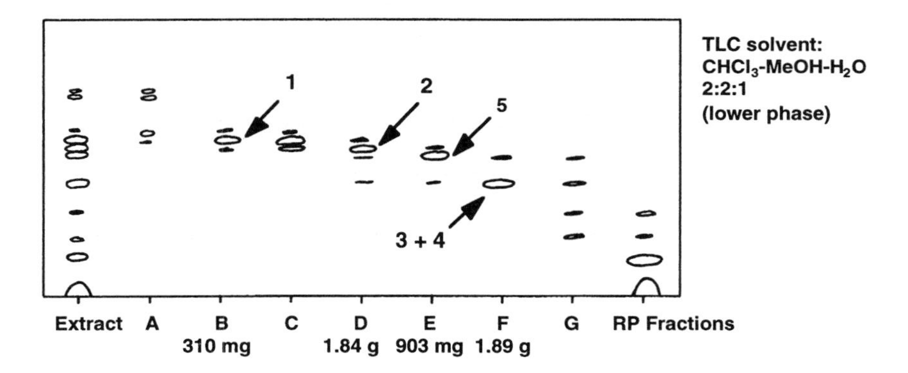

Fig. 16.4. TLC monitoring of fractions from the Pharma-Tech CCC-1000 CPC separation of *B. zanguebarica* stem bark methanol extract.

CPC separation conditions: Coil capacity 660 ml; 1000 rpm; flow-rate 3 ml/min; solvent system cyclohexane-ethyl acetate-methanol-water 8:8:6:6 (mobile phase = upper layer); sample 20 g.

In addition, another polyphenol (**5**) was isolated by the scheme shown in Fig. 16.5. This novel chalcone-dihydrobenzofuran derivative was a spiro compound which can be considered as deriving from an intramolecular cyclization of dihydrobenzofuran derivative **2**. The structure was determined by a combination of ^{13}C-NMR, heteronuclear multiple bond correlation (HMBC) and mass spectrometry. Compound **5** has a structure similar to that of daphnodorin C, which is a spiro biflavonoid isolated from *Daphne odora* (Thymelaeaceae) (Baba et al. 1987)

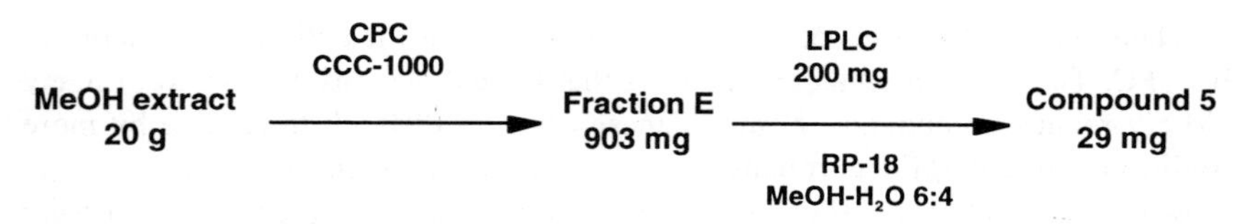

Fig. 16.5. Separation of brackenridgenol (**5**) from *B. zanguebarica*.

All five polyphenols isolated were yellow or orange pigments and responsible at least in part for the coloration of the inner bark of *B. zanguebarica*. Compounds **1-4** exhibited antifungal activity against *Cladosporium cucumerinum*, in the TLC bioautographic assay (Table 16.1.). In fact, the benzofuran derivative **1** had the same activity as the reference compound amphotericin B on the TLC plate. Polyphenols **1-4** were also active against the bacterium *Bacillus subtilis* in a TLC bioassay using an agar overlay method (Rahalison *et al.* 1991) (Table 16.1.). The spiro compound **5** was, however, inactive against *C. cucumerinum* and only had borderline activity against *B. subtilis*.

Table 16.1. Antifungal and antibacterial activities of polyphenols **1-5** isolated from *B. zanguebarica* stem bark

	Cladosporium cucumerinum[a]	*Bacillus subtilis*[a]
1	10.5	
2	50.5	
3	100.2	
4	100.5	
5	inactive	10
Amphotericin B	1	
Chloramphenicol	0.01	

[a] Minimum amount (μg) of compound required to inhibit growth on TLC plate

In addition to their antimicrobial activities, the polyphenols **1-4** were also effective as protein kinase C (PKC) inhibitors (see Table 16.2.). The most active was cordigone (**3**), whereas **1** was inactive in this bioassay.

Table 16.2. PKC inhibition by polyphenols **1-4**

	% inhibition at 100 μg/ml	IC_{50} (μg/ml)
1	22%	inactive
2	60%	67
3	93%	27
4	70%	45

Conclusion

The true identity of the batch of *B. zanguebarica* from Tanzania has now been established and the four polyphenols from the stem bark which are responsible for the antifungal activity have been isolated and characterized. Furthermore, an additional spiro polyphenolic compound has been identified. The antifungal principles also possess antibacterial and protein kinase C inhibitory activities.

During this investigation, the usefulness of liquid-liquid partition chromatography has been shown. Crude extracts of *B. zanguebarica* in up to 20 g quantities could be injected onto the CPC instrument used. Subsequently, a minimum of further purification was required to obtain the pure products. This work thus provides another application of CPC to the isolation of natural products.

Acknowledgements

The Swiss National Science Foundation is gratefully acknowledged for financial support.

References

Baba, K., Takeuchi, K., Doi, M., and Kozawa, M. (1987) *Chemical and Pharmaceutical Bulletin* **35**, 1853-1859.

Haerdi, F. (1964). *Afrikanische Heilpflanzen*. Acta Tropica Supplementum 8. Verlag Recht und Gesellschaft, Basel.

Homans, A.L. and Fuchs, A. (1970). Direct bioautography on thin-layer chromatograms as a method for detecting fungitoxic substances. *Journal of Chromatography* **51**, 325-327.

Marston, A. and Hostettmann, K. (1994). Counter-current chromatography as a preparative tool - applications and perspectives. *Journal of Chromatography A*. **658**, 315-341

Marston, A., Zagorski, M.G., and Hostettmann, K. (1988). Antifungal polyphenols from *Cordia* goetzi Gürke. *Helvetica Chimica Acta* **71**, 1210-1219.

Rahalison, L., Hamburger, M., Hostettmann, K., Monod, M., and Frenk, E. (1991). A bioautographic agar overlay method for the detection of antifungal compounds from higher plants. *Phytochemical Analysis*. **2**, 199-203.

Watt, J.M. and Breyer-Brandwijk, M.G. (1962). *Medicinal and poisonous plants of southern and eastern Africa*. E. and S. Livingstone, Edinburgh.

17. Constituents of muthi plants of Southern Africa: magical and molluscicidal properties

S.E. DREWES[1], M. HORN[1], B.M. SEHLAPELO[1], L. KAYONGA[1], C.C. APPLETON[2], T. CLARK[1] AND T. BRACKENBURY[2]

[1]*Department of Chemistry and Chemical Technology, University of Natal, Pietermaritzburg 3209, South Africa and* [2]*Department of Zoology and Entomology, University of Natal, Pietermaritzburg 3209, South Africa*

Introduction

Southern Africa has a variety of muthi plants which are used for magical and other purposes. The focus in this contribution is on four such plants, namely, *Brackenridgea zanguebarica*, *Cassipourea gerrardii*, *Hypoxis hemerocallidea* (formerly *rooperi*) and *Ocotea bullata*. Finally there is a contribution on a muthi tree which has potential as a Third World molluscicide, *Apodytes dimidiata*.

Results and discussion

The small tree *Brackenridgea zanguebarica* (Ochnaceae; see also chapter 16), which grows in the Northern Province of South Africa and Zambia, has a bright yellow bark. Various magical powers are ascribed to it by indigenous Venda people who call it "*mutavhatsindi*". It is supposed to protect from lightning strikes and also no adult may dig up the roots of the tree as this can lead to sterility. From the bark two new compounds (**1**, **2**) were isolated and fully characterized (Drewes and Hudson 1984, Drewes *et al.* 1987).

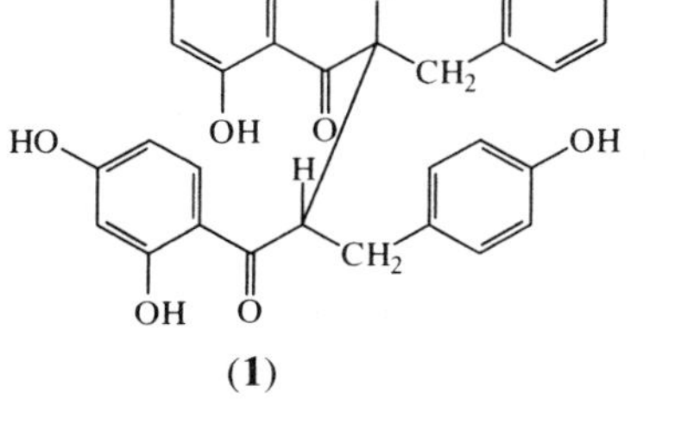

(**1**) 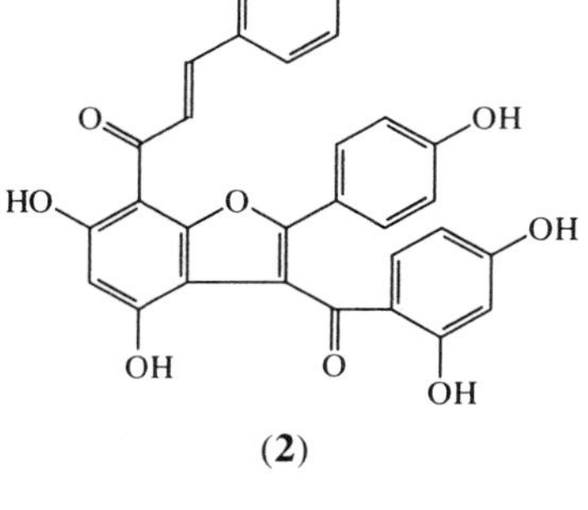 (**2**)

These are a dimeric dihydrochalcone (**1**) and a dimeric chalcone (**2**). Other compounds were present in lower concentration.

The tree known as onionwood, *Cassipourea gerrardii* (Macarisiae), is believed to have many magical properties but it is used as a skin lightener, particularly by the indigenous people of Kwazulu-Natal. The finely ground bark is mixed with sodium carbonate and milk and then applied as a face-pack by young maidens.

Since hydroquinone was used in commercial preparations of skin lighteners (now no longer on the market) and since *Cassipourea* was known to contain phenolic compounds, this investigation was undertaken. The related plant *C. gummiflua* ("*umqonga*") was also examined. Both plants contained novel dimeric A-type proanthocyanidins (Drewes *et al.* 1992, Drewes and Taylor 1994). The structures (**3** and **4**) are shown below.

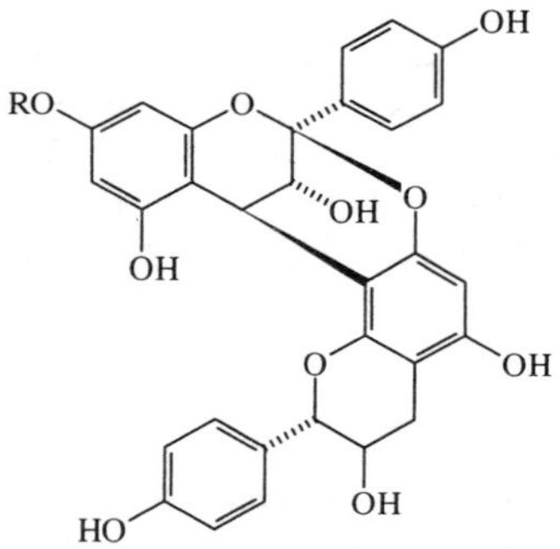

(**3**) R = H

(**4**) R = CH_3

The compound **4**, obtained from *C. Gummiflua* undergoes an interesting rearrangement, which involves a hydroquinone intermediate, when treated with $NaHCO_3$.

The commonly occurring bulbous plant (Kwazulu-Natal) *Hypoxis rooperi* (now *hemerocallidea*) has certain magical properties but is also used mainly for treating urinary ailments (Drewes and Liebenberg 1982, Marini-Bettolo *et al.* 1982, Drewes *et al.* 1984). Our initial investigation was prompted by the observations of a representative of a pharmaceutical company that it cured certain cancers. From the plant we obtained as major product, the new compound (**5**), named hypoxoside by Marini-Bettolo (Marini-Bettolo *et al.* 1982).

Hydrolysis of compound **5** with β-glucosidase leads to the aglycone which we named rooperol. In this form the compound certainly inhibits the growths of certain cancer cells. The tests carried out on rooperol form the basis of chapter 21.

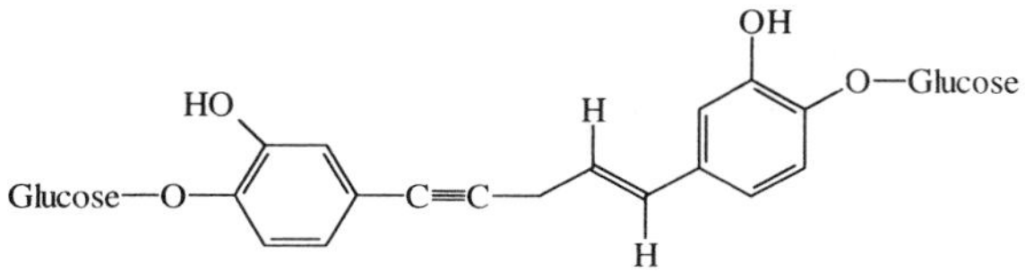

(**5**)

The black stinkwood (*Ocotea bullata*; Lauraceae) which occurs in the Eastern Cape Province and Kwazulu-Natal is one of the top-selling "muthi" plants. Reserves are now very low and the tree is specially protected by Government Ordinance. The bark is used by traditional healers to cure headaches, but its main use relates to its magical powers. It is a favorite love potion and it is alleged to promote a wide variety of magical effects. Our interest here was twofold:

i) to establish the structure of the main constituents in the bark
ii) to examine the chemical relationship between *Ocotea* bark and *Cryptocarya* bark (also belonging to the Lauraceae family). This was particularly important as the scarcity of *Ocotea* bark has led to the use (by traditional healers) of *Cryptocarya* bark as substitute.

The bark of *O. bullata* gave as major product the neolignan **6** which we have named ocobullenone (Sehlapelo *et al.* 1993).

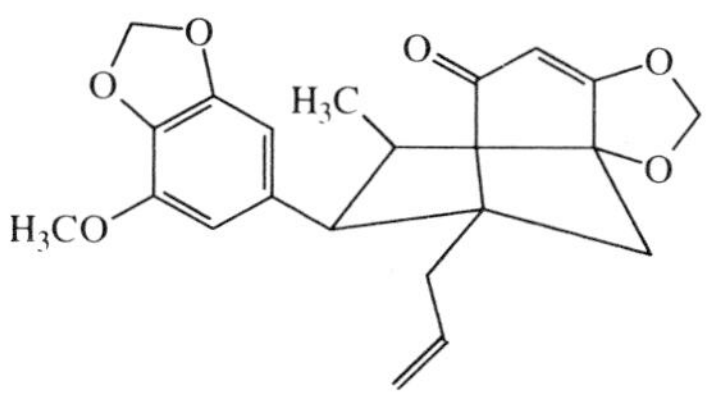

(**6**)

From four *Cryptocarya* species we were able to isolate a variety of new α-pyrones, for example cryptofolione (**7**), but no ocobullenone. (Drewes *et al.* 1995).

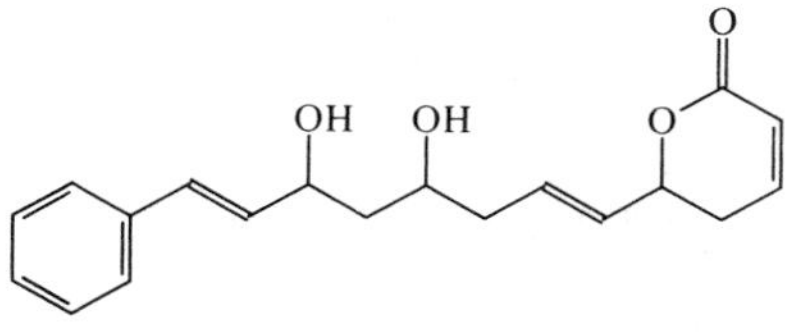

(**7**)

In one species of *Cryptocarya, C. liebertiana*, growing in the northern part of South Africa and in Zimbabwe, we were however, finally able to demonstrate that ocobullenone also occurred, thus supplying the "missing link".

Molluscicides

From *Apodytes dimidiata* (Icacinaceae) the iridoid genipin (**8**) and its 10-acetyl derivatives (**9**) were isolated (Drewes, S.E:, Kayonga, L., Appleton, C.C., Clark, T., and Brackenbury, T., *Journal of Natural Products*, in press). These two compounds were shown to have molluscicidal properties as shown by tests on *Bulinus africanus* the intermediate host of *Schistosoma haematobium*.

Results are shown below:

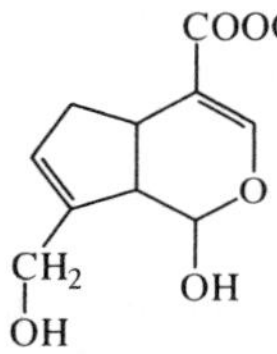

Genipin (**8**)	Genipin acetate (**9**)
LD_{50} 25.27 ppm	LD_{50} 21.72 ppm
LD_{90} 32.57 ppm	LD_{90} 39.40 ppm

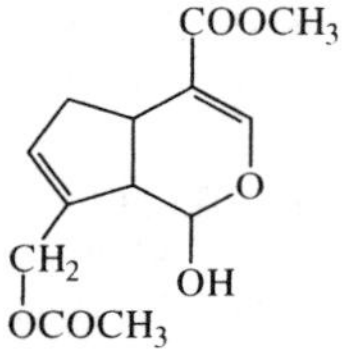

The use of this plant as a Third World molluscicide is being investigated.

Acknowledgments

Thanks are due to the University Research Fund for financial assistance.

References

Drewes, S.E. and Hudson, N.A. (1984). Novel dimeric chalcone-based pigments from *Brakenridgea zanguebarica*. *Phytochemistry* **22**, 105-108.

Drewes, S.E. and Liebenberg, R.W. (1982). Novel pentenyne derivative as potential anticancer agent. *British Patent* No 8211294 (19 April 982).

Drewes, S.E. and Taylor, C.W. (1994). Methylated A-type proanthocyanidins from *Cassipourea gerrardii*. *Phytochemistry* **37**, 551-555.

Drewes, S.E., Hall, A.J., Learmonth, R.A., and Upfold, U.J. (1984). Isolation of hypoxoside from *Hypoxis rooperi* and synthesis of its aglycone. *Phytochemistry* **23**, 1313-1316.

Drewes, S.E., Hudson, N.A., Bates, R.B., and Linz, G.S. (1987). Medicinal plants of Southern Africa. *Journal of the Chemical Society, Perkin Transactions I*, 2809-2813.

Drewes, S.E., Taylor, C.W., Cunningham, A.B., Ferreira D., Steenkamp, J.A., and Mouton, C.H.L. (1992). Epiafzelechin from *Cassipourea gerrardii*. *Phytochemistry* **31**, 2491-2494.

Drewes, S.E., Sehlapelo, B.M., Horn, M.M., Scott-Shaw, R., and Sandor, P. (1995). 5,6-Dihydro-α-pyrones from *Cryptocarya latifolia*. *Phytochemistry* **38**, 1427-1430.

Marini-Bettolo, G.B., Patamia, M., Nicoletti, M, Galeffi, C., and Messana, I. (1982). Hypoxoside, a new glycoside from *Hypoxis obtusa*. *Tetrahedron* **38**, 1683-1687.

Sehlapelo, B.M., Drewes, S.E., and Sandor, P. (1993). Ocobullenone from *Ocotea bullata*, *Phytochemistry* **32** 1352-1353.

18. Unusual metabolites from Tanzanian annonaceous plants: the genus *Uvaria*

M.H.H. NKUNYA

Department of Chemistry, University of Dar es Salaam, P.O. Box 35061, Dar es Salaam, Tanzania

Introduction

The co-existence of organisms is undoubtedly very strongly controlled by ecological conditions, *i.e.* intra- and inter-species interactions. These are regulated by the chemicals that are metabolized by the organisms. Such chemical compounds may be secreted by a particular organism as defensive agents against predation of that organism by another, as reproductive or growth regulators within and outside the source organism (*e.g.* plant and insect juvenile hormones, pheromones, etc.), or just as products of the plants' own metabolic processes. Such plant metabolites may be compounds of no apparent benefit to the former's prosperity. Usually, such compounds are formed in the plants through enzyme mediated metabolic processes, thereby resulting into stable compounds with well defined chemical structures. However, some of these compounds may be formed through non-enzymatic pathways. Non-enzymatic biosynthesis of plant natural products can be appreciated from the fact that certain plant metabolites occur in their racemic forms, which is an uncommon phenomenon among enzyme mediated biochemical transformations. This has prompted phytochemists to envisage a new outlook in explaining the biogenesis of certain naturally occurring compounds. Such proposals include the idea of chemical evolutionary processes to be occurring within plants, whereby enzymatic as well as non-enzymatic processes are taking place. These may eventually lead to transformations of originally formed unstable and therefore chemically very reactive compounds, into new and stable natural products. Such synthetic processes may therefore proceed alongside the plants' evolutionary transformations, which would ultimately yield compounds with rather complex chemical structures. In other words, some naturally occurring compounds may actually be substrates of enzymatic biosynthetic processes that eventually yield compounds which, in one way or another, would be needed by the plants.

Many postulates have been proposed in order to relate the diversity of plant natural products and their potential socio-economic importance, as well as on the chemistry of the plants' ecosystems and prosperity. One such proposal links the metabolism of biologically active plant metabolites to the latter's defensive mechanisms. It is thus argued that since plants are immobile, they are highly vulnerable to attack by other predatory organisms. Specific attention is drawn to tropical plants, since these are faced with challenging survival conditions due to the presence of many organisms that thrive under warm and usually humid tropical climatic conditions. Thus, through evolutionary transformations, plants growing in tropical rain forests must have evolved metabolic processes that enable them to secrete chemotoxins which will interfere with the basic biochemistry of predatory or harmful organisms. The chemotoxins therefore are able to deter predators, hence promoting the survival and prosperity of these plants. Such compounds would then be suitable candidates for the development of pest control agents for agricultural and/or household applications. Coincidentally, some of the plant toxins may also interfere with the biochemical processes in other organisms. Indeed such compounds will be potentially useful in the development of pharmaceuticals for bacterial, fungal, viral and other infectious diseases, or as agrochemicals. Such and other similar proposals have therefore prompted efforts to investigate plants, especially those found in tropical forests, for their biomedically and other useful constituents, studies in which we are actively involved for over a decade now. In our endeavors we combine this scientific philosophy with the fact that certain plant materials are known for their curative effects from the point of view of traditional medical practices. Nonetheless we have been investigating even those plants which are not reported to be used in traditional remedies. Such plants may in fact offer interesting chemicals.

Natural products research and African plant resources

The abundance and diversity of African flora, especially in tropical rain forests, has recently prompted enormous efforts to investigate these plant resources for lead compounds in the development of new pharmaceutical agents and other compounds with economic potential. Despite the threats to biological diversity Africa is facing at the moment due to, among other factors, man-induced environmental degradation activities, the African continent, especially the Sub-Saharan region, still boasts of a wide variety of indigenous plant species. Countries that exhibit the greatest variation in the plant habitats, and are thus internationally important for plant diversity and therefore, for the inherent natural product resources include Cameroon, Kenya, Nigeria, Tanzania, Uganda, South Africa and Zaire (Stuart *et al.* 1990). In this context, the overall plant diversity has been recorded to be at the highest level in Southern African countries (Angola, Botswana, Lesotho, Mozambique, Namibia, South Africa, Swaziland, Tanzania, Zambia and Zimbabwe), including Madagascar, in comparison with other equally sized regions in Africa (Fuggle and Rabie 1992). The available knowledge on the use of plant preparations in traditional medicines in these countries allows a direct search for such compounds following scientifically established norms. In other cases, the medicinal usefulness of the plants may not be known, but still chemical investigations might reveal some interesting bioactive compounds. After all, there are many African plant species which have even not been taxonomically described and therefore, such plants are still unknown to science.

Tanzanian forests are estimated to constitute at least 10'000 vascular plant species, among which about 10% are endemic (Tesha 1991). Some of the endemic species are considered to be occurring only in Tanzania, the majority being in remote forests and therefore have not been investigated scientifically for their biology, ecology and for their bioactive and other chemical constituents. From this point of view therefore, during the past ten years we investigated Tanzanian plants, especially those which are used traditionally as herbal remedies, as well as other plant species which are endemic to Tanzania, for their antimalarial and other chemical constituents. Recently, we also included in our studies, investigations of Tanzanian plants for trypanocidal agents in a hope to obtain compounds which could be promising in the development of drugs for sleeping sickness, a tropical disease which is transmitted by tse-tse flies and is prevalent in many Sub-Sahara African countries including Tanzania.

Most of the plant species which we have investigated so far belong to the family Annonaceae. This plant family is considered to comprise of 130 genera with more than 2300 species (Colman-Saizarbitoria *et al.* 1994). Many species in this family are used in traditional medicines and are found in the tropical regions. This family is known to be a source of natural products which have unique chemical structures, some of which are not known to be occurring in other plant families. Most compounds from the family Annonaceae have been shown to possess potent pharmacological and other bioactivities. The compounds include acetogenins, indole terpenoids, cyclohexene epoxides, C-benzylflavonoids, etc. Such phytochemical results and the fact that several annonaceous plants which grow in Tanzania have actually not been phytochemically investigated prompted our interests to investigate Tanzanian annonaceous plants for their bioactive and other phytochemicals. Our results are reviewed in this article.

The genus *Uvaria*

The genus *Uvaria* consists of climbing, straggling or erect shrubs or small trees. Over 100 species are known to comprise this genus (Verdcourt 1971). These species are distributed in tropical and sub-tropical regions, where some of them are deployed in indigenous therapies. In Tanzania *Uvaria* species are found in coastal evergreen forests and in mountain forests upcountry. So far 13 *Uvaria* species which occur in Tanzania have been described (Verdcourt 1971; Vollesen 1980). Two undescribed species have recently been spotted in Pande and Pugu forests near Dar es Salaam. The described *Uvaria* species which are found in Tanzania include *U. accuminata* Oliv., *U. angolensis* Oliv., *U. decidua* Diels, *U. dependens* Engl. & Diels, *U. faulknerae* Verdc., *U. kirkii* Hook. f., *U. leptocladon* Oliv., *U. lucida* Benth., *U. lungonyana* Vollesen, *U. pandensis* Verdc., *U. scheffleri* Diels, *U. tanzaniae* Verdc. and *U. welwitschii* (Hiern) Engl. & Diels.

Nine *Uvaria* species were included in our investigations of Tanzanian antimalarial plants which we initiated some ten years ago (Nkunya *et al.* 1991). The plants included *U. dependens* Engl. & Diels, *U faulknerae* Verdc., *U. kirkii* Hook. f., *U. leptocladon* Oliv., *U. lucida* ssp. *lucida* Benth., *U. pandensis* Verdc., *U. scheffleri* Diels, *Uvaria* sp. (Pande) and *U. tanzaniae* Verdc. In the course of these investigations we isolated several unusual metabolites, including pairs of mono-, di-, and tri-(*o*-hydroxy)benzyl flavanones and dihydrochalcones, oxygenated cyclohexene epoxides, condensed chalcones and benzopyranyl sesquiterpenes.

C-Benzylflavonoids

C-Benzyldihydrochalcones and flavanones are natural products in which one or more *o*-hydroxybenzyl groups are coupled to a dihydrochalcone or a flavanone nucleus. The C-benzylation occurs at either C-3' or C-5' or C-6 and C-8 of the dihydrochalcone or flavanone nuclei respectively, with further benzylation taking place at the original *o*-hydroxybenzyl groups *para*- to the hydroxyl functionality. These compounds are restricted to the family Annonaceae, being found almost exclusively from the genus *Uvaria* save for the recent isolation of such compounds from a *Xylopia* species, *viz. X. africana* (Ekpa *et al.* 1993; Anam 1994a and 1994b). One of the simplest members of C-benzyldihydrochalcones is uvaretin (**1**), which was isolated from an East African medicinal plant, *Uvaria accuminata* (Leboeuf 1992). The positional isomer of uvaretin, isouvaretin (**2**) occurs in *U. angolensis* and *U. chamae* from West Africa (Lasswell and Hufford 1977; Hufford and Oguntimein 1982). The dibenzyl dihydrochalcone diuvaretin (**3**) has been isolated from *U. angolensis* (Hufford and Oguntimein 1982), *U. chamae* (Lasswell and Hufford 1977), *U. lucida* ssp. *lucida* (Weenen *et al.* 1990a), *U. tanzaniae* (Weenen *et al.* 1991) and U. *kirkii* (Nkunya 1985). The other three dibenzyl dihydrochalcone angoluvarin (**4**), chamuvaritin (**5**) and isochamuvaritin (**6**) are constituents of *U. leptocladon* and *Uvaria* sp. (Pande) (Nkunya *et al.* 1991; Nkunya *et al.* 1993a).

Chamuvaritin (**5**) and isochamuvaritin (**6**) are the only C-benzyldihydrochalcone occurring in *Uvaria* species in which an *o*-hydroxybenzyl group forms a benzopyranyl unit fused to ring A of the dihydrochalcone unit.

We recently isolated the two hitherto unreported tri-(*o*-hydroxy)benzyl dihydrochalcones triuvaretin (**7**) and isotriuvaretin (**8**) from the ethanolic extract of the root bark of *U. leptocladon* and their structures were established by spectroscopic methods, particularly homo- and heteronuclear 2D-NMR (Nkunya *et al.* 1993a). These compounds were obtained together with pairs of the isomeric mono-, di- and tri-(*o*-hydroxy)benzyl dihydrochalcones **1** and **2**, **3** and **4**.

As described for the dihydrochalcones **1-5** (Hufford and Lasswell 1978; Hufford *et al.* 1981), we did establish that triuvaretin (**7**) and isotriuvaretin (**8**) possess antimicrobial activity (Nkunya *et al.* 1993a). Uvaretin (**1**) and isouvaretin (**2**) also exhibit cytotoxicity (Lasswell and Hufford 1977).

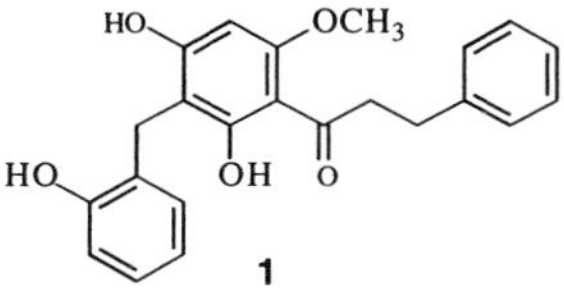

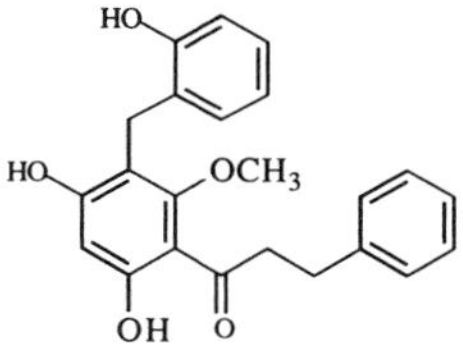

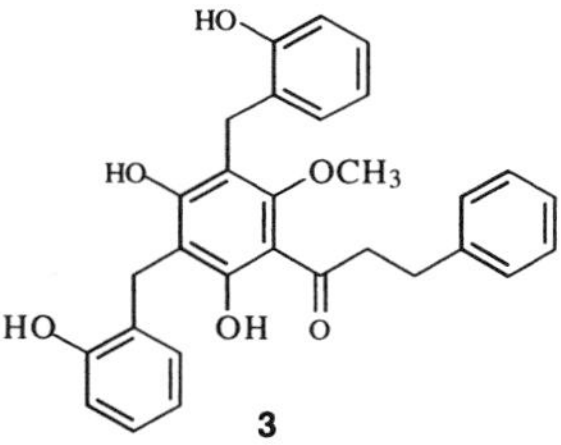

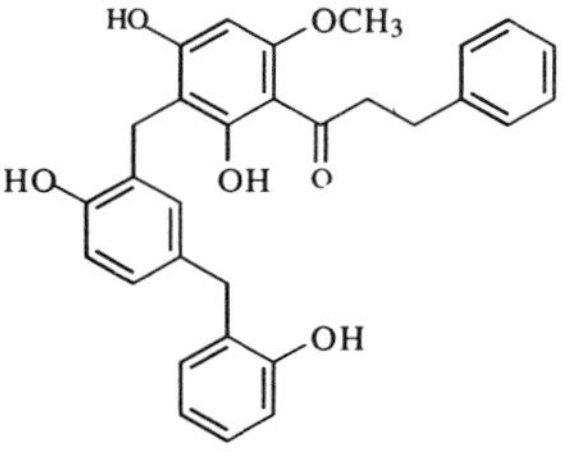

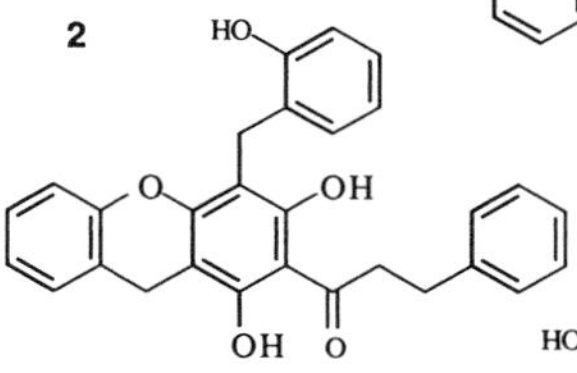

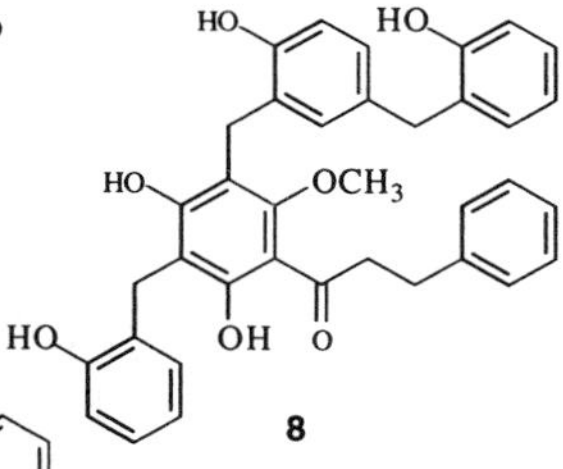

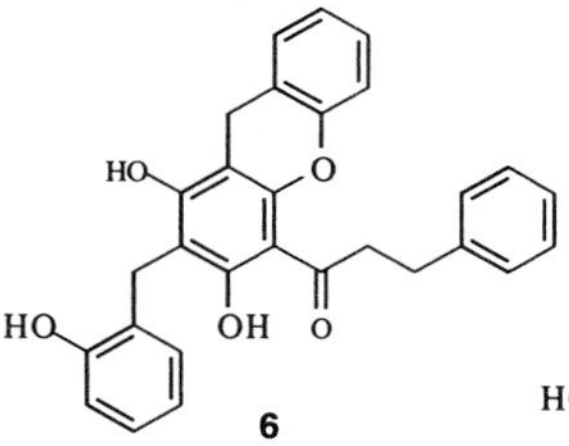

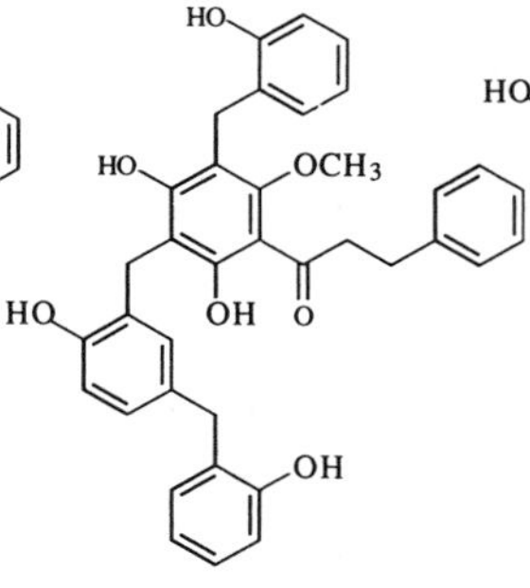

The (*o*-hydroxy)benzyl flavanones that occur in plants of the genus *Uvaria* include chamanetin (**9**), isochamanetin (**10**) and dichamanetin (**11**), all of which were obtained from *U. chamae* and other *Uvaria* species (Lasswell and Hufford 1977). We recently isolated the three flavanones from the antimalarially active stem bark extract of *U. lucida* ssp. *lucida*, together with the tri-(*o*-hydroxy)benzyl flavanones uvarinol (**12**) and isouvarinol (**13**) and the hitherto unknown oxygenated pyrenes, 2,7-dihydroxy-1,8-dimethoxypyrene (**14**) and 2-hydroxy-1,7,8-trimethoxypyrene (**15**) (Achenbach *et al.*, *Phytochemistry* in press).

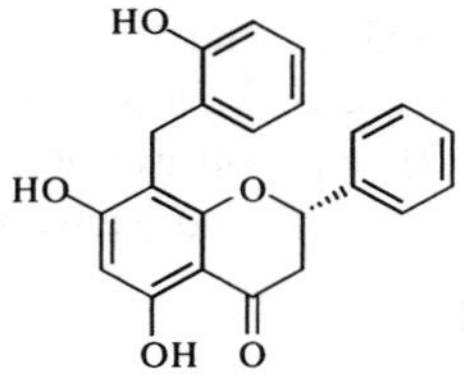

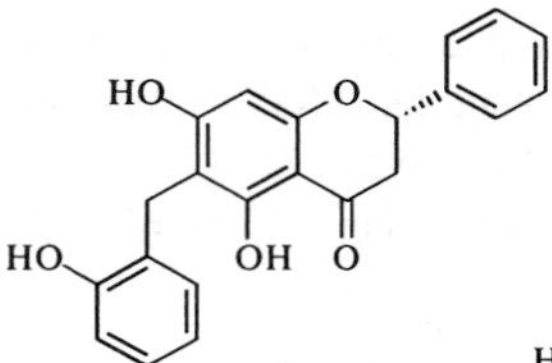

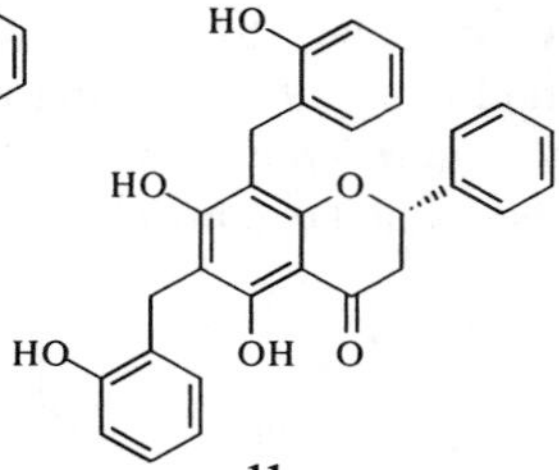

Again in these investigations the occurrence of pairs of (*o*-hydroxy)benzylated natural products from plants of the genus *Uvaria* was revealed. This, and our earlier results from *U. leptocladon* (Nkunya *et al.* 1993a) thus demonstrate the uniqueness of plants of the genus *Uvaria* in their ability to accumulate C-benzylated natural products whose biosynthesis is still not well understood.

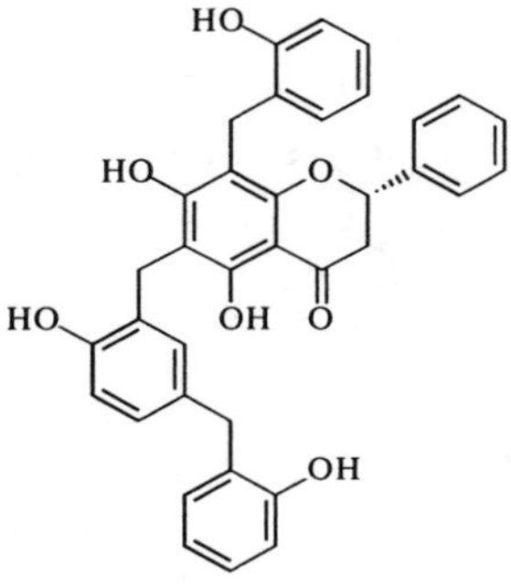

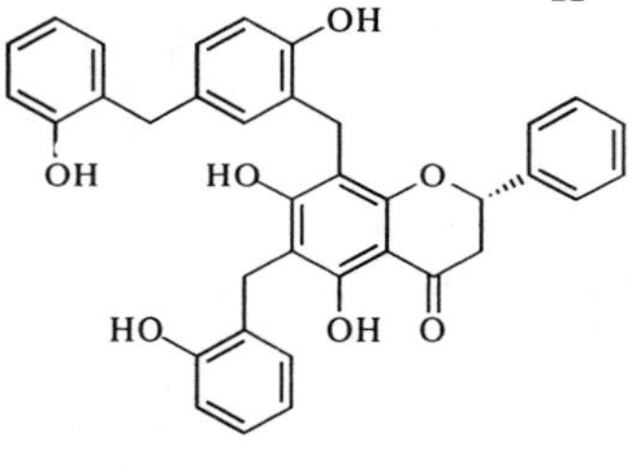

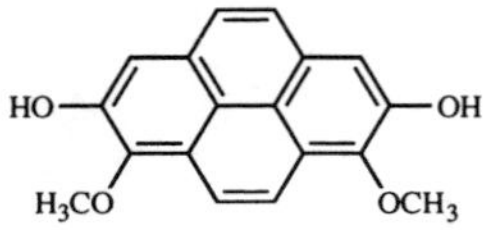

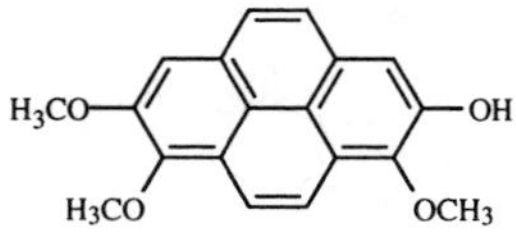

The two pyrenes (**14**) and (**15**) were also isolated from a West African *Uvaria* species, *U. doeringii*, together with a potential biogenetic precursor of the pyrenes, 1,2-dihydro-2,9-dihydroxy-10-methylfura[2,3-a]phenanthrene (**16**)

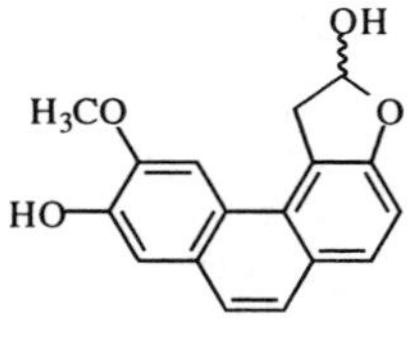

(Achenbach *et al.*, *Phytochemistry* in press). Pyrene **15** was found to have an activity which is comparable to that of podophyllotoxin in the brine shrimp test. However, pyrene **14** was distinctively less active.

Condensed flavonoids

The genus *Uvaria* is also known to contain natural products which may be envisaged to originate from condensation reactions involving flavonoid units and terpenoids, or flavonoid units alone. These compounds include the 'condensed' chalcones schefflerin (**17**) and isoschefflerin (**18**) which we isolated from the antimalarial stem bark of *U. scheffleri*, together with their apparent biogenetic precursor, 2',6'-dihydroxy-3,4'-dimethoxychalcone, the rare triterpene D: β–friedo olean-5-en-3β-ol (glutin-5-en-3β-ol), 3-farnesylindole (**19**), 2'-hydroxy-3',4',6'-trimethoxychalcone, β–sitosterol and benzyl benzoate, which is widespread in *Uvaria* species (Nkunya *et al.* 1990).

The hitherto unreported compounds **17** and **18** are structurally related to the panduratins, which have been isolated from the rhizomes of *Boesenbergia* species (Zingiberaceae) (Pancharoen *et al.* 1984). The MS of schefflerin (**1**) and isoschefflerin (**2**) exhibited comparatively weak molecular ions and highly abundant fragments at m/z 197, which are due to the substituted benzoylium ion. A base peak ion at m/z 301 was observed only in the MS of **1**, the formation of which obviously involves a McLafferty-type rearrangement in which the carbonyl group takes part. Therefore, the formation of this key fragment ion established the neighborhood of the benzoyl and the prenyl groups in **1**. NMR studies using (S)-2,2,2-trifluoro-1-(9-anthryl)-ethanol as a chiral shift reagent demonstrated **17** and **18** to be racemic mixtures.

Biosynthetically schefflerin and isoschefflerin as well as the panduratins can be regarded as resulting from a Diels-Alder-like cyclization of 2',6'-dihydroxy-3',4'-dimethoxychalcone and the acyclic monoterpene β–ocimene (3,7-dimethyl-1,3,6-octatriene).

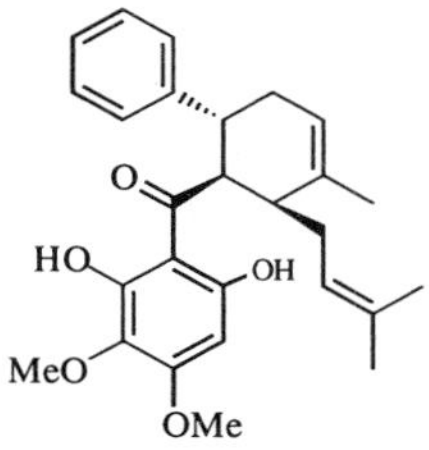

HO OH MeO OMe O

17

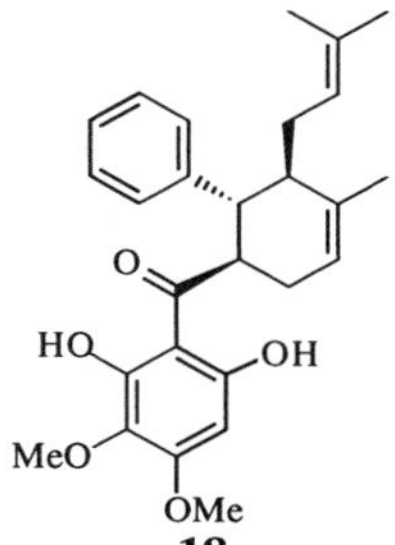

HO OH MeO OMe O

18

OH OH N H

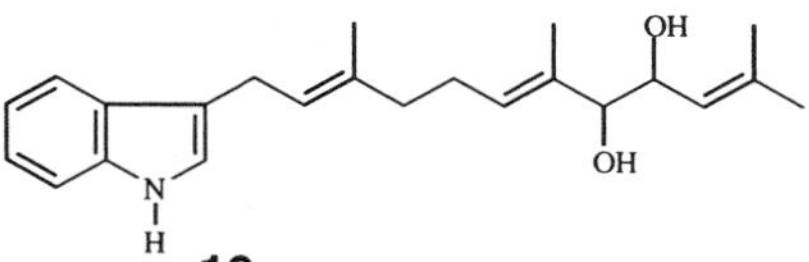

19

The crude extract of the root bark of *U. dependens* was found to be active against *P. falciparum* malaria parasite *in vitro* (Nkunya *et al.* 1991). In attempts to obtain the active constituents we isolated 5,7,8-trimethoxyflav-3-ene (**20**), 2-hydroxy-3,4,6-trimethoxychalcone (**21**) and the dimeric benzopyranoid, dependensin (**22**), (-)-pipoxide and a mixture of β–sitosterol and stigmasterol

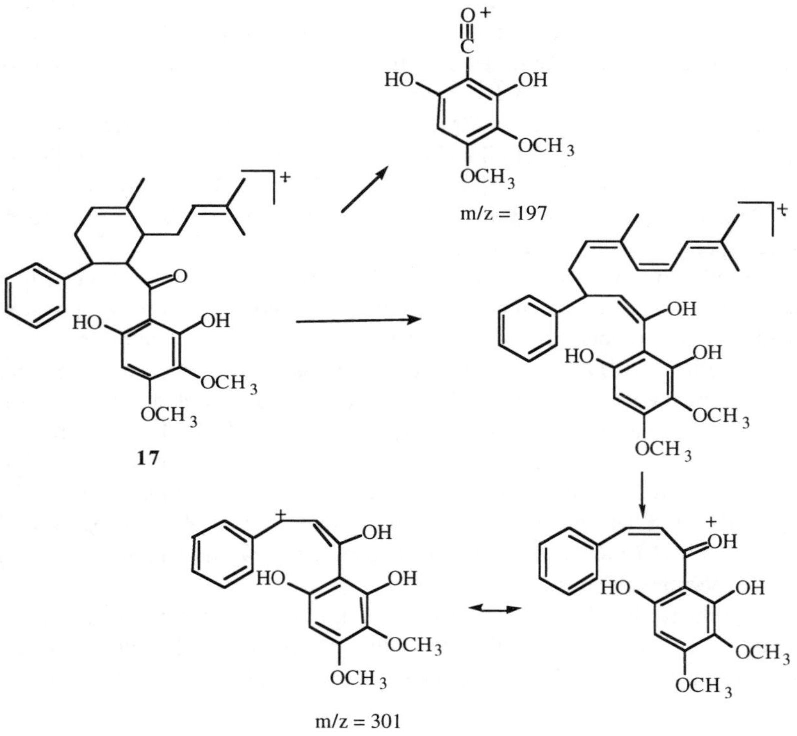

HO OH OCH 3 OCH 3 m/z = 197

HO OH OCH 3 OCH 3

17

OH HO OH OCH 3 OCH 3

OH HO OH OCH 3 OCH 3

m/z = 301

+ OH HO OH OCH 3 OCH 3

(Nkunya *et al.* 1993b). Both compounds **20** and **22** were isolated as racemic mixtures, as particularly revealed by the separation of all ^{1}H NMR resonances into double sets of signals when measured in the presence of the optically active NMR shift reagent, tris[3-(heptafluoropropyl-hydroxymethylene)-(+)-camphorato]europium (III).

Flavenes have been rarely found as natural products. This might be due to their sensitivity to air. We in fact established that flavene **20** is decomposed in the presence of air, especially under acidic conditions, the main decomposition product being the 'reversed' chalcone **21**. The latter compound as isolated from *U. dependens* may therefore be regarded as an artifact.

Hypothetically, dependensin would originate by proton catalyzed dimerisation of **20**, or through dimerisation of chalcone **23**. Since dependensin was isolated as a racemic mixture we immediately suspected that the compound might be an artifact, formed from either **20** or **23**. However, preparative decomposition of **20** did not yield dependensin, even in trace amounts. Furthermore, the chalcone **23** was not detected in the extract which was obtained from fresh plant materials. We therefore regarded **22** as a genuine natural product.

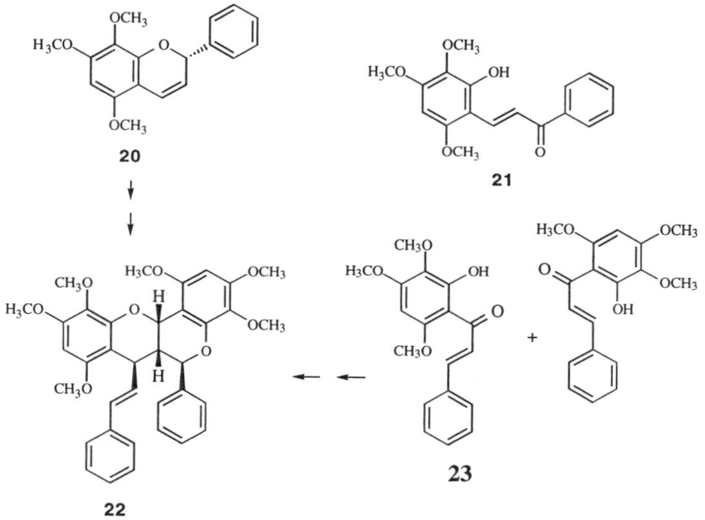

Cyclohexene epoxides

Monocyclic cyclohexene derivatives occur widely in nature and usually possess a wide variety of interesting biological activities. The compounds are found in plants (mainly in the genus *Uvaria*), micro- and marine organisms. They possess a unique chemical skeleton, in which the cyclohexene nucleus is highly oxygenated. Typical examples of such compounds are (+)-β–senepoxide (**24**), (+)-pipoxide (**26**), (+)-pandoxide (**27**), (-)-tingtanoxide (**25**) and (-)-pipoxide (**28**) (Nkunya *et al.* 1986a; Thebtaranonth and Thebtaranonth 1989).

Uvaria pandensis Verdc. which grows in Pugu and Pande forests near Dar es Salaam, was only recently taxonomically described (Verdcourt and Mwasumbi 1988). We isolated from the stem and root barks of this plant the two hitherto unreported cyclohexene epoxide (+)-pandoxide (**27**) and (-)-pipoxide (**28**), in addition to (+)-β–senepoxide (**24**) and farnesylindoles (Nkunya *et al.* 1987a; 1987b; 1989).

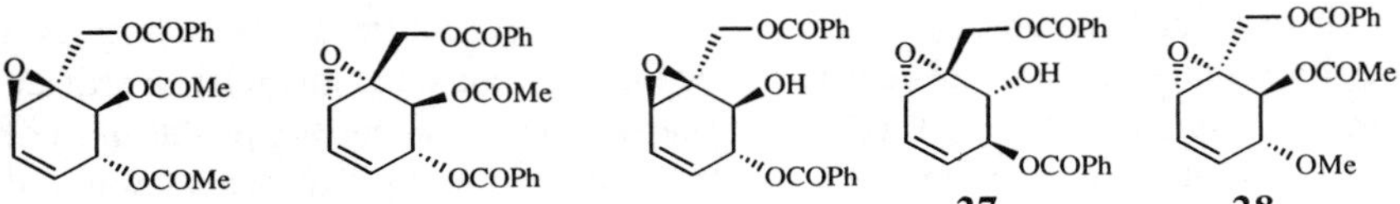

The three epoxides **24**, **27** and **28** showed antibacterial, antifungal and antitumour activities *in vitro*. (-)-Pipoxide was also mildly active against *P. falciparum* malaria parasite *in vitro* ($IC_{50} = 8.37$ μgml^{-1}) (Nkunya *et al.* 1991)

The isolation of cyclohexene epoxides from *Uvaria* species has stimulated interests to propose their biosynthetic pathways. The most plausible proposal involves an arene oxide as the biogenetic intermediate. The compound, which is conceivably obtained from epoxidation of benzoic acid or its ester (a shikimic acid derivative) would undergo a spontaneous and therefore non-stereoselective nucleophilic ring opening reaction at C-3 (ß-attack) to give enantiomeric cyclohexadienes A and B. Epoxidation of cyclohexadiene A would lead to the formation of two diastereomers having the stereogenic centers found in either (+)-β–senepoxide or (-)-senepoxide. This epoxidation process is probably enzymatically controlled and therefore rationalizes the formation of (+)-ß-senepoxide (**24**) and (-)-senepoxide or its derivative, *e.g.* in *U. ferruginea* (Kodpinid *et al.* 1983). Cyclohexadiene B would undergo a similar epoxidation process to give epoxides having the stereochemical configuration of (-)-pipoxide or the still unknown enantiomeric form of (-)-senepoxide (Fig. 18. 1.).

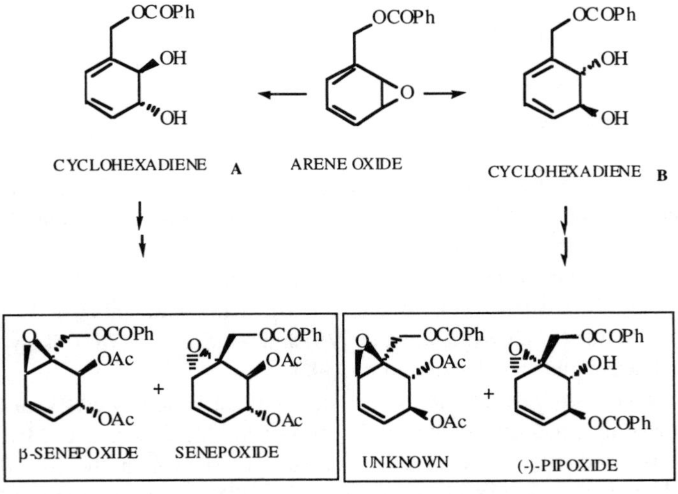

Fig. 18.1. Proposal of a biosynthetic pathway for the cyclohexene epoxides isolated from *Uvaria* sp.

There are indications that the group of Professor Achenbach may have recently isolated the unknown enantiomer from *Uvaria doeningii* (Achenbach, personal communication 1996). If this is the case therefore, all stereoisomers of cyclohexene epoxides which are predictable on the basis of the biosynthesis discussed above will now have been obtained from *Uvaria* species, this thus further supporting the envisaged biosynthesis of these compounds.

Benzopyranyl sesquiterpenes

Sesquiterpenes containing *o*-hydroxybenzyl groups have been isolated from *Uvaria* species. Five such compounds have thus far been reported. These include lucidene (**29**), a bis (benzopyranyl) sesquiterpene which we recently isolated from the root bark of *Uvaria lucida* ssp. *lucida,* and its structure was determined by spectroscopic methods and single-crystal X-ray crystallography (Weenen *et al.* 1990a). The compound was isolated as a racemate and high-resolution ^{1}H NMR spectroscopy (400 and 600 MHz) allowed the assignment of all protons and indicated the existence of relatively slowly interconverting conformations.

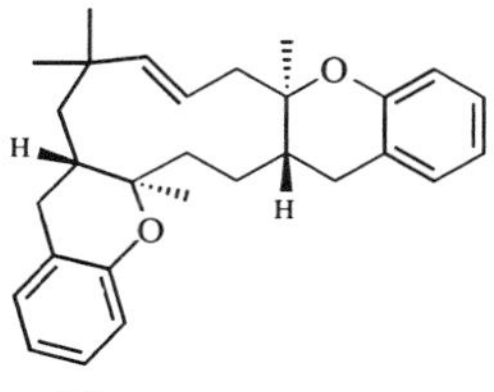

29 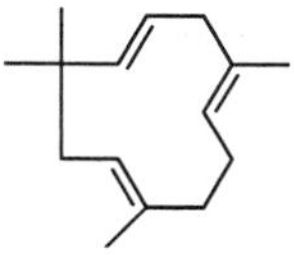 **30** 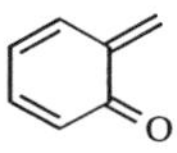 **31**

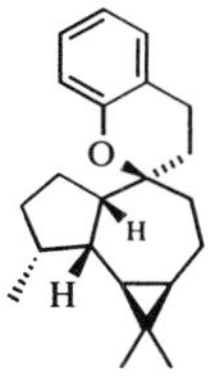

32 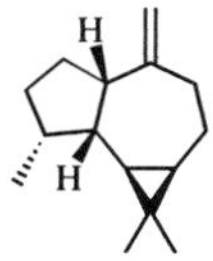 **33** 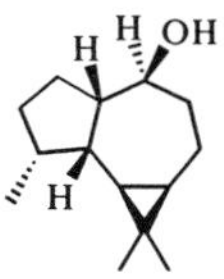 **34** 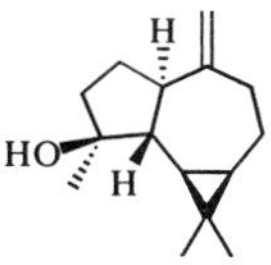 **35**

Conceivably, lucidene (**29**) can be regarded as an addition product of two *o*-hydroxybenzyl groups to humulene (**30**). The latter sesquiterpene was in fact present in the least polar fractions of the root bark extract. The fact that lucidene was isolated as a racemic mixture suggests a biosynthetic formation by a non-enzymatic pathway. In fact the formation of lucidene would involve a double Diels-Alder reaction of humulene (**30**) with two molecules of *o*-benzoquino methide (**31**). The *o*-benzoquinomethide (can also undergo Michael-type conjugate addition reactions to give *o*-hydroxybenzyl moieties. Considering this biogenesis one would expect to obtain other diastereomers of lucidene (ring junction methyls *trans*) and regioisomers (oxygen of *o*-quinomethide bonded to less substituted carbon in the cycloadduct). However, during our investigations such compounds were not detected.

GC-MS analysis of the least polar chromatography fractions of the hexane extract of *U. lucida* ssp. *lucida* showed the presence of the known sesquiterpenes cyperene (39.3% of the combined nonpolar fractions, according to peak integration by GC), β–selinene (17.9%), α–selinene (5.5%), humulene (‹1%) and several other constituents which were either present in very small concentration or could not be identified. Four unidentified sesquiterpenes (1.3%) showed the characteristic fragmentation pattern of sesquiterpenes with an oxybenzyl moiety (m/z 310 and 107).

Another benzopyranyl sesquiterpene, tanzanene (**32**), was isolated from the root bark of *Uvaria tanzaniae* Verdc. and its structure was determined by high-resolution NMR (Weenen *et al.* 1991). The molecular structure of tanzanene can be thought of as a cycloaddition product of alloaromadendrene (**33**) and the quinomethide **31**. Tanzanene has some features in common with other C-benzylated sesquiterpenes. However, unlike any other C-benzylated natural products reported so far, the C-benzyl substituent in **32** forms a spiro connection with the sesquiterpene part of the molecule. The *cis*-fused ring system in tanzanene was demonstrated by decoupling experiments as well as by a phase-sensitive COSY study, which showed the coupling constant between H-1 and H-5 to be 3.5 Hz, the same value as measured for viridiflorol (**34**), whereas in the corresponding *trans*-fused system of spathulenol (**35**), the coupling constant between these bridgehead protons is about 9 ppm.

GC and GC-MS analysis of the apolar sesquiterpene fraction of the roots of *U. tanzaniae* indicated that alloaromadendrene is one of the major sesquiterpene constituents (19.7% of all apolar sesquiterpenes present). This observation may be considered to offer support to the hypothesized biosynthesis of tanzanene as discussed above.

Antimalarial activity

From time immemorial malaria caused by the *Plasmodium falciparum* parasite, the so-called malaria tropical parasite, has claimed countless lives. During the last 25 years, malaria, which is now regarded as a tropical disease, has regained its earlier reputation as being one of the greatest threats to mankind, especially in Sub-Saharan Africa. The emergence of insecticide resistant *Anopheles* mosquitoes and malaria parasites resistant to standard drugs have underlined the necessity to develop additional methods for the control of mosquito vectors and the elimination of the parasites in their human hosts once the latter is infected. For the second point, two approaches may be considered: The development of new chemotherapeutic agents and/or the development of malaria vaccines. Development of effective malaria vaccines as an urgent response to the appearance of drug resistant *P. falciparum* strains is in progress. Thus, recently, clinical trials of the vaccine SPf 66, which has been synthesized by Dr. Patarroyo from Colombia, were conducted in Ifakara, Tanzania. Phase II of these trials which ended in 1994 revealed that the risk of developing malaria in children who received the vaccine was lowered by 31% (Alonso 1994). These results are still not impressive. Therefore, there is still an urgent need to develop new chemotherapeutic agents for malaria, since even if a vaccine is developed, slip out patients will need to be treated. In this respect, plant resources are potential targets for research and development of alternative antimalarial drugs.

Since 1985 we have screened almost 120 plant species for their antimalarial activities *in vitro* against *Plasmodium falciparum* malaria parasite (Weenen *et al.* 1990b; Nkunya *et al.* 1991). These investigations included *Uvaria* species and several other plant species of the family Annonaceae. Thus, petroleum ether, dichloromethane, and methanol extracts of leaves, stem, and root barks of nine *Uvaria* species: *U. dependens, U. faulknerae, U. kirkii, U. leptocladon, U. lucida* ssp. *lucida, Uvaria* sp. (Pande), *U. scheffleri,* and *U. tanzaniae* were tested for their *in vitro* activity against the multidrug resistant K1 strain of *Plasmodium falciparum*. The IC_{50} values of the extracts varied between 5 and 500 μgml^{-1}. The most active extracts were obtained from the stem and root barks of *U. lucida* ssp. *lucida* and *Uvaria* sp. (Pande) and the root bark of *U. scheffleri*, all of which had IC_{50} values between 5 and 9 μgml^{-1}. Among the compounds isolated, uvaretin, diuvaretin, and (8',9'-dihydroxy)-3-farnesylindole were the most active (IC_{50} = 3.49, 4.20, and 2.86 μgml^{-1}, respectively). The activity of the uvaretins decreased with increasing molecular size. Thus, there was a progressive decrease in activity from uvaretin (**1**), diuvaretin (**3**), and triuvaretin (**7**) (IC_{50} = 3.49, 4.20, and 46.02 μgml^{-1}, respectively). The phenolic hydroxyls also somewhat enhance the activity, as shown by the decrease in activity in uvaretin trimethyl ether (IC_{50} = 8.31 μgml^{-1}) as compared to the unmethylated compound (IC_{50} = 3.49 μgml^{-1}).

Conclusion

This article has summarized our investigations of plants of the genus *Uvaria* (Annonaceae) for their antimalarial and other natural products, several of which possess unique chemical structures whose biogenesis is quite intriguing and thus needs further elaboration. Some of the isolated compounds were obtained in their racemic forms despite the fact that these compounds were shown to be genuine natural products and not decomposition products or derivatives of the latter. The occurrence of the compounds as racemates can be rationalized by considering a chemical evolutionary process which might be occurring with plant natural products, as it has been briefly explain in this article. However, these proposals need further experimental verifications.

Acknowledgments

These investigations were supported by grants from the Netherlands Organization for International Cooperation in Higher Education, through the Organic Chemistry Project (a project of academic cooperation between the Universities of Dar es Salaam and Nijmegen), the Norwegian Agency for International Development (NORAD), through the NORAD Chemistry Project at the University of Dar es Salaam, the International Foundation for Science (IFS), the University of Dar es Salaam and the German Academic Exchange Services (DAAD). I am very grateful to these institutions for this support. I also thank Mr. L.B. Mwasumbi of the Herbarium, Department of Botany, University of Dar es Salaam for locating and identifying all the plants which we have investigated and to Ms Monika C. Gessler of the Swiss Tropical Institute, Basle in Switzerland for antimalarial assays.

References

Alonso, P.L. 1994. Randomized trial of efficacy of SPf66 vaccine against *Plasmodium falciparum* malaria in children in Southern Tanzania. *Lancet* **344**, 1175-1181.

Anam, E.M. (1994a). 2''',2'''',2'''''-Trihydroxy-5'',3''',5''''-tribenzyldiuvaretin and 2''',2'''',2'''''-trihydroxy-5''',3''',5''''-tribenzylisodiuvaretin: Two novel tri-C-benzylated dihydrochalcones from the root extract of *Xylopia africana* (Benth.) Oliver (Annonaceae). *Indian Journal of Chemistry* **33B**, 204-205.

Anam, E.M. (1994b). 2''''-Hydroxy-3'''-benzyluvarinol, 2''''-hydroxy-5'''-benzylisouvarinol-A, and 2''-hydroxy-5''-benzylisouvarinol-B: Three novel tetra-C-benzylated flavanones from the root extract of *Xylopia africana* (Benth.) Oliver (Annonaceae). *Indian Journal of Chemistry* **33B**, 1009-1011.

Colman-Saizarbitoria, T., Zambrano, J., Ferrigni, N.R., Gu, Z.-M., Ng, J.H., Smith D.L., and Mc Laughlin, J.L. (1994). Bioactive annonaceous acetogenins from the bark of *Xylopia aromatica*. *Journal of Natural Products* **57**, 486-493.

Ekpa, O.D., Anam, E.M., and Gariboldi, P.V. (1993). Uvarinol and novel iso-uvarinol: Two C-benzylated flavanones from *Xylopia africana* (Annonaceae). *Indian Journal of Chemistry* **32B**, 1295-1297.

Fuggle, R.F. and Rabie, M.A. (1992). Environmental management in South Africa. Juca and Company, Ltd, Cape Town.

Hufford, C.D. and Oguntimein, B.O. (1982). New dihydrochalcones and flavanones from *Uvaria angolensis*. *Journal of Natural Products* **45**, 337-342.

Hufford, C.D., Oguntimein, B.O., and Baker, J.K.(1981). New flavonoids and coumarin derivatives of *Uvaria afzelii*. *Journal of Organic Chemistry* **46**, 3073.

Hufford, C.D. and Lasswell, M.L., Jr. (1978). Antimicrobial activity of constituents of *Uvaria chamae*. *Lloydia* **41**, 156-160.

Kodpinid, M., Sadavongvivad, C., Thebtaranonth, C., and Thebtaranonth, Y. (1983). Structures of β–senepoxide, tingtanoxide and their diene precursors. Constituents of *Uvaria ferruginea. Tetrahedron Letters* **24**, 2019-2022.

Lasswell, M.L. (Jr.) and Hufford, C.D. (1977). Cytotoxic C-benzylated flavonoids from *Uvaria chamae. Journal of Organic Chemistry* **42**, 1295-1302.

Leboeuf, M., Cave, A., Bhaumik, P.K. Mukherjee, B., and Mukherjee, R. (1982). The Phytochemistry of the Annonaceae. *Phytochemistry* **21**, 2783-2813.

Nkunya, M.H.H. (1985). 7-Methyljuglone, diuvaretin and benzyl benzoates from the root bark of *Uvaria kirkii. Journal of Natural Products* **48**, 999-1000.

Nkunya, M.H.H. and Weenen, H. (1989). Two indolosesquiterpenes from *Uvaria pandensis. Phytochemistry* **28**, 2217-2218.

Nkunya, M.H.H., Weenen, H., Koyi, N.J., Thijs, L., and Zwanenburg, B. (1987a). Cyclohexene epoxides, (+)-pandoxide, (+)-β–senepoxide and (-)-pipoxide, from *Uvaria pandensis. Phytochemistry* **26**, 2563-2565.

Nkunya, M.H.H., Weenen, H., and Koyi, N.J. (1987b). 3-Farnesylindole from *Uvaria pandensis* Verdc. *Phytochemistry* **26**, 2402-2403.

Nkunya, M.H.H., Achenbach, H., Renner, C., Waibel, R., and Weenen, H. (1990). Schefflerin and isoschefflerin: Prenylated chalcones and other constituents of *Uvaria scheffleri. Phytochemistry* **29**, 1261-1264.

Nkunya, M.H.H., Weenen, H., Bray, D.H., Mgani, Q.A., and Mwasumbi, L.B. (1991). Antimalarial activity of Tanzanian plants and their active constituents: The genus *Uvaria. Planta Medica* **57**, 341-343.

Nkunya, M.H.H., Weenen, H., Renner, C., Waibel R., and H. Achenbach (1993a). Benzylated dihydrochalcones from *Uvaria leptocladon. Phytochemistry* **32**, 1297-1300.

Nkunya, M.H.H., Waibel, R., and Achenbach, H. (1993b). Three flavonoids from the stem bark of the antimalarial *Uvaria dependens. Phytochemistry* **34**, 853-856.

Pancharoen, O., Picker, K., Reutrakul, V., Skelton, B.W., Taylor, W.C., Tuntiwachuttikul, P., and White, A.H. (1984). Crystal structure of panduratin A: (1'RS, 2'SR,6'RS)-(2,6-dihydroxy-4-methoxyphenyl)-[3'-methy-2'-(3"-methybut-2"-enyl)-6'-phenylcyclohex-3'-enyl]methanone. *Australian Journal of Chemistry* **37**, 2589-2592.

Stuart, S.N., Adams, R.J., and Jenkins, M.D. (1990). Biodiversity in Sub-Saharan Africa and its islands: Conservation, management and sustainable use. IUCN, Gland, Switzerland.

Tesha, A.J. (1991). Effect of rise in population on the endangered plant species in Tanzania. In *Proceedings, 4th NAPRECA Symposium on Natural Products*, pp.67-69. 16-21 Dec., 1991, Addis Ababa.

Thebtaranonth, C. and Thebtaranonth, Y. (1989). Naturally occurring cyclohexene oxides. *Accounts of Chemical Research* **19**, 84-90.

Verdcourt, B. (1971). *Flora of Tropical and East Africa - Annonaceae.* Royal Botanic Garden, Kew, Crown Agents For Overseas Government and Administration, London.

Verdcourt, B. and Mwasumbi, L.B. (1988). A new species of *Uvaria* (Annonaceae) from Tanzania. *Kew Bulletin* **43**, 99-101.

Vollesen, K. (1980). Notes on Annonaceae from Tanzania. *Bot. Notiser* **133**, 53-62.

Weenen, H., Nkunya, M.H.H., Abdul El-Fadl, A., Harkema, S., and Zwanenburg, B. (1990a). Lucidene, a bis(benzopyranyl) sesquiterpene from *Uvaria lucida* ssp. *lucida. Journal of Organic Chemistry* **55**, 5107-5109.

Weenen, H., Nkunya, M.H.H., Bray, D.H., Mwasumbi, L.B., Kinabo, L.S., and Kilimali, V.A.E.B. (1990b). Antimalarial activity of Tanzanian medicinal plants. *Planta Medica* **56**, 368-370.

Weenen, H., Nkunya, M.H.H., Mgani, Q.A., Achenbach, H., Posthumus, M.A., and Waibel, R. (1991). Tanzanene, a spiro benzopyranyl sesquiterpene from *Uvaria tanzaniae. Journal of Organic Chemistry* **56**, 5865-5867.

19. Quantitative and qualitative analysis of the saponins from berries of cultivated *Phytolacca dodecandra* by LC/UV and LC/MS

S. RODRIGUEZ[1], J.-L. WOLFENDER[1], J.D. MSONTHI[2] AND K. HOSTETTMANN[1]

[1] *Institut de Pharmacognosie et Phytochimie, BEP, Université de Lausanne, CH-1015 Lausanne, Switzerland and* [2] *Department of Chemistry, University of Swaziland, Kwaluseni, Swaziland*

Introduction

The dried berries of Endod, *Phytolacca dodecandra* L'Hérit (Phytolaccaceae), are used in Ethiopia as a soap substitute. The molluscicidal properties of their constituents were discovered by Lemma in 1965 and this plant became rapidly of great importance for the local control of bilharzia or schistosomiasis (Lemma, 1970). This parasitic disease affects more than 200 million people in over 70 countries in Africa, South America and in the Far East. As shown in Fig. 19.1. (Marston and Hostettmann 1985), the use of molluscicides affects dramatically the life cycle of the parasitic nematode *Schistosoma* species.

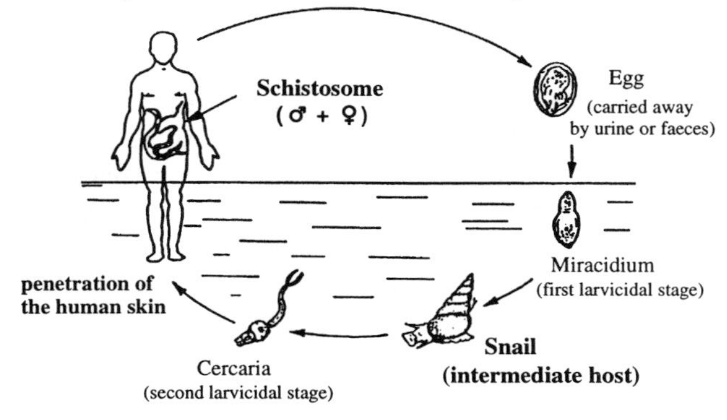

Fig. 19.1. Life cycle of *Schistosoma* species.

Eggs from infected individuals are carried away with urine and produce miracidia which locate snails as intermediate host. They multiply into thousands of cercariae which can penetrate the intact skin of humans who come into contact with the infested water. Once in the body they gradually change into schistosomes and pass into veins and bladder.

The destruction of the snails - intermediate hosts of the parasit - represents a very interesting alternative in the struggle against this tropical disease, and the use of snail-killing compounds of plant origin for the local control of schistosomiasis is attractive due to the economic advantage of cultivating the plants locally, instead of importing costly synthetic chemicals.

However, *Phytolacca dodecandra* is not grown in Swaziland and there was the need to develop a large scale cultivation. To improve the yields of berries and their saponin content, an experimental cultivation plan was undertaken. The growth of three phenotypes was tested under four conditions of field treatment, including the use of a new polymer developed in Slovenia called Eco-agrogel (see Fig. 19.2.).

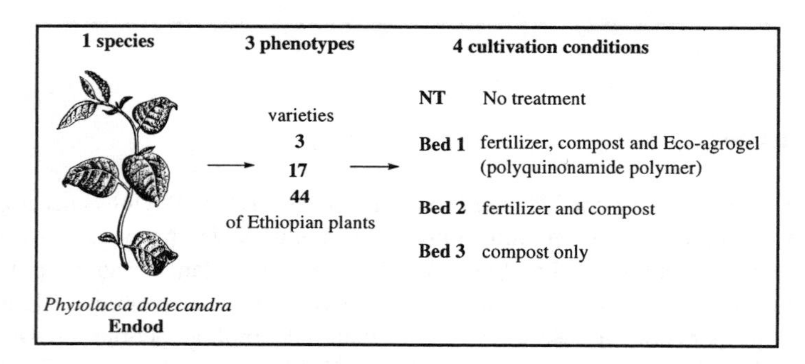

Fig. 19.2. Cultivation plan undertaken for the study of the growth of *Phytolacca dodecandra.*

The number of leaves and average height of each cultivated plant were recorded at fortnightly intervals and the berries harvested. These measurements have shown that a higher growth rate could be obtained with the first field treatment (Bed 1). The effect of the addition of Eco-agrogel to the soil is nevertheless variable for the different phenotypes (Makhubu *et al.* University of Swaziland, unpublished resuls). But the aim of the cultivation was not only to obtain a high yield of berries at maturity stage, but also to maintain a high saponin content. Thus, the study of the growth had to be completed by a quantitation of the saponin content of the berries to evaluate the effect of cultivation conditions on molluscicidal potency.

Extraction procedure

The molluscicidal activity of saponin containing extracts is strongly dependant of the extraction solvents used. Indeed, extraction with methanol provides principally inactive bidesmosidic saponins (glycosylated at positions C-3 and C-28), while water gave active monodesmosidic saponins (glycosylated at position C-3 only) (Slacanin *et al*, 1988). Temperature and water quality also influence the extraction process and often, differences between field and laboratory extracts can be observed. Thus, in order to evaluate the real content of the berries, the extractions were carried out with methanol as follows: the dried berries were powdered and 10 g of powder were extracted with 200 ml of methanol during 24 hours at 20°C. After filtration, the solvent was evaporated and the extracts were lyophilised and weighed.

In Fig. 19.3., the extraction yields of berries produced in different cultivation conditions are compared. It can be observed that these yields vary between 20 and 30% and that non-treated berries give higher extraction yields than those grown with special field treatment.

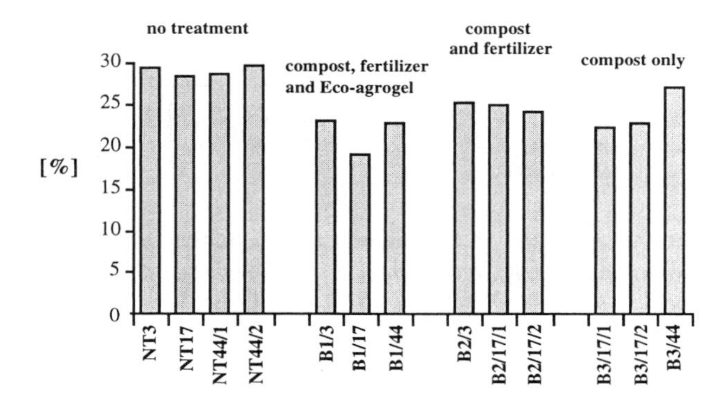

Fig. 19.3. Comparison of methanolic extract yields.

Quantitative determination

The quantitation of the saponin content of all these extracts was then realised by HPLC-UV analysis, with five pure bidesmosidic saponins as standard mixture. These saponins were previously isolated from the plant and were chosen for the quantitative determination because they are representative of the more abundant aglycones present in *P. dodecandra.* These aglycones are oleanolic acid, bayogenin and hederagenin. The structures of the five standards are presented in Fig. 19.4. They present all glucosyl moieties at position C-28 and different sugar chains at C-3.

In Fig. 19.5. are shown the two UV chromatograms of the standard mixture and of a methanolic extract of dried berries. These HPLC analyses were performed on a Waters NovaPak C_{18} (4 µm, 150 x 3.9 mm i.d.) column under the following conditions: MeCN-H_2O 15:85 to 50:50 in 30 min at a flow rate of 1 ml/min. Detection was realised at 206 nm.

R^2	R^1	Oleanolic acid
Rha - ²Glc - ²Glc - (4 | Glc)	Glc -	**saponin 3** (MW=1250)
Glc - ²Glc - (4 | Glc)	Glc -	**saponin 4** (MW=1104)

R_1O ... 3 ... $COOR_2$ 28

HO ... R_2O 3 ... CH_2OH ... $COOR_1$ 28

Bayogenin	R^1	R^2
saponin 5 (MW=974)	Glc -	Gal - ²Glc -
saponin 1 (MW=1136)	Glc -	Gal - ²Glc - (4 | Glc)
Hederagenin		
saponin 2 (MW=1120)	Glc -	Glc - ²Glc - (4 | Glc)

R_2O 3 ... CH_2OH ... $COOR_1$ 28

Fig. 19.4. Structures of the five standards representative of the more abundant aglycones present in *P. dodecandra*: oleanolic acid, bayogenin and hederagenin.

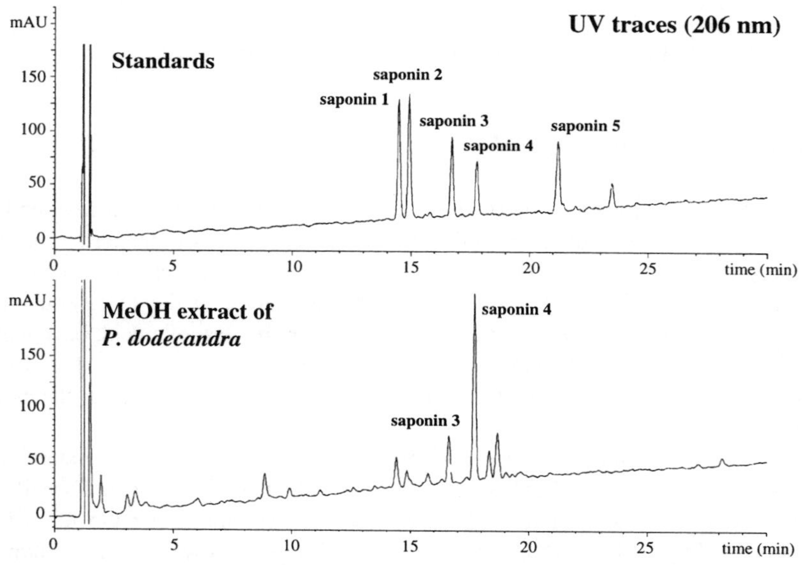

Fig. 19.5. LC/UV analyses of the standard mixture and of a methanolic extract of dried berries.

It can be observed that the methanolic extract is mainly composed of saponins. Of these, glycosides of oleanolic acid (compounds **3** and **4**) represent the major components.

These LC/UV analyses showed that the composition of the extracts did not vary significantly from one sample to another, but differences were observed in the total saponin content of the extracts as shown in Fig. 19.6.

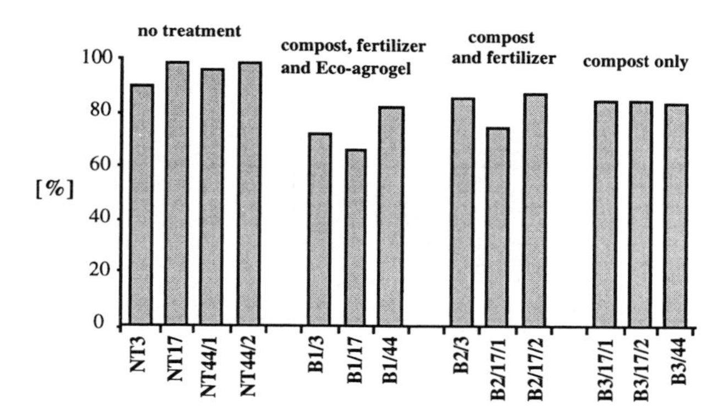

Fig. 19.6. Comparison of the saponin content of the methanolic extracts.

Again, extracts of non-treated berries present higher saponin contents than those obtained with berries grown with special field treatment. Thus, it seem that the increased growth rate obtained by the use of compost, fertiliser or Eco-agrogel is accompanied by a diminution in the saponin content in the berries. Field and laboratory measurements have now to be carefully analysed to determine if these two phenomena are linked.

It can also be observed that the saponin content in the extract is directly proportional to the mass of the extracts. Thus, by simply weighing the standardised methanol extracts, it is possible to get a good idea of the saponin content of the dried berries. This is an important point for future cultivation experiments, because it means that the measurements could be realised without sophisticated material.

Qualitative determination

While LC/UV is an adequate technique for quantitative determination, it does not give any structural information as to the type of the aglycones or the sugar chain sequence of the saponins. Some qualitative experiment have been performed by the hyphenated technique LC/MS with two different ionisation modes, thermospray (TSP) and electrospray (ES).

In each case, the analyses were performed under the same chromatographic conditions as with the methanolic extract NT44/1. For each ionisation mode, the characteristic MS spectrum of the pentaglycosylated saponin **3** obtained on-line will be presented. This bidesmosidic saponin is one of the five standards and is present in all the extracts. Its molecular weight is 1250 u and its aglycone is oleanolic acid (MW: 456 u). This molecule is an interesting example due to its branched sugar chain at position C-3 and its terminal rhamosyl unit. Its structure is shown in Fig. 19.7.

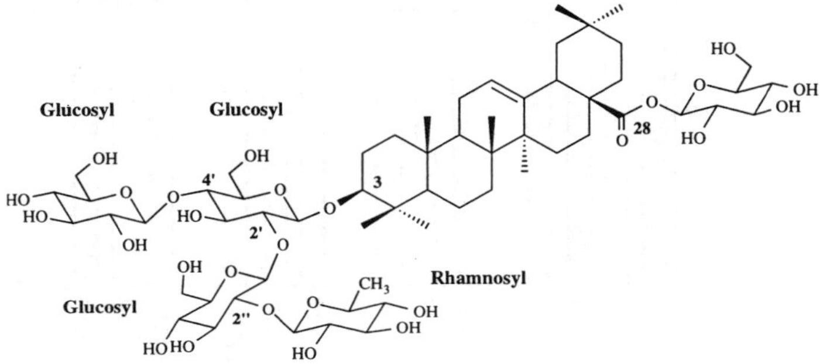

Fig. 19.7. Structure of the pentaglycosylated saponin **3.**

LC/TSP-MS analysis

The LC/TSP-MS analyses were performed on a Finnigan MAT TSQ-700 instrument with a TSP 2 interface.

In Fig. 19.8., are presented the LC/TSP-MS traces of the standard mixture, together with the on-line TSP-MS spectrum of saponin **3**.

A buffer (NH_4OAc 0.5 M, 0.2 ml/min) was added post column and the conditions were as follow: source 280-310°C, vaporiser 90, aerosol 280-360°C, positive ion mode, filament off.

LC/TSP-MS analyses permitted the identification of the aglycones, due to strong and characteristic dehydrated aglycone ions in the TSP spectra. Oleanolic acid presented only one $[(M-H_2O)+H]^+$ at *m/z* 439, whereas bayogenin and hederagenin exhibited $[(M-H_2O)+H]^+$ and $[(M-2H_2O)+H]^+$ ions at *m/z* 455, 437 and *m/z* 471, 453, respectively, due to the presence of more than one hydroxyl group. By displaying the single ion traces corresponding to each dehydrated aglycone moiety, it is possible to obtain a chromatogram for each type of aglycone, in this case the three aglycones oleanolic acid, bayogenin and hederagenin. However, due to the variations of the thermospray response, quantitative analyses are not highly reproducible. Weak sodium adducts were also observable and allowed the determination of the molecular weights of the most abundant saponins in the extracts. Indeed, the $[M+Na]^+$ ions at *m/z* 1273 are clearly visible in the case of saponin **3**. Nevertheless, TSP analyses do not give supplementary information about the sugar chain.

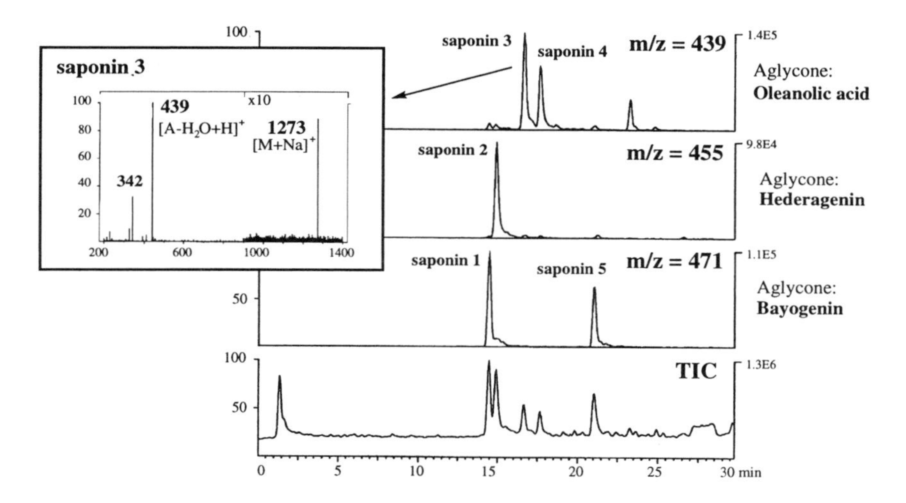

Fig. 19.8. LC/TSP-MS analysis of the saponin standards.

LC/ES-MS analysis

The electrospray interface (ES) was also tested for the analysis of the methanolic extracts. These analyses were performed in the negative ion mode on a Finnigan MAT TSQ-7000 instrument with a capillary temperature of 220°C. The electrospray interface appeared to be the most suitable technique for the analysis of extracts of *P. dodecandra*. In addition to its relatively easy use, electrospray allowed a good ionisation of saponins and was much more sensitive than TSP. Indeed, intense pseudomolecular ions $[M-H]^-$ were exhibited in the ES-MS spectra of all the saponins present in the extracts (see Fig. 19.9.).

With this interface, some information of the sugar sequence were obtained. Indeed, the fragmentation level can be adjusted by choosing an adequate voltage in the entrance of the optic system. The ions accelerated this way are fragmented by collision induced dissociation (up-front CID). It was thus possible to observe differences of 162 or 146 u between fragmentation peaks in the ES-MS spectra. These were characteristic for the loss a hexosyl moiety such as glucosyl or a desoxyhexosyl moiety such as rhamnosyl.

Thus, in the case of saponin **3**, the molecular weight was deduced from the $[M-H]^-$ ion at *m/z* 1249. The strong $[(M-Glc)-H]^-$ ion at *m/z* 1087 was due to the loss of the glucosyl moiety attached at position C-28 with an ester linkage. Peaks at *m/z* 941, 925 and 779 were also observable, suggesting the loss of the glucosyl and rhamnosyl units of the sugar chain.

However, due to the number of peaks, the electrospray spectra may be sometimes difficult to interpret.

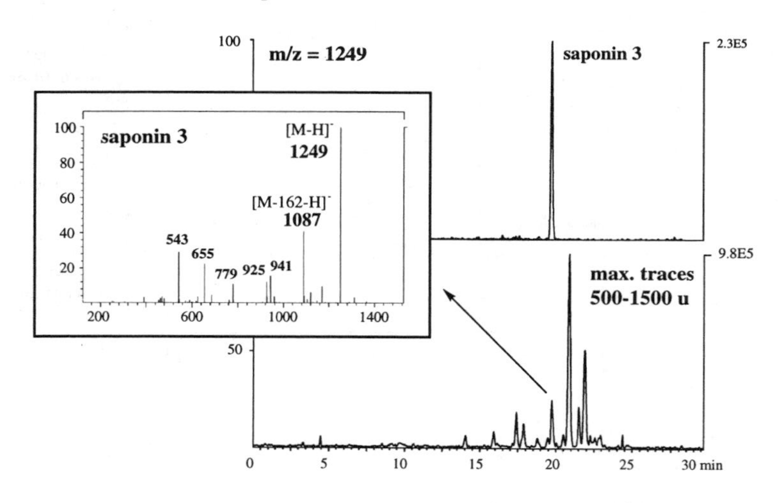

Fig. 19.9. LC/ES-MS analysis of the methanolic extract of *P. dodecandra* berries.

MS^n electrospray analysis

An early prototype of a new ion trap mass spectrometer was tested for the determination of sugar sequence. With this MS-MS instrument, it was possible to isolate and excite only one ion of interest and thus, to decrease the amount of consecutive reactions (Wolfender *et al.* 1995). The sugar sequence information was obtained by successive decomposition of the main ion, as shown in Fig. 19.10.

The first step was the fragmentation of the strong TFA adduct at *m/z* 1363 (Fig. 19.10a.), giving a deprotonated molecular ion $[M-H]^-$ (Fig. 19.10b.). This latter ion yielded a first fragment at *m/z* 1087 (Fig. 19.10c.). This first loss of 162 u corresponded to the departure of the glucosyl moiety at position C-28. This sugar was particularly sensitive due to the ester linkage. Then, the $[(M-Glc)-H]^-$ ion cleaved into two fragments at *m/z* 941 and *m/z* 925 (Fig. 19.10d.), showing the simultaneous loss of a rhamnosyl or a glucosyl unit, respectively. These losses were characteristic for a branched sugar chain. In Fig. 19.10e., the ion at *m/z* 779 issuing from the fragmentation of *m/z* 941 (-Rha) or *m/z* 925 (-Glc) was observed. This ion corresponded to the diglucoside moiety, which gave finally the monoglucoside $[(A+Glc)-H]^-$ at *m/z* 617 (Fig. 19.10f.) and the aglycone ion characteristic of oleanolic acid at *m/z* 455.

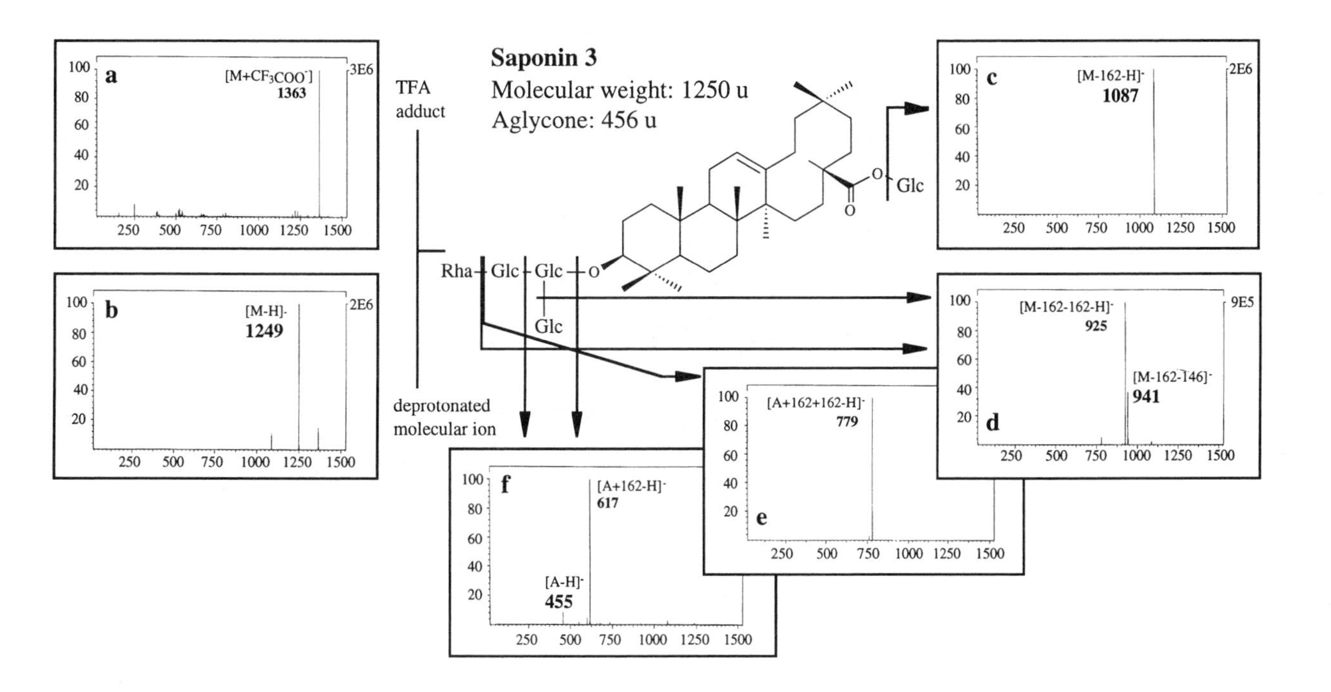

Fig. 10. MSn electrospray experiment on the pentaglycosylated saponin **3**

Thus, this MS^n analysis showed that it was possible to cleave only one sugar at a time by adjusting the collision energy, making the interpretation simpler. This type of experiment was found to be very useful for clarifying the sugar sequence of saponins.

Conclusion

The HPLC/UV analyses carried out on the different methanolic extracts of the *Phytolacca dodecandra* berries harvested showed that the cultivation conditions had only a weak influence on the qualitative composition of the extracts. However, extraction with methanol demonstrated that these conditions can affect significantly the total saponin content of the berries. It seems that the increased growth rate obtained by the use of compost, fertiliser or Eco-agrogel is accompanied by a diminution in the saponin content in the berries. Other measurements have now to be done to determine if these two phenomena are effectively linked. For example parameters such as the maturity or the time of collection of the berries have to be considered (Ndamba *et al*, 1994).

LC/UV remains a rapid and efficient technique for the quantitation of the saponin content of the different samples of cultivated *P. dodecandra.* However, weighing the standardised methanolic extracts is sufficient for comparing the saponin content of dried berries. Indeed, as the extracts are mainly constituted of saponins, this simple method gives a good idea of the yield in saponin in the dried berries and could thus be used for the further measurements.

Extraction with water under the same conditions as those encountered in the field have now to be performed to obtain active extracts. These will have to be tested in order to evaluate and compare their molluscicidal properties.

LC/TSP-MS analyses gave supplementary information on the composition of the extracts, or more precisely, the type of the aglycones and the molecular weights of the major peaks. However, due to the variation of the thermospray response, this method is not ideal for the precise quantitation of molecules as large as bidesmosidic saponins.

LC/ES-MS, combined with MS-MS experiments appears to be perfectly adapted for qualitative analyses of triterpene glycosides. Indeed, it allows the determination of the molecular weight, even for minor peaks and gives essential structural information on the sugar chain. The use of an electrospray interface for quantitation purposes has still to be tested, but might be promising.

Acknowledgements

This work was supported by the *Swiss National Science Foundation*. Thanks are due to *Dr. W. Wagner-Redeker* from *Spectronex AG*, Basel, for the use of the LC/ES-MS equipment.

References

Lemma, A. (1970). *Bulletin WHO* **42**, 597.

Marston, A. and Hostettmann, K. (1984). Plant Molluscicides. *Phytochemistry* **24**, 639-652.

Ndamba, J., Lemmich, E. and Mølgaard, P. (1994). Investigation of the diurnal, ontogenic and seasonal variation in the molluscicidal saponin content of *Phytolacca dodecandra* aqueous berry extracts. *Phytochemistry* **35**, 95-99.

Slacanin, I., Marston, A., and Hostettmann, K. (1988). High performance liquid chromatography determination of molluscicidal saponins from *Phytolacca dodecandra* (Phytolaccaceae). *Journal of Chromatography* **448**, 265-274.

Wolfender, J.-L., Rodriguez, S., Wagner-Redeker, W., and Hostettmann, K. (1995). Comparison of liquid chromatography/electrospray, atmospheric pressure chemical ionization, thermospray and continuous-flow fast atom bombardment mass spectrometry for the determination of secondary metabolites in crude plant extracts. *Journal of Mass Spectrometry and Rapid Communications in Mass Spectrometry*, **S35-S46**.

20. Isolation and characterization of two saponins from *Becium grandiflorum* var. *obovatum*

I. BURGER[1], C.F. ALBRECHT[2], H.S.C. SPIES[1] AND B.V. BURGER[1]

[1]*Department of Chemistry, Laboratory for Ecological Chemistry, University of Stellenbosch, Stellenbosch, 7600, South Africa and* [2]*Department of Pharmacology, University of Stellenbosch, Tygerberg, 7505, South Africa*

Introduction

Hot water extracts of five plants, indigenous to Natal, South Africa, has reportedly been used by a traditional herbalist in the successful treatment of two cancer patients during the first half of this century. No modern chemo- or radiotherapy was available at that time.

Conventional chemotherapy relies heavily on a combination of cytotoxic chemotherapeutic substances. Such therapy often fails due to the development of drug resistance in the cancer cells and the patient usually suffers from cytotoxic side-effects. However neither of the two patients experienced side-effects and the one patient, a woman with breast cancer, still bore 3 children after the treatment. This lead to the idea that one of the plants in the combination might possess unique cancer-curing properties.

Using different cancer cell lines the biological effects of hot water extracts of all the plants were investigated. An extract of one of the plants in the combination, umatanjane, identified as *Becium grandiflorum* var. *obovatum* (Lamiaceae) by the National Botanical Institute, produced a significant *in-vitro* response in H522 human lung carcinoma cells.

It was consistently found that a 1:500 dilution of a hot water extract (1g/10ml) of the powdered root bark of this plant induces the formation of numerous intra-cytoplasmic vesicles within 48 hours. These vesicles tend to increase in size and eventually inhibit cell division and cell growth. Different combinations of hot water extracts of the five plants were also tested on melanoma cells and the combination of *B. grandiflorum* and a *Crinum* sp. was found to be cytotoxic although neither of the two extracts showed cytotoxicity on its own.

This paper deals with the isolation and characterization of the biologically active compounds in *Becium grandiflorum* var. *obovatum*.

Experimental

Analytical methods

HPLC analyzes were carried out with a Hewlett Packard HP 1092 liquid chromatograph equipped with a HP 1040 diode-array detector. Analytical and semipreparative separations in reverse phase mode were performed with a 250 mm x 4.6 mm i.d. Phenomenex Partisil C_8 analytical column (5 μm particle size) under step gradient conditions and a constant flow (1.5 ml/min) (see Table 20.1.).

Table 20.1. Solvent systems used in HPLC analyses of *Becium grandiflorum*

gradient 1		gradient 2	
Time (min)	Eluent A (%)	Time (min)	Eluent A (%)
0.01	90	0.01	90
1.00	90	1.00	90
16.00	30	5.00	72
20.00	30	20.00	72
		25.00	30
		30.00	30

Eluent A: aq. solution of KH_2PO_4 (0.05 M). Eluent B: acetonitrile - isopropanol (4:1). Better separation for collection of K2 and K3 was achieved by using solvent gradient 2.

Electron impact (EI) mass spectra were recorded at 70 eV on a Carlo Erba QMD 1000 gas chromatograph-mass spectrometer (GC-MS) system equipped with a flame ionization detector and a glass column coated by the Laboratory for Ecological Chemistry with a 0.25 μm film of the apolar stationary phase PS-089, a silanol-terminated, 95%-dimethyl-5%-diphenylsiloxane copolymer. The flame ionization detector was operated at 280°C and the injector was used at 250°C. Samples were injected in the split mode and analyzed using a temperature program of 2°C/min from 160°C to 250°C (hold).

Time of flight matrix assisted laser desorption/ionization (TOF-MALDI) analyzes were carried out with a Kratos KOMPACT MALDI III instrument using 2,5-dihydroxybenzoic acid as matrix solution and bovine insulin as an external calibrant.

^{1}H and ^{13}C nuclear magnetic resonance (NMR) spectra (in DMSO-d6) were recorded with a Varian Unity Plus 500 NMR spectrometer (^{1}H frequency:

499.868 MHz, ^{13}C frequency: 125.703 MHz) at 30°C with TMS as internal standard.

Preparation of derivatives of monosaccharides

Alditol peracetates - K2 and K3 (200 μg of each) were hydrolyzed with trifluoroacetic acid (TFA) (2 M, 40 μl) at 120°C for 1 hr. The soluble portion of the hydrolysates were evaporated to dryness at 50°C with nitrogen. The resulting monosaccharides were reduced to their respective alditols with 40 μl of a solution of sodium borohydride (10 mg) in ammonia (1 M, 0.5 ml). After 1 hr at room temperature the excess of borohydride was decomposed by dropwise addition of glacial acetic acid, until effervescence had ceased. Boric acid, produced on decomposition of the borohydride, was removed by adding methanol (40 μl) to the reaction mixture followed by evaporation to dryness at 50°C with nitrogen. This drying process was repeated 4 times with methanol (40 μl) and 4 times with dichloromethane (40 μl). Acetylation of the alditols were performed in a sealed tube with acetic anhydride (40 μl) heated for 3 hr at 120°C. The acetylated products were dissolved in dichloromethane and analyzed by GC-MS. Internal standards were prepared similarly.

Trimethylsilylethers - K2 and K3 (200 μg of each) were hydrolyzed with TFA (2 M, 40 μl) at 120°C for 1 hr. The soluble portion of the hydrolysates were evaporated to dryness at 50°C with nitrogen and dried an additional 3 times with dichloromethane. The oximes of the resulting monosaccharides were obtained by adding a solution of hydroxylamine hydrochloride in pyridine (25 mg/ml, 20 μl) and heating the reaction mixtures for 30 min at 70-75°C. After cooling to room temperature, hexamethyldisilazane (20 μl) and TFA (2 μl) was added. The reaction mixtures were shaken well and allowed to stand at room temperature for 30 min. The silylated products were diluted with dichloromethane and analyzed by GC-MS. Internal standards were prepared similarly.

Determination of mole ratios of monosaccharides

K2 and K3 were hydrolyzed as in the preparation of the alditol peracetates. The hydrolyzed material (20 μl) was fractionated on a Dionex Carbopac PA1 column (250 mm x 4 mm) at a flow speed of 0.75 ml/min using pulsed amperometric detection. Sodium hydroxide (150 mM) was used as eluent.
The hydrolyzed material (20 μl) was also fractionated on a HPX 87H column (250 mm x 4 mm) at a flow speed of 0.5 ml/min and temperatures of 35°C and 75°C, using a refractive index detector. Sulfuric acid (0.01 N) was used as eluent.

Results and discussion

The formation of vesicles in H522 human lung carcinoma cells was used as a bioassay in the isolation of the biologically active principles in *B. grandiflorum*. The powdered root bark was extracted with boiling water giving a frothy extract which suggested the presence of saponins.

The extract was fractionated by elution from a short C_{18}-column with different methanol/water concentrations. HPLC analysis of the active fraction (Fig. 20.1.) revealed a peak at 11.8 minutes with an absorbance maximum at 200 nm.

This fraction was further purified on an analytical HPLC column. By adapting the HPLC solvent gradient it was found that the active fraction consists of several constituents (Fig. 20.2.) of which only two, K2 and K3, could be isolated in a relatively pure state.

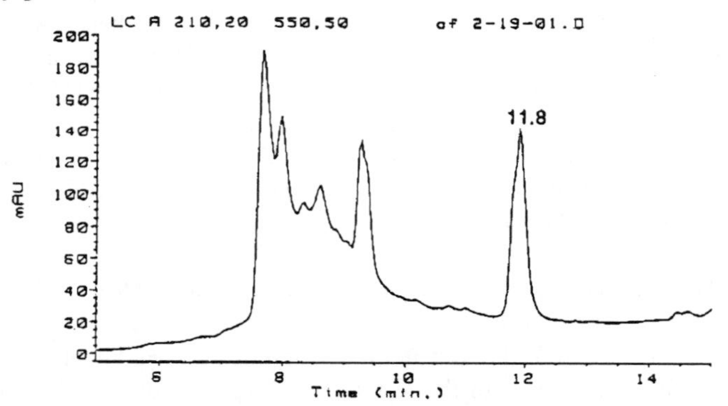

Fig. 20.1. HPLC chromatogram of the 70% methanol/water fraction at 210 nm.

Separations were done on a reversed phase column (250 mm x 4.6 mm, 5 µm C_8, Partisil). Eluent A: 0.05 M KH_2PO_4; Eluent B: acetonitrile- isopropanol (4:1); gradient 1 (see Table 20.1).

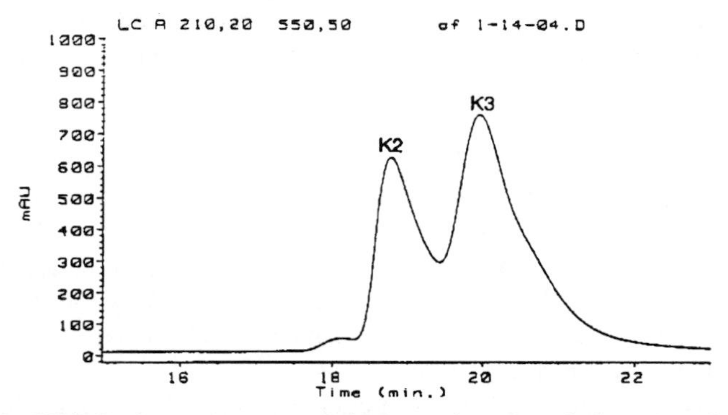

Fig. 20.2. HPLC chromatogram of the constituents of the biologically active fraction.

Conditions as in Fig. 20.1., with gradient 2 (see Table 20.1)

Both of these compounds cause vesicle formation and the same degree of growth inhibition in lung carcinoma cells.

As mentioned before, the extract was frothy and therefore possibly contained saponins. A test with Liebermann-Burchard reagent (Van Atta and Guggolz 1958) revealed that both compounds are triterpenoid saponins and therefore consist of a triterpenoid group with attached sugar groups.

K2 and K3 were hydrolyzed with TFA and the resulting monosaccharides were converted to alditol peracetates (Albersheim *et al.* 1967). Several sugar derivatives were identified using GC-MS analysis of the derivatized hydrolysates. Components 1067, 1109 and 1655 in the total ion chromatogram of the derivatized hydrolysate of K2 (Fig. 20.3.) were identified as arabinose, xylose and glucose, respectively, by using authentic reference compounds.

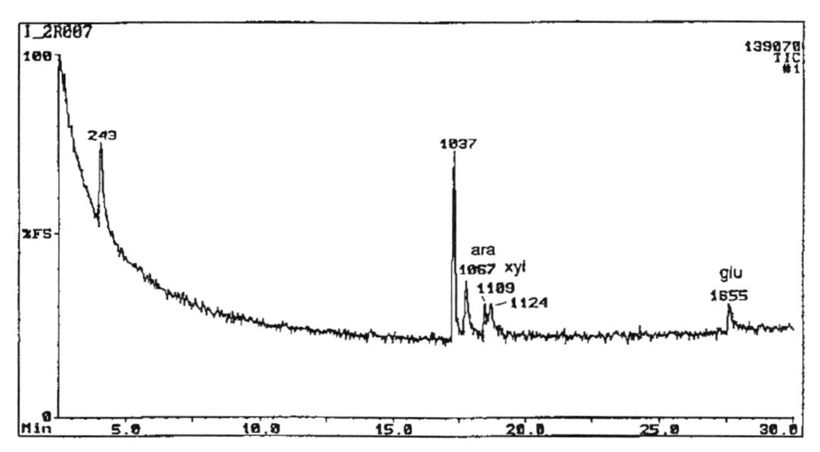

Fig. 20.3. GC-MS analysis of the alditol peracetates derived from the monosaccharides in K2.

Capillary glass column coated with PS-089 (40 m x 0.3 mm, film thickness 0.25 μm). Temperature program 160°C - 250°C at 2°C/min.

These three sugars were identified in K3 in a similar manner. The results were confirmed by analyzing the trimethylsilylderivatives of the monosaccharides.

The mole ratios of the monosaccharides in the hydrolyzed sugar chains of K2 and K3 were determined by HPLC (Hardy *et al.* 1988). Using pulsed amperometric detection as well as refractive index detection, glucose, arabinose and xylose were found to be present in equimolar amounts in both compounds.

Further research was concentrated on the elucidation of the structure of K2, the purest of the two samples. The complete structure and stereochemistry of K2 were determined by a series of 500-MHz NMR experiments.

The presence of anomeric and olefinic, sugar and aliphatic funtions, appearing in discrete regions, is discernable in the 2D HSQC spectrum of K2 (Fig. 20.4.).

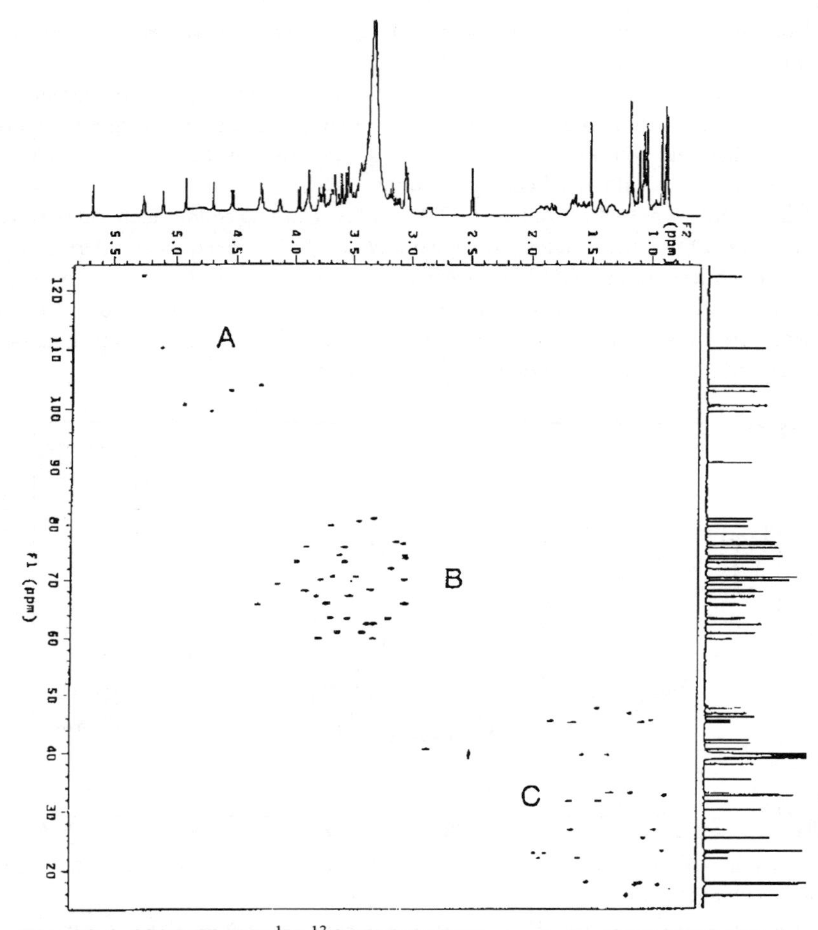

Fig. 20.4. 500 MHz 2D ^{1}H-^{13}C HSQC NMR spectrum of K2 in DMSO-d_6. TMS as internal standard. Discernable functions: (A) anomeric and olefinic, (B) sugars, and (C) alifatic including methyl groups.

The triplet at δ 5.27 in the ^{1}H-spectrum (Fig. 20.5.) indicates the presence of an olefinic proton at C-12. The six three-proton singlets in the ^{1}H-spectrum can be assigned to the six methyl groups of the aglycone. The singlet at δ 175.50 in the ^{13}C NMR spectrum is due to an ester carbonyl group at C-28. A correlation between C-28 and H-1 of sugar R_1 (see formula of beciumecine I) in the 2D HMBC spectrum indicates a glycosidic ester linkage between the aglycone and sugar R_1. The correlation between H and C-3 and H and C-1 of sugar R_6 indicates a glycosidic bond between the aglycone and sugar R_6.

1D TOCSY experiments were used to identify six individual sugar groups. The ^{1}H-^{1}H coupling constants of the anomeric protons were used to determine the anomeric configuration of these sugars. Interglycosidic linkages were determined by 2D HMBC and 1D TOCSY and the molecular conformation of the structure was determined by 2D NOESY.

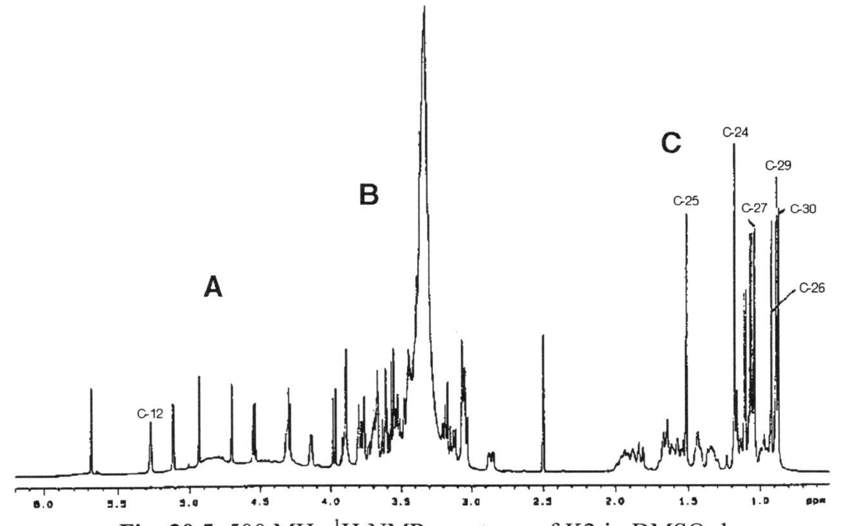

Fig. 20.5. 500 MHz ^{1}H-NMR spectrum of K2 in DMSO-d_6

The proposed structure for K2 is 3-O-β-D-glucopyranosyl terminolic acid 28-O-α-L-rhamnopyranosyl-(1-3)-β-D-xylopyranosyl-(1-4)-[apiofuranosyl-(1-3)]-α-L-rhamnopyranosyl-(1-2)-α-L-arabinopyranosyl ester. Beciumecine I is proposed as a name for this saponin.

Based on the structure determined by NMR, the molecular mass of K2 was calculated as 1417.9 Da, which differs by the mass of two watermolecules from the mass determined by TOF-MALDI, possibly due to the loss of water on laser desorption.

The structure of K3 is very similar to that of K2 and further work will entail the full structure determination of K3 using NMR and, if required, chemical degradation.

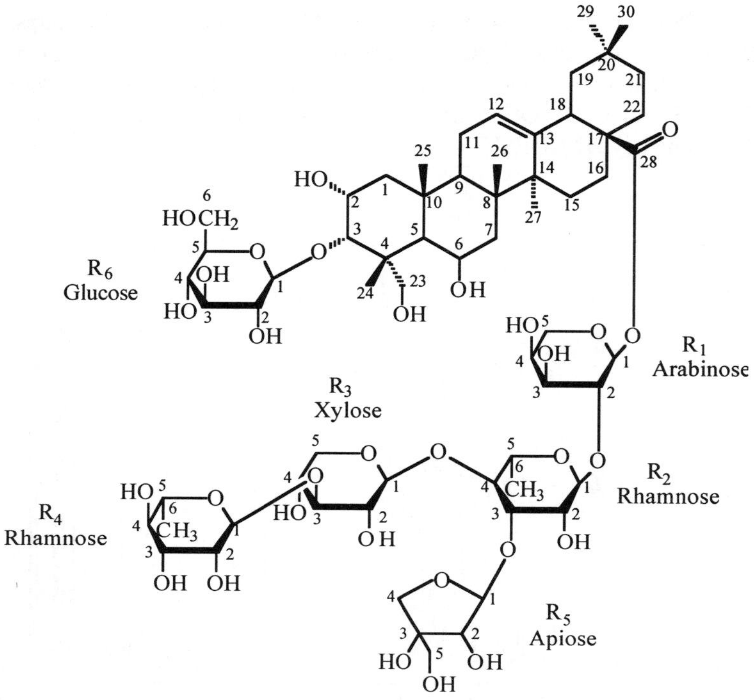

Beciumecine I

Acknowledgments

Thanks are due to the Foundation for Research Development and SASOL, for financial support, and to Dr. P. Sándor at the applications laboratory of Varian in Darmstadt, Germany, for NMR analyzes.

References

Albersheim, P., Nevins, D.J., English, P.D., and Karr, A. (1967). A method for the analysis of sugars in plant cell-wall polysaccharides by gas-liquid chromatography. *Carbohydrate Research* **5**, 340-345.

Hardy, M.R., Townsend, R.R., and Lee, Y.C. (1988). Monosaccharide analysis of glycoconjugates by anion exchange chromatography with pulsed amperometric detection. *Analytical Biochemistry* **170**, 54-62.

Van Atta, G.R. and Guggolz, J. (1958). Detection of saponins and sapogenins on paper chromatograms by Liebermann-Burchard reagent. *Journal of Agriculture and Food Chemistry* **6**, 849-850.

21. Hypoxoside: a putative, non-toxic prodrug for the possible treatment of certain malignancies, HIV-infection and inflammatory conditions

C.F. ALBRECHT

Department of Pharmacology, Faculty of Medicine, University of Stellenbosch, Tygerberg 7505, South Africa

---

Introduction

One of the best studied phytochemicals from an African plant is hypoxoside ((*E*)-1,5-bis(4'-β-D-glucopyranosyloxy-3'-hydroxyphenyl)pent-4-en-1-yne). This unique norlignan diglucoside was first described in the early eighties (Marini-Bettolo *et al.* 1982; Drewes *et al.* 1984). It is the first phytochemical found to contain a pent-1-en-4-yne structure. Hypoxoside is a major component of the corms of Hypoxidaceae and can readily be converted to a more lipophylic aglucone, rooperol, by β-glucosidase deconjugation (Fig. 21.1.) (Theron *et al.* 1994)

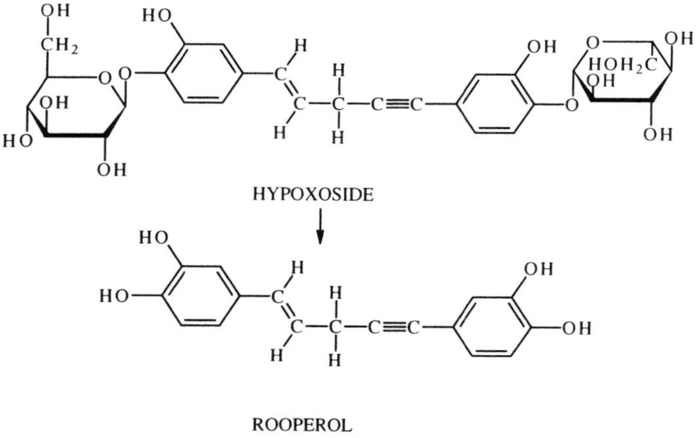

HYPOXOSIDE

ROOPEROL

Fig. 21.1. The inactive diglucoside, hypoxoside, is deconjugated by β-glucosidase to form the lipophylic biologically active aglucone, rooperol.

Of considerable interest over the past 14 years has been the question whether hypoxoside has any therapeutic properties and if so, what the molecular mode(s) of action could be. According to reports, extracts of various members of the Hypoxidaceae have been used *inter alia* as a tonic, rejuvenator, treatment for testicular tumors, internal cancers, hypertrophy of the prostate and urinary diseases (Nicoletti *et al.* 1992). Indeed patents were granted for the use of extracts of the family Hypoxidaceae and rooperol and derivatives for the treatment of cancer (Drewes and Liebenberg 1987a, 1987b).

Research conducted over the past 14 years has shown that rooperol has potent pharmacological activity whereas hypoxoside is mainly inert. Nevertheless a hypothesis has been formulated that hypoxoside could act as a non-toxic, multi-functional prodrug (Albrecht *et al.* 1995a).

Pharmacological properties of rooperol

Rooperol is a potent inhibitor of lipoxygenase in the synthesis of leukotrienes at concentrations as low as 1 μM (Van der Merwe *et al.* 1993). It is also active against a wide panel of rodent and human cancer cells lines giving 50% growth inhibition at about 10 μg/ml. Some cell lines such as the NCI-H522 and ATCC HTB 53; A-427 derived from human non-small-cell-lung cancers, were found to be 10 times more sensitive, suggesting specific inhibitory mechanisms (Albrecht *et al.* 1995a). Recent work in our laboratory has clearly shown that rooperol can induce apoptosis in HL60 human promyelocytic leukemia cells (Theron, E., University of Stellenbosch, unpublished data). We have also found that rooperol can actively inhibit mutagenesis in the Ames test and scavenge free radicals 10 times more actively than ascorbate (Albrecht, C. and Bester, C., University of Stellenbosch, unpublished data). Using a cell line transfected with HIV-LTR-Luc-reporter gene (Israel *et al.* 1992) we also found that rooperol can inhibit the phorbol ester induction of this gene (Theron, E. and Albrecht, C., University of Stellenbosch, unpublished data).

Hypoxoside as a prodrug

When hypoxoside is taken orally by man it is not absorbed as such, but is first deconjugated by colonic bacterial β-glucosidase to form rooperol, which can be found in the faeces (Kruger *et al.* 1994). Subsequently, rooperol metabolites are found in the serum and urine. These metabolites were shown to be mainly glucuronides, sulfates and mixed glucuronide/sulfates (Kruger *et al.* 1994). A similar situation is found in the mouse except that the metabolites are only detected in the bile and not in the serum or urine (Albrecht *et al.* 1995a). Direct injection of rooperol into the bloodstream of baboons showed that it was conjugated to form metabolites within a few minutes (Coetzee *et al.*, *Drug*

Research, in press). Injection of hypoxoside showed no evidence of any metabolism.

When rooperol metabolites (220 μg/ml) were added to cancer cells in culture no inhibition of growth was found. However when β-glucuronidase (100 μg/ml) was added, some of the metabolites were deconjugated to form rooperol and the cells were killed (Albrecht *et al.* 1995a). Furthermore when rooperol metabolites were incubated with activated macrophages for a day, it was found that rooperol formed (Gabrielse *et al.*, University of Stellenbosch, unpublished data). These data lead to the hypothesis that rooperol metabolites could act as prodrugs which are activated by enzymatic deconjugation to form pharmacologically active rooperol. Such conversion of the prodrug to the active drug could conceivably take place in pathological tissue where the lysosomal deconjugases such as β-glucuronidase were present outside the cells. It is known that many tumors contain necrotic areas in which lysosomal enzymes are present. Furthermore release of such enzymes also occurs in inflammatory sites due to macrophage activation.

These considerations motivated us to conduct clinical studies of cancer patients and patients with HIV. Motivation for the later was based on the premise that HIV-infected lymph nodes could also contain active macrophages, releasing enzymes that could cause rooperol to form and that the rooperol could inhibit the expression of the HIV genome. It was also found that the di-sulfate metabolite of rooperol inhibited the *in vitro* proliferation of HIV-1 (Albrecht, C. and Kruger, P. , University of Stellenbosch, unpublished data).

Clinical evaluation of hypoxoside as a prodrug

Ethical permission was granted by the University of Stellenbosch and the South African Medicines Control Council to conduct Phase 1 studies with hypoxoside for the putative treatment of cancer patients (for whom no other therapy was indicated) and patients with HIV.

Patients received 1200 -3200 mg of dried methanolic extract of the dried corms of *Hypoxis rooperi* in capsule form (200 mg/capsule), in three daily doses. Total rooperol metabolite concentrations were in the order of 100 μg/ml during these studies. All patients were subjected to frequent and extensive laboratory and clinical examinations including computed tomography of the cancer patients.

Nineteen non-small-cell lung cancer patients on hypoxoside therapy survived for an average of 4 months with progression of their primary tumors and metastases, while 5 survived for more than one year and one survived for 5 years before dying of tuberculosis. Histological examination of a lesion in the lung showed that it only contained connective tissue and no cancer cells.

No toxic effects, in clinical examinations or biochemical or hematological measurements, were found that could be ascribed to hypoxoside (Smit *et al.* 1995).

A further Phase 2 clinical study was conducted on more than 100 cancer patients for whom no further conventional therapy was available. No toxic effects were detected in these patients and it was found that the median survival of a cohort of 16 pancreas patients was increased from an expected 3 months to 10 months.(Smit, B. and Albrecht, C., University of Stellenbosch, unpublished data).

Extensive pharmacokinetic analyses of these patients were conducted. Main findings were the absence of hypoxoside or rooperol in the serum and the presence of the rooperol metabolites which had half lives ranging from 20 to 50 hours following first order kinetics. (Albrecht *et al.* 1995b).

Patients with HIV have taken the methanolic extract of *Hypoxis rooperi* corms now for more than two years and their CD4 lymphocyte counts have remained remarkably stable while a decrease in serum p24 HIV antigen has been found and a decreased expression of the HLA-DR CD8 lymphocyte activation marker (Bouic, P., University of Stellenbosch, unpublished data).

Hypoxoside and NDGA

During our studies we became aware of similar studies on the phytochemical nordihydroguaiaretic acid (NDGA) derived from the creosote bush (*Larrea* sp.). This molecule is very similar to rooperol, *i.e.* containing two catechols suspended on a carbon bridge, however, NDGA lacks the pent-1-en-4-yne configuration. NDGA has been shown to inhibit leukotriene synthesis (Van der Merwe *et al.* 1993), act as an anti-mutagen (Wang *et al.* 1991), inhibit the transcription of the HIV gene (Gnabre *et al.* 1995) and induce apoptosis in HL60 cells (Theron, E., University of Stellenbosch, unpublished data).

Conclusions

These studies have demonstrated that rooperol has potent, diverse and important pharmacological properties relevant to cancer, inflammation and HIV. Furthermore extensive clinical experience has shown that oral dosing of hypoxoside in man is safe. The reason for this, is the rapid metabolism of rooperol to form conjugated metabolites that are inert. It is postulated that these metabolites can act as prodrugs and can be activated to form rooperol in targeted, pathological sites where lysosomal deconjugases are present. Although not conclusive, some of our clinical data suggests that such a process may be occurring *in vivo.* Direct proof of this is now required.

Acknowledgments

These studies were initiated and sponsored by Essential Sterolin Products according to an agreement entered into with the University of Stellenbosch.

References

Albrecht, C. F., Theron, E. J., and Kruger, P.B. (1995a). Morphological characterization of the cell-growth inhibitory activity of rooperol and pharmacokinetic aspects of hypoxoside as an oral prodrug for cancer therapy. *South African Medical Journal* **85**, 853-860.

Albrecht, C.F., Kruger, P.B., Smit, B.J., Freestone, M., Gouws, L., Miller, R., and Van Jaarsveld, P.P. (1995b). The pharmacokinetic behaviour of hypoxoside taken orally by patients with lung cancer in a phase I trial. *South African Medical Journal* **85**, 861-865.

Drewes, S. and Liebenberg, R.W. (1987a). Rooperol and its derivatives. U.S.Patent No. 4,644,085.

Drewes, S. and Liebenberg, R.W. (1987a). Pentene-diphenyl-diglucoside containing compound. U.S.Patent No. 4,652,636.

Drewes, S.E., Hall, A.J., Learmonth, R.A., and Upfold, U.J. (1984). Isolation of hypoxoside from *Hypoxis rooperi* and synthesis of (*E*)-1,5-bis(3',4'-dimethoxyphenyl)pent-4-en-1-yne. *Phytochemistry* **23**, 1313-1316.

Gnabre, J.N., Brady, J.N., Clanton, D.J., Ito, Y., Dittmer, J., Bates, R.B., and Huang, R.C. (1995). Inhibition of human immunodeficiency virus type 1 transcription and replication by DNA sequence-selective plant lignans. *Proceedings of the National Academy of Science, USA* **92**, 11239-11243.

Israel, N., Gougerot-Pocidalo, M-A., Aillet, F., and Virelizier, J-L. (1992). Redox status of cells influences constitutive or induced NF-KB translocation and HIV long terminal repeat activity in human lymphocytes T and monocytic cell lines. *Journal of Immunolog*, **149**, 3386-3393.

Kruger, P.B., Albrecht, C.F., Liebenberg, R.W., and Van Jaarsveld, P.P. (1994). Studies on hypoxoside and rooperol analogues from *Hypoxis rooperi* and *Hypoxis latifolia* and their biotransformation in man using high-performance liquid chromatography with in-line sorption enrichment and diode-array detection. *Journal of Chromatography* **662**, 71-78.

Marini-Bettolo, G.B., Patamia, M., and Nicoletti, M. (1982). Hypoxoside, a new glycoside of uncommon structure from *Hypoxis obtusa* Busch. *Tetrahedron* **38,** 1683-1687.

Nicoletti, M., Galeffi, C., Messana, I., and Marini-Bettolo, G.B. (1992). Hypoxidaceae. Medicinal uses and the norlignan constituents. *Journal of Ethnopharmacology* **36**, 95-101.

Smit, B.J., Albrecht, C.F., Liebenberg, R.W., Kruger, P.B., Freestone, M., Gouws, L., Theron, E., Bouic, P.J.D., Etsebeth, S., and Van Jaarsveld, P.P. (1995). A phase I trial of hypoxoside as an oral prodrug for cancer therapy - absence of toxicity. *South African Medical Journal* **85**, 865-870.

Theron, E.J., Albrecht, C.F., Kruger, P.B., Jenkins, K., and Van der Merwe, M.J. (1994). β-Glucosidase activity in fetal bovine serum renders the plant glucoside, hypoxoside, cytotoxic toward B16-F10-Bl-6 mouse melanoma cells. *In Vitro Cellular and Developmental Biology* **30**, 115-119.

Wang, Z.Y., Agarwal, R., Zhou, Z.C., Bickers, D.R., and Hasan, M. (1991). Antimutagenic and antitumorigenic activities of nordihydroguaiaretic acid. *Mutation Research* **261**, 153-162.

22. Antimalarial active principles of *Spathodea campanulata*

O.O.G. AMUSAN[1], E.K. ADESOGAN[2], J.M. MAKINDE[2]

[1]*Chemistry Department, University of Swaziland, Private Bag 4, Kwaluzeni, Swaziland and* [2]*University of Ibadan, Ibadan, Nigeria*

Introduction

Spathodea campanulata P. BEAUVAIS (Bignoniaceae) is one of the important plants used in traditional medicine, whose chemical analysis has been recommended (Oliver-Bever 1960). This plant is used in traditional medicine for the management of malaria and the blood schizontocidal action of the alcoholic extract of its leaves against *Plasmodium berghei berghei* in mice has been described (Makinde *et al.* 1987). Extracts of the stem bark of the tree also demonstrated antimalarial activity against *P. berghei berghei* in mice both in early and established infections (Makinde *et al.* 1988).

Column chromatography was effective for the isolation of three fractions of the stem bark which demonstrated antimalarial properties (Makinde *et al.* 1990). Two of which, fractions B and C were obtained from the chloroform extract while one fraction (Z) was obtained from the hexane extract of the stem bark (Makinde *et al.* 1990). Phytochemical investigation has led to the characterization of the antimalarial active principles in the leaves and in the three fractions of the stem bark of *Spathodea campanulata* using spectroscopic methods and chemical transformations. The isolation of these antimalarial compounds from the stem bark of *S. campanulata* is noteworthy in the current search for new antimalarial drugs since these compounds have never been reported to have antimalarial action.

Results

Fresh leaves and stem bark of *S. campanulata* were collected and sun-dried. Details of the method of extraction and biological screening for antimalarial property have been published elsewhere (Makinde *et al.* 1988).

The aqueous methanol extract of the leaves (8 g) which was active in biological screening was fractionated on a silica gel column chromatography with ethyl acetate-methanol (3:1) as solvent to afford an orange crystalline solid, recrystallized in aqueous alcohol and identified as caffeic acid.

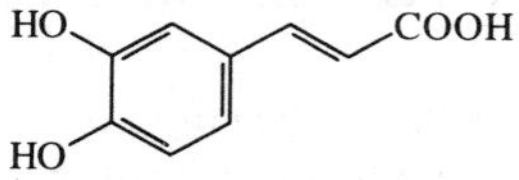

Caffeic acid

The chloroform extract of the stem bark of *S. campanulata* (15 g) has been separated in different fractions by column chromatography on silicagel. Elution with hexane-ethylacetate (2:3) yielded fraction B, which afforded after further purification on a short column of alumina and recrystallization twice in methanol a white amorphous powder identified as 3,20β-dihydroxyurs-12-ene-28-oic acid (20β-hydroxyursolic acid, **1**).

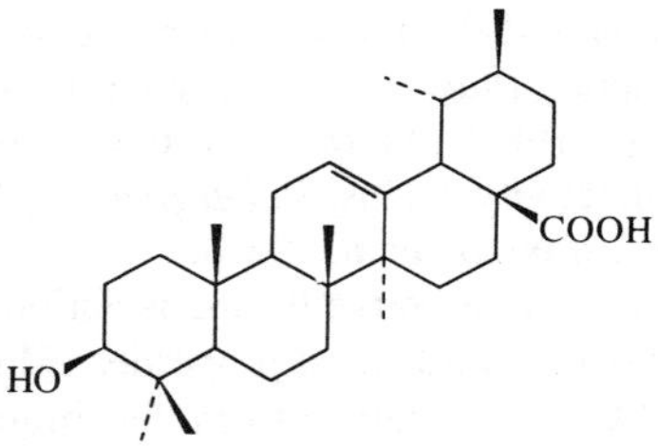

(1)

Fraction C was eluted by ethylacetate-methanol (9:1) from the same column than fraction B. Further purification on a short column of alumina, preparative TLC and crystallization in ethanol afforded white crystals characterized as 3β-hydroxyurs-12-19-dien-28-oic acid (tomentosolic acid, **2**).

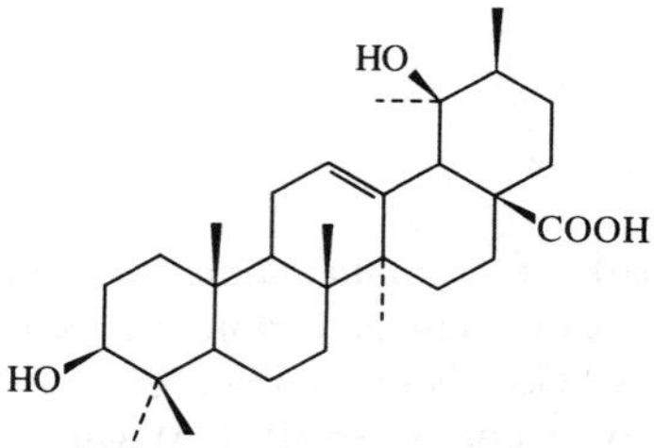

(2)

The hexane extract (12 g) of *S. campanulata*, eluted on silicagel column chromatography with ethylacetate-methanol (9:1) afforded fraction Z. Purification on a short column of alumina and recrystallization in ethanol gave white crystals identified as ursolic acid (**3**).

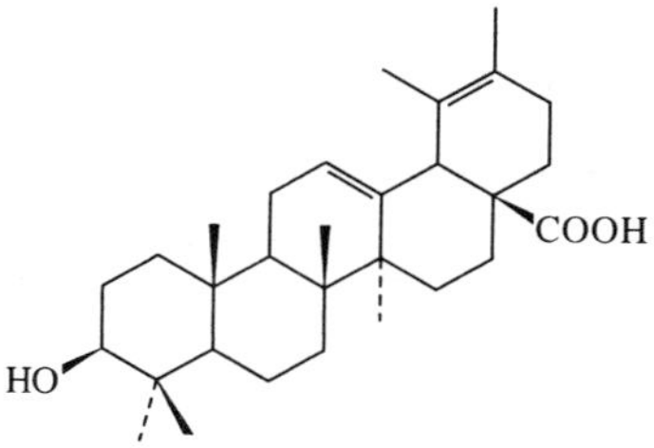

HO COOH

(**3**)

The three triterpenes isolated from the stem bark of *Spathodea campanulata* exhibited antimalarial activity in different assays and Table 22.1. shows the results of the action of these three compounds against *Plasmodium berghei berghei* in Fink and Kretschmar's test (Makinde *et al.* 1990). Each triterpene demonstrated a marked dose-dependent suppressive effect and high mean survival times.

Table 22.1. Blood schizontocidal action in mice of compounds isolated from the stem bark of *S. campanulata* on *P. berghei berghei* using Fink and Kretschmar's tests

	Dose ($mgKg^{-1}$ day^{-1})	**% suppression ± SE**	**% suppression of parasitaemia**	**Mean survival time (days)**
Control (2.5 % Tween 80)	-	25.7 ± 2.9		7.0 ± 1.4
20ß-hydroxyursolic acid (**1**)	20	23.0 ± 3.8	10.5	8.0 ± 1.2
	40	14.9 ±3.6	42.0	12.6 ± 2.8
	80	i2.2 ± 0.8	52.5	12.2 ± 1.3
Tomentosolic acid (**2**)	5	25.8 ± 0.7	-0.4	7.8 ± 1.2
	10	16.8 ± 2.8	34.6	10.0 ± 1.3
	20	8.7 ± 2.9	66.1	12.4 ± 1.7
	40	4.7 ± 1.9	81.7	19.2 ± 1.8
	80	6.1 ± 1.4	76.3	1.8 ± 1.1
Ursolic acid (**3**)	15	16.9 ± 3.3	34.2	12.8 ± 3.3
	30	8.5 ± 2.0	66.9	17.8 ± 5.1
	60	0.8 ± 0.7	96.9	25.3 ± 3.5
Chloroquine	10	0.4 ± 0.3	98.4	25.8 ± 1.2

Similar results were obtained from the blood schizontocidal action of the isolated compounds in an established infection test (Rane test). There was a fall in parasitaemia in groups of mice treated with these products while the control (blank) group showed increase in parasitaemia. The mean survival time produced by the control and triterpenes **1**, **2** and **3** are as shown in Table 22.2.

Table 22.2. Mean survival time in mice produced by triterpenes of *S. campanulata* stem bark extracts on *P. berghei berghei* in Rane test

	Dose ($mgKg^{-1} day^{-1}$)	Mean survival time (days)
Control (Tween 80)	-	7.2 ± 1.0
20ß-hydroxyursolic acid (**1**)	20	5.6 ± 0.3
	40	14.4 ± 4.1
	80	16.4 ± 4.2
Tomentosolic acid (**2**)	5	8.8 ± 2.0
	10	13.2 ± 4.1
	20	17.0 ± 4.8
	40	18.4 ± 4.2
Ursolic acid (**3**)	15	8.6 ± 0.5
	30	19.8 ± 4.5
	60	24.0 ± 4.0
Chloroquine	5	26.0 ± 2.0

Discussion

The isolation of caffeic acid as the antimalarial principle in the leaves of *S. campanulata* is noteworthy because caffeic acid is already known to have some antimalarial properties. It demonstrated antipyretic property and suppressed malaria in chicks (Helbecque *et al.* 1963). The use of the alcoholic decoction of the leaves of *S. campanulata* in the treatment of malaria in traditional medical practice has therefore some scientific basis.

Purification of fraction B, C and Z of the stem bark of *S. campanulata* yielded 3,20β-dihydroxyurs-12-en-28-oic acid, 3-hydroxyurs-12,19-dien-28-oic acid and 3-hydroxyurs-12-en-28-oic acid respectively. These compounds are structural analogues and are biogeneticallly related (Drake and Duvall 1936, Barton *et al.* 1962). The isolation of ursolic acid and two of its derivatives as antimalarial agents is a major step in the search for new antimalarials because ursolic acid or any other triterpenoids has never been reported as antimalarial agent. The only terpenoids reported to be active against *Plasmodium* sp. are the sesquiterpenoid artemisinin or qinghaosu and its structural analogues which have been found effective for the treatment of some drug-resistant strains of the malarial parasite (Klayman 1985).

Ursolic acid is well tolerated in the body. It is non-toxic when fed to rats, guinea pigs, chickens, rabbits at levels of 1000 - 5000 mg/kg body weight and to humans at a dose of 20 mg/kg/day (Lubitz and Fellers 1941).

It is also important to note that the isolation of the three triterpenoids of the urs-12-ene series from *S. campanulata* has never been reported.

References

Barton, D.H.R., Cheung, H.T, Daniells, P.J.L., Lewis, K.G., and McGhie, J.F. (1962). Triterpenoids, Part XXVI The triterpenoids of *Vangueria tomentosa*. *Journal of the Chemical Society*, 5163.

Drake, N.L. and Duvall, H.M. (1936). The dehydrogenation of ursolic acid by selenium. *Journal of the American Chemical Society* **58**, 1682-1688.

Helbecque, C., Juilliand, A.M. Herold, M., and Cahn, J. (1963). Action of different inhibitors of L-dopa decarboxylase on a provoked hypothermia of central origin. *Comptes-Rendus de la Société de Biologie et de ses Filiales* **157**, 996-999.

Klayman, D.L. (1985). Qinghaosu (Artemisinin): An antimalarial drug from China. *Science* **228**, 1049-1055.

Lubitz, J.A. and Fellers, C.R. (1941). Nontoxic character of ursolic acid. Preliminary study. *Journal of the American Pharmaceutical Association* **30**, 207-208.

Makinde, J.M., Adesogan, E.K., and Amusan, O.O.G. (1987). The schizontocidal activity of *Spathodea campanulata* leaf extract on *Plasmodium berghei berghei* in mice. *Phytotherapy Research* **1**, 122-125.

Makinde, J.A., Adesogan, E.K., and Amusan, O.O.G. (1988). The schizontocidal activity of *Spathodea campanulata* leaf extract on *Plasmodium berghei berghei* in mice. *Planta Medica* **54**, 122-125.

Makinde, J.M., Amusan, O.O.G., and Adesogan, E.K. (1990). The antimalarial activity of chromatographic fractions of *Spathodea campanulata* stem bark extracts against *Plasmodium berghei berghei* in mice. *Phytotherapy Research* **4**, 53 -56.

Oliver-Bever, B. (1960). *Spathodea campanulata* in Medicinal plants in Nigeria, p. 83,84. Nigeria College of Arts, Science and Technology, Ibadan.

23. Antimalarial activity of *Artemisia afra* against *Plasmodium falciparum in vitro*

M.A. ABRAHAMS[1], B.M. SEHLAPELO[2], W.E. CAMPBELL[2], D.W. GAMMON[2], P.J. SMITH[1] AND P.I. FOLB[1]

[1]Department of Pharmacology, University of Cape Town, Observatory, 7925, South Africa and [2]Department of Chemistry, University of Cape Town, Observatory, 7925, South Africa

---

Introduction

With the increase in recent years in the prevalence of malaria, and in drug resistance of *Plasmodium falciparum*, there has been much interest in natural plant products as a source of new antimalarials with novel modes of action against *Plasmodium*. Artemisinin or Qinghaosu (**1**) is one such antimalarial isolated from a Chinese herb, *Artemisia annua* L. (*Asteraceae*) and is currently undergoing phase I and II clinical trials (White 1994; Meshnick *et al.* 1996). The Southern African species, *Artemisia afra* Jacq. (African wormwood, wildeals, lengana) is commonly used by local traditional healers for the treatment of fevers and for symptoms of malaria. In a recent survey (Williams 1996), *A. afra* was rated the most commonly stocked plant from more than 500 species being harvested and traded on the Witwatersrand, South Africa. There has been one report of a crude extract of a Tanzanian variety having antimalarial activity (Weenen *et al.* 1990) but as far as we are aware there have been no further reports of any active principle isolated. A number of compounds have been isolated and identified from *A. afra* (Jakupovic *et al.* 1988), but none of these have been screened for antimalarial activity.

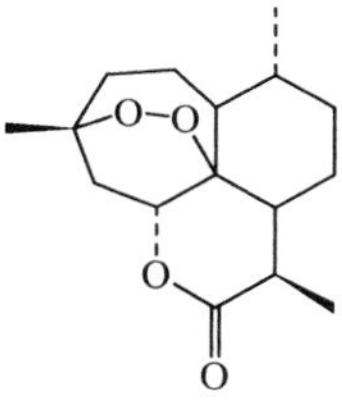

(**1**) Artemisinin

Therefore, the objectives of this study were to investigate *A. afra* for antimalarial activity and to isolate the active compound(s). It was also of interest to establish whether artemisinin present in *A. afra* could be responsible for any activity demonstrated.

Overview of Strategy

The strategy that we adopted is similar to that reported by Hamburger and Hostettmann (1991). It involved the following steps:

(1) positive identification of the plant by a botanist, collection and drying of the plant material;

(2) preparation of crude extracts and preliminary chromatographic analysis by TLC;

(3) bioassay of crude extracts;

(4) several consecutive steps of chromatographic separation with each fraction obtained submitted for bioassay in order to follow the activity;

(5) verification of the purity and activity of the isolated compounds;

(6) structure elucidation of isolated pure compounds;

(7) toxicological testing of isolated pure compounds.

Experimental Techniques

Plant collection

Artemisia afra was collected at the National Botanical Institute, Kirstenbosch in Cape Town, South Africa and allowed to air dry at room temperature, away from direct sunlight, for 6 weeks.

There were some observations concerning this plant that are worth noting. *A. afra* was harvested in its flowering stage. This may be relevant in that some reports (Woerdenbag *et al.* 1990) have indicated that the highest yield of artemisinin from *A. annua* can be obtained from plants harvested just before flowering occurs, although more recent reports (Laughlin 1994) indicate the contrary *i.e.* the greatest amount of artemisinin is isolated at the time of flowering.

In addition, there is evidence (Graven, E.H., Grassroots Natural Products cc., unpublished data) that the composition of the essential oils of individual *A. afra* plants in South Africa varies widely. Furthermore, we have observed by TLC and HPLC distinct chemical differences in crude extracts from 2 different bushes of *A. afra* not more than 5 metres apart at Kirstenbosch Gardens.

Extraction

Air-dried plant material was extracted for 14 hours in petroleum ether (60-80°C) using a soxhlet apparatus.

In our initial study of *A. afra*, we investigated the activity of crude extracts from different parts of the plant, namely:

(1) stems, leaves and flowers;

(2) stems only;

(3) leaves and flowers.

The TLC profiles of these three extracts were found to be the same and there was comparable potency against *P. falciparum*. Subsequent investigations were conducted on the extract of leaves and flowers.

Fractionation and purification of crude extracts

The crude petroleum ether extract of leaves and flowers was chromatographed over silica gel eluting with petroleum ether, petroleum ether/ethyl acetate mixtures and methanol. Fractions (20 ml each) were collected and pooled on the basis of their TLC profiles. The resulting fractions, P1 to P7 were then screened for antimalarial activity. The most active fraction, P6 was subjected to further repeated flash chromatography to yield several further sub-fractions (Fig. 23.1.). After screening these fractions for antimalarial activity, sub-fraction B was found to be the most active and was subsequently purified.

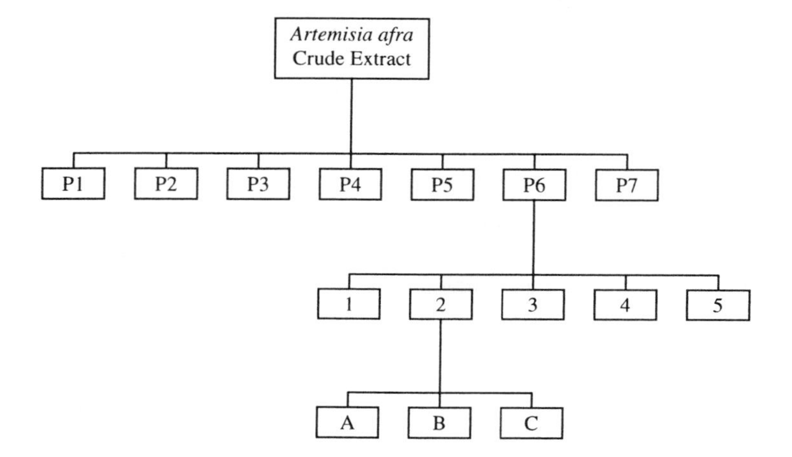

Fig. 23.1. Fractionation sequence.

Maintenance of parasite cultures

Two strains of *P. falciparum* were grown continuously in stock cultures by a modification of the methods of Trager and Jensen (1976). The strains used in this investigation were D10 (chloroquine-sensitive) and FAC8 (chloroquine-resistant) which were donated by Dr. A.F. Cowman at the Walter Eliza Hall Institute, Melbourne, Australia. Type O^+ human erthyrocytes infected with *P. falciparum* were cultured at 5% haematocrit in HEPES buffered RPMI 1640 culture medium supplemented with glutamine, glucose, hypoxanthine, gentamycin, $NaHCO_3$ and 10% human A^+ serum. The parasites were grown in tissue culture flasks flushed with filtered gas containing 3% O_2, 4% CO_2 and 93% N_2 and incubated at $38^{\circ}C$.

Bioassay: antimalarial screening

The method of Desjardins *et al.* (1979) was used to determine IC_{50} values of chloroquine, artemisinin and extracts of *A. afra.* Asynchronous cultures (1% parasitemia, 1.5% haematocrit) were exposed to increasing concentration of drug or extract. After 24 hours in culture, [^{3}H]-hypoxanthine was added (5μCi/ml final), followed by an additional 21 hours incubation period. Cells were harvested in triplicate using a cell harvester and filters were counted in a scintillation counter. The IC_{50} value which is reported is the inhibitory concentration at which there is 50% parasite growth.

Careful consideration had to be given to the preparation of these plant extracts in a form suitable for addition to the parasites at a known concentration. The extracts were insoluble in water and had to be dissolved in an organic solvent such as acetonitrile. The samples were then serially diluted with culture medium to achieve the desired concentrations in the 96 well flat-bottom microtitration plate. The final concentration range to which the parasites were exposed ranged from 0.1 ng/ml to 100 μg/ml. The highest percentage of acetonitrile used in the final concentration was 3% which had no adverse effect on the parasites.

Results and discussion

The activities of the crude petroleum ether extract of the leaves and flowers of *A. afra* against 2 strains of *P. falciparum* are summarized in Table 23.1. The activities of chloroquine and artemisinin against the two strains were also determined (Table 23.1.). These values are in good agreement with literature values (Foote *et al.* 1990; Barnes *et al.* 1992), thus confirming the reliability of the assay system. The appreciably high activity of the crude extract against both the chloroquine-sensitive strain, D10, and the chloroquine-resistant strain, FAC8, led to a series of bioassay guided fractionations as shown in Fig. 23.1., with priority given to investigation of fraction **P6** which showed the highest activity against both strains (Table 23.1.). Fractions **P2, P4** and **P7** in particular, with activities in the range 1 - 4 μg/ml, are also under investigation.

Fraction **P6** was separated further using flash chromatography to yield ultimately a pure compound **P6.2B** as a clear, colourless oil. **P6.2B** migrated as a single spot on TLC in a variety of solvents, and eluted as a single peak on HPLC. It was established by TLC and HPLC-UV that fraction **P6** did not contain artemisinin, and preliminary structural information on **P6.2B** confirms that the activity in this fraction is not due to artemisinin. The detailed chemical structure of **P6.2B** is currently under investigation.

Compound **P6.2B** has an IC_{50} value of 0.95 μg/ml against D10. Reasons for the apparent loss of activity in the fractionation of **P6.2**, which had an IC_{50} of 0.5 μg/ml, have not been established. A synergistic mechanism cannot be ruled out, and it is recognized that it may be of interest to reconstitute fraction **P6.2**, or prepare mixtures of pure compounds derived therefrom, and investigate their activities. Preliminary cytotoxicity screening against a normal fibroblast cell line has been carried out on **P6.2B**, showing that at a concentration of 1 μg/ml of this compound is not toxic.

Table 23.1. IC_{50} values (μg/ml) of chloroquine, artemisinin and samples of *A. afra* against two strains of *P. falciparum in vitro*

Sample	D10	FAC8
Chloroquine	0.014	0.085
Artemisinin	0.004	0.019
Crude	1.03	1.50
P1	5.4	>100
P2	1.3	3.7
P3	3.1	6.4
P4	1.7	2.9
P5	3.2	3.1
P6	0.5	0.5
P7	1.3	1.9
P6.1	21.0	nd
P6.2A	16.0	nd
P6.2B*	0.95	nd
P6.2C	17.4	nd
P6.3	5.9	nd
P6.4	7.2	nd
P6.5	15.3	nd

P6.2B* : isolated pure compound; nd : not determined

Conclusion

The petroleum ether extract of *A. afra* exhibits *in vitro* activity against both chloroquine-sensitive (D10) and chloroquine-resistant (FAC8) strains of *P. falciparum*. A bioassay guided fractionation of the extract resulted in the isolation of at least one pure active compound whose complete structure will be reported separately, together with the structures of other compounds with lower activity obtained from the extract.

The significant activity of the crude extracts, the association of some activity with at least one compound, and the low cytotoxicity suggest roles for this plant in phytomedicine and the treatment of malaria which are worth investigating further. The demonstration of activity in the aerial parts has implications for the sustainable agriculture and harvesting of the plants.

Acknowledgments

Thanks are due to the botanists at Kirstenbosch Gardens in assisting us in the collection of plant material. This research was supported by funds obtained from the Medical Research Council of South Africa and the University of Cape Town Research Committee.

References

Barnes, D.A., Foote, S.J., Galatis, D., Kemp, D.J., and Cowman, A.F. (1992). Selection for high-level chloroquine resistance results in deamplification of the *pfmdr*1 gene and increased sensitivity to mefloquine in *Plasmodium falciparum*. *EMBO Journal* **11,** 3067-3075.

Desjardins, R.E., Canfield, C.J., Haynes, D., and Chulay, J.D. (1979). Quantitative assessment of antimalarial activity *in vitro* by a semiautomated microdilution technique. *Antimicrobial Agents and Chemotherapy* **16,** 710-718.

Foote, S.J., Kyle, D.E., Martin, R.K., Oduola, A.M.J., Forsyth, K., Kemp, D.J., and Cowman, A.F. (1990). Several alleles of the multidrug-resistance gene are closely linked to chloroquine resistance in *Plasmodium falciparum*. *Nature* **345**, 255-258.

Hamburger, M. and Hostettmann, K., (1991). Bioactivity in plants: the link between phytochemistry and medicine. *Phytochemistry* **30**, 3864-3874.

Jakupovic, J., Klemeyer, H., Bohlmann, F., and Graven, E.H. (1988). Glaucolides and guaianolides from *Artemisia afra*. *Phytochemistry* **27,** 1129-1133.

Laughlin, J.C. (1994). Agricultural production of artemisinin - a review. *Transactions of the Royal Society of Tropical Medicine and Hygiene* **88**, Supplement 1, 21-22.

Meshnick, S.R., Taylor, T.E., and Kamchonwongpaisan, S. (1996). Artemisinin and the antimalarial endoperoxides: from herbal remedy to targeted chemotherapy. *Microbiological Reviews* **60**, 301.

Trager, W. and Jensen, J.B. (1976). Human malaria parasites in continuous culture. *Science* **193**, 673-675.

Weenen, H., Nkunya, M.H.H., Bray, D.H., Mwasumbi, L.B., Kinabo, L.S., and Kilimali, V.A.E.B. (1990). Antimalarial activity of Tanzanian medicinal plants. *Planta Medica* **56**, 368-370.

White, N.J. (1994). Artemisinin: current status. *Transactions of the Royal Society of Tropical Medicine and Hygiene* **88**, Supplement 1, 3-4.

Williams, V.L. (1996). The Witwatersrand muti trade. *Veld & Flora* **82**, 12-14.

Woerdenbag, H.J., Lugt, C.B., and Pras, N., (1990). *Artemisia annua* L.: a source of novel antimalarial drugs. *Pharmaceutisch Weekblad Scientific edition* **12**, 169-181.

24. Efficacy and duration of activity of *Lippia javanica* SPRENG, *Ocimum canum* SIMS and a commercial repellent against *Aedes aegypti*

N. LUKWA[1], C. MASEDZA[1], N.Z. NYAZEMA[2], C.F. CURTIS[3] AND G.L. MWAIKO[4]

[1]Blair Research Laboratory, P.O. Box CY573 Harare, Zimbabwe; [2]Department of Clinical Pharmacology, P.O. Box A178 Avondale, Harare, Zimbabwe; [3]London School of Tropical Medicine and Hygiene, Keppel Street, London, United Kingdom and [4]NIMR, P.O. Box 4, Amani, Tanzania.

Introduction

Essential oils of a multitude of plants have been shown to have mosquito-repellent properties. For example, repellency has been ascribed to the oils of cassia, camphor, citronella, lemongrass, clove, thyme, geranium, bergamot, pine, wintergreen, pennyroyal and eucalyptus. These oils and mixtures of them, were the basis of most commercial repellents, and many different varieties were produced and tested until diethyl toluamide (DEET) took over most of the market (Curtis *et al.* 1990).

Information and data on traditional uses of plants against pests all over the world have been accumulated. For example, the ancient Chinese classics contain many prescriptions for repellents against mosquitoes and other blood sucking flies (Roark 1947). However, little attention has been paid to the question of whether traditional preparations do, in fact, work when used in the traditional way and hence to the possibility that they may show us how to use plants directly, without the need for petroleum products or industrial processes.

Personal protection by use of mosquito repellents is, potentially an important component of integrated disease vector control. It has been documented, for example, that the Ainu people of Hokkaido, Japan and the Micmac Indians of Newfoundland both wore leggings of sedge, bark or cloth to reduce insect biting nuisance which tended to concentrate around the lower legs (Curtis *et al.* 1990). There is therefore need to develop appropriate tools and methods for vector control at individual as well as community level. N,N,-diethyl-3-methylbenzamide, DEET, for example, formulated with permethrin as a soap has been demonstrated to be effective as an insect repellent in Malaysia (Yapp 1986) and Australia (Frances 1986). Use of home-grown plant repellents could also avoid the problems which often bedevil conventional vector control programs.

Work on home grown plant repellents has tended to fall into two categories: the analytical and ethnobotanical. The latter has documented the use of plants by different cultures (Secoy and Smith 1983). In Zimbabwe, for example, *Lippia javanica* SPRENG (Verbenaceae) and *Ocimum canum* SIMS (Lamiaceae) have been reported to be commonly used as mosquito repellents (Lukwa 1994).

We have compared the efficacy and duration of effect of preparations made from dried *Lippia javanica* "Zumbani" and *Ocimum canum* "Rukovhi" leaves with an alcoholic commercial insect repellent which contains 11% DEET as the active ingredient against *Aedes aegypti.*

Materials and methods

A commercial preparation which contained 11% N,N,-diethyl-3-methylbenzamide, DEET, in alcohol was kindly donated S. Mharakurwa, (Blair Research Laboratory, Harare, Zimbabwe) *Lippia javanica* and *Ocimum canum*, correctly identified with assistance of a botanist, were collected from Gokwe and the leaves air dried. Fifty grams of dried leaves from either *Lippia javanica* or *Ocimum canum* were ground to fine powder and dissolved in 70% ethanol to give final concentration of 250 mg/ml. Laboratory bred *Aedes aegypti* mosquitoes (Blair Research Laboratory Insectary, Harare) were used in the experiment.

Following the test method previously described by Curtis et al.(1987) repellency properties of *Lippia javanica* and *Ocimum canum* were evaluated. Briefly human palms onto which about 2 ml of either plant extract, DEET or alcohol as the control were exposed to biting by previously starved 250 *Aedes aegypti* caged mosquitoes. Each palm was exposed for 1.5 minutes and the number bites/landings recorded. Retesting with each preparation was then done at an hourly interval making sure that the treated palms did not come into contact with water. Participants were also discouraged from doing any manual work that would encourage any sweating. The procedure was carried out for a period of 5 hours using the same batch of mosquitoes. Each hand was exposed an average of 4 times per each time period, in other words eight replicates were done for each test and controls. Percentage repellency was calculated, following the method previously described by Mehr *et al.* (1985). The calculation was carried out as follows:

$$\frac{\overline{B_c} - \overline{B_t}}{\overline{B_c}} * 100$$

where $\overline{B_c}$ = mean number of bites on control and $\overline{B_t}$ = mean number of bites on treated

To investigate insecticidal properties, filter paper discs, 8.5 cm in diameter, that had previously been soaked in either alcoholic *Lippia javanica* or *Ocimum canum* extract solution, 1 mg/ml, were introduced into perforated petri dishes. Four petri dishes were used for each extract solution. Ten blood-fed mosquitoes were introduced into the dishes through a hole. The mosquitoes were allowed to be in contact with the soaked discs for 20 minutes. A separate dish which contained discs soaked in 70% alcohol was also prepared to serve as the control. The mosquitoes were transferred to medium sized mosquito cage with sugar water and observed for 24h. The paired Student's *t*-test was applied to the data.

Results

Efficacy and residual effects of the plant preparations and DEET in providing up to 5h protection, against mosquitoes are shown in Table 24.1. None of the plant preparation provided complete protection throughout the 5 hours period. There was no significant differences in terms of reduction of mosquito rate of landing when the two plants were compared. Table 24.1. and Fig. 24.1. also shows that, in general the activities of each repellent drastically dropped 3 hours after application of the test solutions. Peak activity for all the tests including that of the over-the-counter preparation was observed at 2 hours. The graph on Fig. 24.1. shows the trend, as time went by. There appeared to be rapid decline in protection 2 hours post application. Alcoholic extract of *Lippia javanica* appeared to be more effective than that of *Ocimum canum*, 100 and 70% repellency at 2 hours respectively.

Table 24.1. Repellent activity of *L. javanica* and *O. canum* at different time period

	percentage of repellency after					
	1	2	3	4	5	[h. post treatment]
CONTROL	0	0	0	0	0	
Lippia javanica	100	100	40	13	0	
Ocimum canum	77	70	20	25	0	
DEET	92	80	0	0	0	

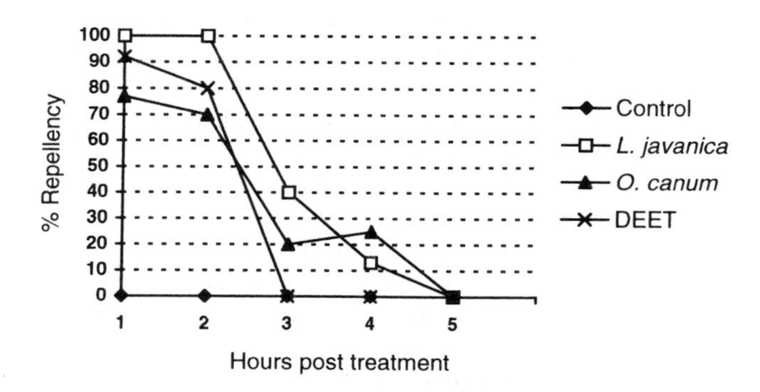

Fig. 24.1. Duration of repellency.

There appeared to be no significant difference between the effectiveness of the insecticidal properties of the two plants investigated. As shown in Table 24.2. the average mortality rate of the mosquitoes observed after 24 hours was 93%.

Table 24.2. Toxicity of *Lippia javanica* and *Ocimum canum* to *Aedes aegypti*

	% Mortality at 24 hours	
expt. no	*Lippia javanica*	*Ocimum canum*
1	100	80
2	100	100
3	70	100
4	100	90

Average mortality at 24h was 93% and reversal of knockdown was observed to 7.5%

Discussion

Ocimum spp. belongs to basil family, Lamiaceae whose repellency has been attributed to constituent eugenol. *Lippia javanica* belongs to Verbenaceae family that has been variously described as having the odor of vanilla or of mint (Watt and Breyer-Brandwijk 1962). The traditional way in Zimbwbe of using these plants, is to smear the juice from the leaves onto the skin and, fresh application is always thought to be more effective and would be affected by sweating.

In the present experiment alcoholic extracts were used. The assumption was that, alcohol would make it easy to concentrate any water soluble active constituents. This would also to take care of any seasonal variation that are known to occur in plants affecting concentration of the plant constituents. In other words plants do become senescent and lose much of their active ingredients at certain times of the year. Hence the idea of cultivating those plants, that have been found to have economic value, under controlled environment. Our results with *Ocimum* agreed with those that were obtained by a previous study which used the same species (White 1973). Not much work has been done on *L. javanica* which could be used fro comparison with our own results. However, field trials and laboratory tests using the plant have been carried out and results compared, taking into account all the complex and interacting variables controlling the attraction of insects to hosts. The results appear to be in agreement comparable (Lukwa 1994).

As the aim of the study was to attempt to validate traditional practice of mosquito control in the community, it would appear that there is evidence to support active promotion of such practices. In fact, what this means, therefore, is that people would not need much encouragement to continue using such methods with persuasion from scientific experts. It is not altogether necessary for people to understand all the complex and interacting variables controlling the attraction of biting insects, such as mosquitoes, to hosts: an effective repellent should reduce biting for everyone in a range of conditions. A method of control which is freely available and of small benefit may be more useful than one which is effective but unaffordable. DEET is beyond the reach of most people in the rural areas. In addition DDT, which is still the mainstay of malaria control program in Zimbabwe, has come under heavy criticism as an environmental pollutant.

In conclusion, it would appear that both *Lippia javanica* and *Ocimum canum* have the potential of being widely used as mosquito repellents to compliment other control methods. Studies into the possibility of cultivating these plants under controlled conditions need to be carried out.

References

Curtis, C.F., Lines, J.D., Baolin, L., and Renz, A. (1989). Natural and synthetic repellents. In *Appropriate Technology in Vector Control.* (ed. C.F. Curtis), pp 75-92. CRC Press, Bacu Raton, FL., USA.

Frances, S.P. (1986). Effectiveness of DEET and permethrin alone and in a soap formulation as skin and clothing protectants against mosquitoes in Australia. *Journal of the American Mosquito Control Association* **3**, 643-650.

Granett, P. (1940). Studies of mosquito repellent II. Relative performance of certain chemicals and commercially available mixtures as mosquito repellents. *Journal of Economic Entomology* **33**, 566.

Lukwa, N. (1994). Do traditional mosquito repellent plants work as mosquito larvicides. *Central African Journal of Medicine* **40**, 306-309.

Mehr, Z.A., Rutledge, L.C., Morales, E.L., Meixsall, V.E., and Korte, D.W. (1985). Laboratory evaluation of controlled release insect repellent formulations. . *Journal of the American Mosquito Control Association* **1**, 143-147.

Roark, R.C. (1947). Some promising insecticidal plants. *Journal of Economic Botany* **1**, 437.

Secoy, D.M. and Smith, A.E. (1983). Use of plants in control of agricultural and domestic pests. *Journal of Economic Botany* **37**, 28.

Watt, J.M. and Breyer-Brandwijk, M.G. (1962). *Medicinal and Poisonous Plants of Southern and Eastern Africa*, p. 1051. E. and S. Livingstone LTD. Edinburgh and London.

White, G.B. (1973). The insect repellent value of *Ocimum* spp. (Labiatae) traditional anti-mosquito plants. *East African Medicinal Journal* **50**, 248.

Yapp, H.H. (1986). Effectiveness of soap formulations containing DEET and permethrin as personal protection against outdoor mosquitoes in Malaysia. . *Journal of the American Mosquito Control Association* **2**, 63-67.

Index